FLIGHT IN AMERICA

THE JOHNS HOPKINS UNIVERSITY PRESS
Baltimore and London

FLIGHT
IN AMERICA

From the Wrights
to the Astronauts

Third Edition

ROGER E. BILSTEIN

The Johns Hopkins University Press
2715 North Charles Street
Baltimore, Maryland 21218-4363
www.press.jhu.edu

Library of Congress Cataloging in Publication Data

Bilstein, Roger E.
    Flight in America : from the Wrights to the astronauts / Roger E. Bilstein — 3rd ed.
        p. cm.
    Includes bibliographical references and index.
    ISBN 0-8018-6685-5 (alk. paper)
    1. Aeronautics—United States—History. 2. Astronautics—United States—History. I. Title.
TL521.B528 2001
629.1′0973—dc21   00-065494

A catalog record for this book is available from the British Library.

Parts of chapter 2, "The Aviation Business: 1918–1930," also appear in *Flight Patterns: Trends of Aeronautical Development in the United States, 1918–1929,* by Roger E. Bilstein, © 1983 by the University of Georgia Press.

Frontispiece: For credit, see illustration on page 118.

For Robert H. Bremner

Contents

Preface to the Third Edition ix

Abbreviations xi

1. The Awkward Years: Early Flight to 1918 3

2. The Aviation Business, 1918–1930 41

3. Adventure, Airways, and Innovation, 1930–1940 83

4. Air Power at War, 1930–1945 125

5. Air-Age Realities, 1945–1955 167

6. Higher Horizons, 1955–1965 205

7. From the Earth to the Moon, 1965–1975 247

8. Aerospace Perspectives, 1975–1983 285

9. Turmoil and Transition, 1983–2000 327

Notes 371

Index 395

Preface to the Third Edition

In a thoughtful study, *A Social History of Engineering* (1961), English social historian Walter H. G. Armytage observed that "it is not too much to say that aviation has had a more comprehensive and far-reaching effect on all fields of material endeavour embarked upon by humans than any other previous single development." Emphasizing World War II and the postwar era, Armytage noted a broad range of influences, from high-precision machining, metallurgy, and high-test fuels to surveying. Even a cursory review of space flight would reveal a similar catalog of benefits of progressive sophistication. Considering the influence of wartime developments such as radar, jet propulsion, and rocketry, it is clear that postwar as well as prewar innovations from Europe were often significant legacies to American aerospace progress. This overseas influence represents a continuing thread within the pattern of American developments. By the 1980s and after, such international influences represented a substantial share of actual hardware marketed by the American aerospace industry and constituted an important element of investment involving airline operations by American companies.

Over the decades following World War II, international tensions generated by the cold war strongly influenced the direction of American aerospace research, development, and air power doctrine. The dramatic collapse of the Soviet Union and Communist bloc alliances across Europe promised fresh prospects for cooperation and at the same time created wrenching readjustments within the aerospace community as defense budgets plummeted and corporations scrambled to achieve stability in an unfamiliar economic environment. Likewise, airlines based in the United States struggled through an era of dramatic change brought about by deregulation and rough spots in both domestic and global economies.

At the close of the twentieth century, it was clear that aviation and aerospace events represented significant milestones in human history. Some events, such as warfare, represented a destructive capability of staggering proportions. On the other hand, flight facilitated rapid disaster relief efforts and presented opportunities for continental and intercontinental travel for unprecedented numbers of the world's citizens. Flight, which had helped generate so much change in global history, continued to experience significant change itself.

For their interest and assistance in the evolution of previous versions of this book, I acknowledge my colleagues at the National Air and Space Museum, Smithsonian Institution. From my tenure as Charles Lindbergh Professor of Aerospace History at the museum during 1992–93, I owe an ongoing debt to Tom Crouch and Dominick Pisano for their support and advice; to Peter Jakab, for his unflagging humor and encouragement; and to R. E. G. Davies, for his encyclopedic knowledge of airline history. At the Johns Hopkins University Press, I again thank Henry Y. K. Tom for his support and also thank Robert Brugger for his knowledgeable guidance and his editorial assistant Melody Herr. I also want to acknowledge Linda Forlifer, Senior Manuscript Editor, for her skill and patience in shepherding the third edition through the editorial process. Any remaining problems and inconsistencies are mine alone.

Dr. Robert Bremner, to whom this book is dedicated, is professor of history, emeritus, at the Ohio State University. As a graduate student in his lecture courses and seminars, I learned to consider a broad range of details from music, art, literature, and popular culture in the process of crafting a historical narrative. As the director of my doctoral dissertation in 1965, he demonstrated the gentle humor, insight, and wisdom of a quintessential mentor. His influence on my approach to teaching, research, and writing has been fundamental. Once again, I want to thank Linda, our family, and patient relatives for sharing my fascination with the world of flight.

Abbreviations

AAF	Army Air Force
ABMA	Army Ballistic Missile Agency
ALPA	Air Line Pilots Association
ATA	Air Transport Association
AWACS	airborne warning and control system
CAB	Civil Aeronautics Board
CSM	command and service module
FAA	Federal Aviation Agency
GALCIT	Guggenheim Aeronautical Laboratory of the California Institute of Technology
IATA	International Air Transport Association
ICAO	International Civil Aviation Organization
ICBM	intercontinental ballistic missile
ILS	instrument landing system
IRBM	intermediate-range ballistic missile
LM	lunar module
MIRV	multiple independently targetable reentry vehicle
NACA	National Advisory Committee for Aeronautics
NASA	National Aeronautics and Space Administration
R&D	research and development
RAF	Royal Air Force
SAC	Strategic Air Command
SAM	surface-to-air missile
SST	supersonic transport
STOL	short takeoff and landing
USAF	United States Air Force
USAFE	United States Air Force in Europe
VfR	*Verein für Raumschiffahrt*
VOR	very high frequency omnidirectional radio

FLIGHT IN AMERICA

1

The Awkward Years:
Early Flight to 1918

During the month of September 1903, few people anywhere had reason to take particular interest in the appearance of the Wright brothers, Orville and Wilbur, at a lonely stretch of beach near Kitty Hawk, North Carolina. The new century had been eventful enough already: American newspapers had followed the Boer War in Africa and the Boxer Uprising in China. Queen Victoria had died and President McKinley had been assassinated. Sigmund Freud had published *The Interpretation of Dreams* and Beatrix Potter was publishing "Peter Rabbit" stories. During 1903 alone, fanatics murdered the king and queen of Serbia; suffragists were joining activist groups like Emmeline Pankhurst's National Women's Social and Political Union in England; an automobile had accomplished the first coast-to-coast crossing of the United States in only 65 days. And the Wright brothers arrived at Kitty Hawk, where they intended to make the world's first successful flight in a powered airplane.

Local citizens knew the brothers from visits the Wrights had made each summer since 1900 in order to fly gliders in the steady currents along the Atlantic coast. During these summer sojourns, the Wrights coped with swarms of insects and blowing sand that continually drifted through the cracks in the crude, wooden shack the brothers used as both shelter and workshop. Winter visits solved the bug problem, but the wooden shack was hardly weatherproof. As Wilbur wrote home, "In addition to . . . 1, 2, 3, and 4 blanket nights, we now have 5 blanket nights and 5 blankets and 2 quilts." They occasionally slept fully clothed, wearing shoes, hats, and overcoats. Along with these discomforts, they struggled by day to assemble their flying machine and tune its engine. During the evenings, Orville wrote to his sister and studied German.

Working carefully and methodically, the Wrights put everything in readiness by Monday, December 14, 1903. Who would be first? The brothers flipped a coin and Wilbur won the toss. He settled himself at the controls as the engine warmed up; gave the nod to start; took off, quickly stalled, and hit the ground, damaging the rudder. For three more days, the brothers made repairs, shivered at night, and studied more

German. But they were confident. As Wilbur wrote his family the same evening, "There is now no question of final success." The Wrights were about to achieve a goal that had captivated dreamers and experimenters throughout recorded history.

AERONAUTS AND BALLOONS

Virtually every epoch of human civilization includes references to flight. From the winged deities of ancient Egypt, to the Greek legend of Icarus, to Germanic Valkyries, to a score of other myths, themes of flight occur again and again. There were undoubtedly sporadic attempts to achieve human flight, probably in imitation of birds. The first credible mention of such efforts appeared in Roger Bacon's *Secrets of Art and Nature,* written by the English monk in 1250, in which he referred to an ornithopter contrived by an acquaintance. The first documented drawings of such experimental flying machines surfaced in the notebooks of the remarkable Florentine genius, Leonardo da Vinci (1452–1519). Based on the flapping motion of a bird's wings, an ornithopter required a good deal of muscular energy from its human operator. As long as there was no other practical means of mechanical power available (such as steam or gas engines), the ornithopter beguiled aspiring aeronauts for generations, with a notable lack of success. Practical heavier-than-air flight finally evolved from fixed-wing aircraft in the form of gliders; with the addition of propellers and engines, airplanes at last became a reality. Gliding itself dated from about the year 1000, when a Benedictine monk, Eilmer of Malmesbury, reportedly launched himself from a tower of Malmesbury Abbey and traveled some 600 feet. A hard landing, so the story goes, broke both of Brother Eilmer's legs and left him crippled. The abbey commemorated his feat in one of its stained-glass windows. Structural and stability problems seem to have frustrated novice gliding enthusiasts until the early nineteenth century. Curiously, the legacy of propellers and power plants came from a different source, because adventurous individuals found other means of exploring the realm of the skies. With the age-old problem of suitable power sources handicapping early experiments, the first person to leave the surface of the earth did so in a balloon. The first balloons were buoyed into the air with air itself—hot air.

The Montgolfier brothers of France had observed that warm air tends to rise, and reasoned that if they could capture it in some sort of lightweight bag, the bag would rise, along with anything attached to it. They experimented with several small linen bags, lined with paper to help retain the hot air, climaxed by a demonstration for Louis XVI and Marie Antoinette at Versailles in September 1783. This large balloon, equipped with a container to hold a blazing fire underneath it, carried a sheep, a rooster, and a duck, to see if air-breathing creatures could stay alive in the unknown atmosphere above the earth. Since the barnyard specimens survived, a young physician named Pilatre de Rozier persuaded the Montgolfiers to let him try several tethered flights. Finally, on November 21, 1783, Pilatre de Rozier and a daring friend (a French infantry major) made the first free flight in a balloon, a 25-minute journey that carried them a total of five miles. En route, the infantry major heroically wielded a wet sponge to douse burning holes in the balloon.

Benjamin Franklin, in Paris to enlist continuing French support for the American Revolution, joined the thousands of Parisians who watched the elegant balloon coast along above the city's skyline. The balloon, colored a striking blue with elaborate gold designs and signs of the zodiac, carried its equally elegant passengers (and an open fire) in a wicker basket hung with red bunting. It was a colorful, sensational event that prompted a wave of balloon mania, with the balloon motif adorning furniture, china, buttons, and snuffboxes. Hot-air balloons of the eighteenth century, however, had evident drawbacks due to the danger of flying sparks from burning straw, generally used as fuel because of its light weight. Gas-filled balloons held the promise of safer, sustained trips, free from the danger of fire, although hot-air balloons could be maneuvered up or down by varying the amount of fuel fed into the fire. Attempting to cross the English Channel in 1785, Pilatre de Rozier tried a hybrid arrangement, combining a hot-air balloon (for altitude control) under a hydrogen-filled gasbag (for sustained flight). Predictably, errant sparks reached the upper balloon, and the contraption exploded, killing de Rozier and a companion.

As long as it was treated respectfully, however, hydrogen successfully provided the lifting medium for most of the leading balloonists of the nineteenth century through the twentieth centuries. Only 10 days after Pilatre de Rozier's first free flight, two more Frenchmen, Noël Robert and J.A.C. Charles, made the first voyage in a hydrogen balloon. A hydrogen balloon piloted by Jean-Pierre Blanchard and a refugee American named John Jeffries (who had supported George III during the American Revolution) first crossed above the English Channel in 1785. Blanchard, minus Jeffries, later demonstrated his balloon in Philadelphia (1793) for a crowd that included John Adams, Thomas Jefferson, James Madison, James Monroe, and the new president of the United States, George Washington. Public displays like these inspired awed enthusiasm for a growing coterie of select aerial adventurers, known appropriately as "aeronauts."

Intensive research in ballooning continued in Europe. Balloon flights became very popular but remained at the mercy of the winds, a handicap that brought the old problem of propulsion back into focus. Henri Giffard, another Frenchman, made an important contribution in this respect when he equipped a cigar-shaped airship (a configuration that enhanced directional control) with a propeller and lightweight steam engine of about 3 HP. On September 24, 1852, this device successfully flew 17 miles, from the Paris Hippodrome to a field outside the city, at an average speed of about 5 MPH. A well-known engineer, Giffard reflected growing interest in the subject among other professionals and mechanically inclined experimenters. Symbolic of this important change in flying activities was the formation of professional groups like the Societé d'Aviation in France in 1863 and the Aeronautical Society of Great Britain (later, the Royal Aeronautical Society) in 1866.

But steam engines of that day proved to be too heavy for truly efficient application in powered balloons. By 1885, when the German engineer, Karl Benz, demonstrated the first practical automobile driven by a gasoline engine, the power plant problem came nearer to solution. Several aerial pioneers experimented with dirigibles and gasoline engines, but the essential catalyst appeared in the person of Alberto Santos-Dumont, a wealthy Brazilian living in Paris. In 1898, he took an engine from one of his small, sputtering motorcars, attached a propeller to the drive shaft, and slung the contraption (along with a seat for the operator) under an elongated gasbag. In spite of many early failures, the dogged Brazilian persisted, designing and flying a dozen different airships during the next five years.

By mid-nineteenth century, aspiring aeronauts in the United States also had conducted innumerable balloon flights, making ballooning a major spectator event. Essentially entertainers, American balloonists provided thrilling free-balloon ascents for enthralled crowds, generated reams of rococo copy for local newspapers, and lifted the country to a state of balloon mania on the eve of the Civil War. By 1859, perhaps 3,000 ascents had been made in the United States, and some 8,000 passengers, men and women, had been airborne. Given the extent of ballooning activities, it was not surprising that someone in the flying fraternity hit upon the use of balloons in warfare. Europeans, in fact, had already done so; in 1849, the Austrians had set loose a number of unmanned hot-air balloons rigged with delayed-action bombs during an attack on Venice. The results were negligible, although portents for the future were ominous.

At the outbreak of the Civil War, both Union and Confederate patriots volunteered their services and made useful reconnaissance ascents in balloons, but the best-organized effort was the Union contingent directed by Thaddeus S. C. Lowe, a well-known balloonist of the prewar era. During 1861, Lowe supervised construction of seven balloons, along with a dozen mobile hydrogen generators. Using 1,600 pounds of acid and 3,300 pounds of iron filing, each hydrogen generator could fill a balloon in three hours; the balloon could remain inflated for about two weeks. Tethered to the ground (or sometimes to a boat), the balloons carried an observer in a wicker basket; the observer telegraphed or dropped messages to the ground. Field officers who relied on hills or the presence of trees to screen troop movements from observation by the enemy were forced to take into account the possibility of aerial reconnaissance. Troops and gun batteries spotted by enemy balloons might have to reposition; troop movements might have to be made at night, certainly an inconvenient and potentially chaotic affair. But despite acknowledged success in directing artillery fire and reporting Confederate troop movements, the Union's balloon service was short lived. Officers remained skeptical, and the balloon service's volunteer civilian status meant that Lowe constantly clashed with the army over necessary horses, wagons, and additional troops needed to transport balloons and equipment in combat zones.

Following the Civil War, free-flight ballooning enjoyed a renewed but brief popularity as the novelty began to wane. Several Americans experimented with navigable dirigibles—characterized by cylindrical gasbags and by pedals geared to a propeller. At the turn of the century, the powered balloon flights of Santos-Dumont spurred fascination with dirigibles, especially since prize money became a powerful incentive. Organizers of the Louisiana Purchase Exposition, held in St. Louis in 1904, announced $150,000 in awards for several categories of airship flights, including altitude, endurance, distance, and time trials over an L-shaped course of 15 miles. The announcement drew 97 potential entrants and created intense interest in navigable

In 1862 an artist from Frank Leslie's Illustrated Newspaper *made this on-the-spot sketch of a Union balloon conducting reconnaissance from a barge operating on the James River. Courtesy of U.S. Navy*

airships. Strict rules caused numerous entrants to withdraw later, and a rash of failures aborted several of the contests, including the time-course trials, but the publicity created a new wave of eagerness to conduct navigable flights.

Among the 1904 contestants was Roy Knabenshue. The following year, he created a sensation by piloting a 62-foot-long dirigible, propelled by a 10-HP engine, over New York City. Knabenshue had to share headlines with a growing number of professional aeronauts, including Lincoln Beachey, who took two airships in 1906–7 on an international junket ranging from Montreal, Canada, across the United States, and down to Mexico City. Knabenshue and Beachey, among others, competed for substantial prize money in a widely publicized airship meet organized a second time by St. Louis promoters in 1907. During the next two years, in the aftermath of the second St. Louis meeting, additional aerial stalwarts, like Charles Hamilton, acquired airships and flitted across the country, riding the crest of a vogue for aerial demonstrations at county fairs or any event that promised a paying crowd. Hamilton eventually made the first Oriental aerial tour and steered his airship across the Bay of Osaka in Japan.

As balloons and airships soared in skies around the globe, human flight had become a phenomenon of some 125 years of practice, and many thought that dirigibles would prove the most practical means of air travel in the future. The dirigible era certainly added much to flight experience in terms of power plants and navigation. The phenomenon of flight became well known to millions of people in the United States and elsewhere in the world. But other aerial experimenters were ready to challenge the suzerainty of lighter-than-air vehicles, flying a strange new contraption with wings instead of a gasbag.

GLIDERS AND AIRPLANES

In the development of their successful airplane, the Wrights had the advantage of over half a century of experimentation involving winged aircraft. Although no practical power plant was available for many years, thousands of short flights were made in gliders, beginning in the early nineteenth century.

As a nine-year-old boy, George Cayley (1774–1857) had been one of hundreds of Englishmen thrilled by accounts of the Montgolfier balloon flights. As an adult, Sir George Cayley became one of Europe's leading scientists and a commanding figure in aeronautical research. He carefully studied the flight of birds and concluded that the ornithopter did not suggest a successful direction of endeavor. He advocated a design that would separate propulsion from the source of lift, which, in his judgment, ought to be fixed wings. By 1809–10, his ideas had crystallized into several model gliders with angled wings and tail assemblies, including vertical and horizontal stablizers. Before he died in 1857, Cayley had designed and flown numerous gliders as well as a full-size craft capable of carrying a person.

During the middle- and late-nineteenth century, various experimenters constructed flying craft based on Cayley's work and tried to find an efficient, lightweight source of power. Like balloonists, these designers found that steam engines were much too cumbersome and heavy for flying craft. Besides, the science of aerodynamics, the art of flying, and control of winged craft in flight were still in infancy, although the work of Otto Lilienthal (1848–96) marked another milestone in the history of flight.

In Germany, Lilienthal designed and built gliders with the care and expertise of the trained engineer that he was. During the 1890s, he made over 2,000 successful gliding flights, and pictures of his gliding feats appeared all over the world. His work inspired unknown numbers of flight enthusiasts, and his basic contribution to the store of knowledge about materials, design, construction, and handling of gliders was of paramount importance. In 1896, Lilienthal died when one of his gliders stalled and crashed, a grim reminder that the conquest of the air held many dangers.

It is significant that so much of the work in flight experiments was being done by professionals in engineering like Cayley and Lilienthal. This meant that their activities were reported and discussed at various scientific and professional meetings, so that researchers in the United States had a chance to keep up with progress overseas. From the time of the formation of the Societé d'Aviation in 1863, heavier-than-air flight had been a prime consideration. At the first meeting of the Aeronautical Society of Great Britain, in 1866, the British researcher F. H. Wenham read a paper on high-aspect-ratio wings, a topic of major significance for later researchers. Cayley and Lilienthal

After a running leap from the top of a hill built for his glider experiments, Otto Lilienthal makes a successful flight during the 1890s. Courtesy of the Smithsonian Institution.

represent only two of the dozens of researchers, like Wenham, who pursued the goal of winged flight step by step, slowly but surely accumulating the essential understanding of lift, control, structure, and propulsion arrangement. The French-born Octave Chanute played a role both as a conduit and as a researcher. Born in 1832, Chanute had immigrated to the United States at an early age, and eventually grew to be one of the country's leading civil engineers. Chanute became fascinated by aviation and in 1894 published *Progress in Flying Machines,* not only the first accurate history but also a valuable practical reference for contemporary researchers. Chanute's own biplane hang-glider of 1896 featured an improved method of rigging the wings. The well-to-do American industrialist James Means accumulated reliable aeronautical information from a variety of sources and reprinted the material in three outstanding collections of aeronautical data, the *Aeronautical Annual,* published annually between 1895 and 1897. With such references available, other researchers had an invaluable compilation of information upon which to proceed.

Such a researcher was Samuel Pierpont Langley (1834–1906), a distinguished astronomer and head of the Smithsonian Institution in Washington, D.C. Langley was determined to be the first to perfect a powered airplane piloted by a human. His design for the *Aerodrome* seemed to be confirmed by a series of gliders and flights by a succession of aircraft powered by small steam engines. The assistant secretary of the navy, Theodore Roosevelt, recognized the possibilities of aircraft for military scouting and helped Langley secure a grant of $50,000 from the War Department at the time of the Spanish-American War. Langley even built and flew a one-quarter-scale version of the full-size airplane in his quest of a successful design. Concluding that steam engines were too heavy for practical use, Langley equipped the full-size *Aerodrome* with a gasoline engine, itself a remarkable piece of machinery for its day. Perfected by a gifted mechanic, Charles M. Manly, the five-cylinder engine weighed only 125 pounds and produced 53 HP. In the autumn of 1903, Manly twice tried to fly Langley's plane, catapulted from a houseboat in the Potomac River. Following Manly's second ducking in the icy Potomac, rescuers cut away his icy garments, administered a shot of whiskey, and heard Manly utter a "most voluble series of blasphemies." The problem was not the engine (its basic design was similar to the modern radial engine) but Langley's catapult system, the plane's structural faults, and controls. Langley's second failure occurred on December 8, 1903. Nine days later, at Kill Devil Hills, near Kitty Hawk, North Carolina, the Wright brothers flew into history.

Sons of a well-known minister and church official in Dayton, Ohio, the Wright brothers, Orville (1871–1948) and Wilbur (1867–1912), grew up in the stimulating atmosphere of a well-read, inquisitive family. The brothers acquired a hometown reputation for indefatigable tinkering and local renown for the design and construction of high-quality bicycles. Although lacking in professional training, they were eventually led by their intelligence, intuitive mechanical aptitude, and catholic reading into the fascinating study of flying machines. They eagerly followed news about Lilienthal's glider flights, learned of Langley's flying models, and wrote to the Smithsonian for available literature on aviation. They also had the benefit of an extensive correspondence with Octave Chanute, who remained in contact with the foremost aeronautical researchers of his day in Europe as well as in the United States. Through letters and discussions with Chanute, the Wrights sharpened their minds on the latest trends, adapting them to the line of development already suggested by their own experiments. Using their own small wind tunnel (6 feet long and 16 inches square), they compiled new data on the relationship of various wing shapes and lift and corrected the errors of their predecessors.

They built their own gliders and began periodic sojourns to Kill Devil Hills in North Carolina, where steady winds off the Atlantic offered attractive conditions for test flights. Each brother gained a wealth of experience in mastering successively advanced designs in the air. The Wright gliders boasted significant new features, including warping (or twisting) the wings slightly so that the glider would bank in flight. They linked the wing-warping controls to a movable rudder in order to make controlled turns in either direction. They installed forward elevators to control the glider's ability to climb or descend. By 1902, the problem of control had been solved, and both brothers became skilled glider pilots, an important attribute that differentiated them from many contemporaries.

Next, they wanted a power plant to keep them aloft for longer periods of time. Knowledgeable mechanics, the Wright brothers built their own 12-HP engine because they could not find a suitable lightweight motor. They began to study propellers, based on additional data from their wind tunnel. Propellers had long been used on steam ships, but the Wrights soon discovered that very little was actually known about the

Samuel P. Langley (right), Secretary of the Smithsonian Institution, stands with Charles Manly, who built the Langley Aerodrome's engine and made two unsuccessful flight attempts. Courtesy of the Smithsonian Institution.

theory of marine propellers. They perfected a new aeronautical propeller, the most efficient ever designed up to that time, then made two of them for their new airplane. In the autumn of 1903, the Wrights crated up the plane and returned to the windswept sand dunes at Kill Devil Hills. Then came the series of frustrating mechanical problems, bad weather, and other difficulties that delayed the inaugural flight until mid-December.

Following Wilbur's mishap on December 14, minor repairs and discouraging weather kept the brothers waiting three days until it was finally Orville's turn. As in the previous attempt, the plane was positioned on a small trolley that ran along a wooden track laid across the soft sand. Orville clambered aboard and warmed up the plane's engine; then Wilbur helped steady the wing tips as the plane rolled down the track, gathered speed, and lifted into the air under its own power. On the morning of December 17, 1903, the Wright brothers achieved the first sustained controlled flight made by a human in a powered airplane: a distance of 120 feet and a duration of 12 seconds before the plane's landing skids touched down on the soft sand. Besides the Wrights, the only other people to witness the event included three helpful men from a lifesaving station down the beach and two curious young onlookers from nearby coastal hamlets. One of them, 18-year-old Johnny Moore, later proclaimed the news in Kitty Hawk: "They did it! They did it! Damned if they didn't fly!" Exhilarated, the Wrights made three more flights, and Wilbur kept the plane up for almost a full minute to cover 852 feet. Even though a gust of wind damaged the plane and stopped further attempts, it had been a great day. The brothers sent their father a telegram proclaiming their success and assured him they would be home for Christmas.

This incredible and historic feat was, even more incredibly, virtually ignored for nearly five years.

THE SKEPTICS AND THE WRIGHTS

In 1908, a curious 14-year-old named Charles Fayette Taylor found a magazine article written by Orville and Wilbur Wright in *Century*. It was the brothers' first account of their own development of the airplane to be published in a popular periodical. Vastly taken by this story, young Charles showed it to his father, who, despite the accompanying photos, refused to accept the notion that human flight was a reality. Undaunted, young Charles persisted in aviation, becoming a prominent aeroengineer and professor at the Massachusetts Institute of Technology. In all probability, Taylor's doubting parent eventually believed that the Wrights had actually achieved what they claimed in 1908. But the son's reaction and his father's skepticism typified the ambivalence toward airplanes that characterized the American public during the pre–World War I years. For all the evidence that accumulated in published reports, most Americans had to see it to believe it. The airplane, the Wrights, and the American public eventually came to terms in the course of a turbulent and flamboyant era of aviation development.

As the world marked its passage from 1899 into the twentieth century, many informed individuals regarded the prospects of piloted aircraft with reasonable con-

The Wright brother's first powered flight took place on the morning of December 17, 1903 with Orville Wright at the controls. After the brothers had their plane set on the launching track and had run up the engine Orville arranged the camera on a tripod and focused it on a point a few feet short of the end of the track. Confident that the flight would be a success, he asked John T. Daniels, one of several men from the Kill Devil (N.C.) life saving station who were present, to snap the shutter as the plane cleared the starting rail. Daniels apparently did as he was told, but the Wrights were not sure they had a picture until Orville developed the film in their darkroom after they had returned to Dayton, Ohio. Courtesy of the Smithsonian Institution.

fidence. Henry Brooks Adams savored the feast of essays celebrating the past and future sent him by his old friend from the Smithsonian, Langley, including the reports and photographs of Langley's aircraft models in flight. These early experiments, along with the glider flights of Lilienthal and others, were known to Adams and scores of Americans who read the Smithsonian publications and journals like *Scientific American*. Thousands of American citizens had witnessed balloon and dirigible flights. Across the country in the late-nineteenth century, many newspapers commented favorably on the prospects of powered flight, and dozens of enthusiasts and tinkerers struggled to develop a feasible airplane. Still, a considerable segment of the American public harbored strong doubts about the notion of human flight in a winged aircraft powered by an engine. Mark Sullivan, one of the eminent journalists of the day, vividly recalled an informal dinner in 1901 with Langley and Alexander Graham Bell at Bell's home in Nova Scotia. Bell strongly believed in airplanes, as opposed to balloons, and remarked to Langley that piloted airplanes would someday be able to fly off with cargoes as heavy as a thousand bricks. Sullivan's reaction was to remind himself that Bell had invented the telephone, and that feat alone entitled him to considerable latitude in predicting the future. Also, Bell was an old man—he had a right to ruminate any way he pleased. But to himself, Sullivan was thinking, "He is talking plain nonsense."

Sullivan was not alone. During 1901, leading popular periodicals like *McClure's* and the *North American Review* still published articles that discussed flying machines in the most condescending terms. One contribution in *McClure's* was written by Simon Newcomb, an internationally known astronomer and mathematician at the Johns Hopkins University. "Man's desire to fly like a bird is inborn in our race," he wrote, "and we can no more be expected to abandon the idea than the ancient mathematician could have been expected to give up the problems of squaring the circle." Other publications were just as negative on the subject, using the flying machine as a vehicle for stock humor in the same genre as "mother-in-law" jokes and the inane yarns that began with the tip-off phrase, "There were two Irishmen. . ." People who attempted to build flying machines fell into the same category as the notorious cranks trying to make perpetual motion machines, tunnel through the earth to China, or perpetuate similar fantasies. As one humorist jibed in the magazine *Puck* (in 1904, nearly a year after the Wrights' success), hopeful airplane inventors would get into the air as soon as the law of gravity could be repealed. It was in this skeptical atmosphere that Langley attempted to fly his full-size airplane, the *Aerodrome*.

The *Washington Post,* which had derisively dubbed Langley's flying machine the *Buzzard,* gleefully reported Langley's lack of success, and other published accounts jeered at the "Smithsonian staff, the stuffers of birds and rabbits. . . ." Langley's soggy fiasco on the Potomac elicited an outpouring of smug criticism from newspapers like the *Chicago Tribune,* which had known all along that God intended mortal men to remain grounded until the roll was called in Heaven. The *Boston Herald* recommended that Langley redirect his efforts toward submarine development, since his handiwork seemed to have more affinity for water than air. The diatribe against the *Aerodrome* poured in from high and low, including a devastating comment from the mordant Ambrose Bierce: "I don't know how much larger Professor Langley's machine is than his flying model was—about large enough, I think, to require an atmosphere a little denser than the intelligence of one scientist and not quite so dense as that of two."

In short, Langley's unhappy efforts to achieve powered flight in an airplane hardened the convictions of most Americans that the notion was totally impractical. If Langley, with his academic background, scientific experience, and prestigious backing, could not do it, then who could? Crazy scientists who wanted to toy with the unremitting law of gravity had been given their comeuppance. Common sense would show up a college degree and other fancy credentials any day of the week. Press and public alike remained skeptical of reports that two bicycle mechanics from Dayton, Ohio, had succeeded where obviously better technicians from the Smithsonian had failed.

For the next five years, the Wrights' Promethean achievement remained virtually unknown, and several contemporary accounts that surfaced were badly garbled. The Wrights themselves had personally issued a statement to the Associated Press in January 1904, attempting to clarify some of the gross misinformation about their plane and the Kitty Hawk flight, but the gesture had little effect. Accurate stories by acquaintances and first-hand observers were, surprisingly, ignored. In remarks to a meeting of the American Association for the Advancement of Science in early 1904, Octave Chanute had reported the Wrights' success at Kitty Hawk and used this address as a basis for an article appearing in the March 1904 issue of *Popular Science Monthly.*

In September, Amos I. Root, publisher of a well-known apiarian journal in Medina, Ohio, came to Dayton out of curiosity, saw the Wrights fly (a performance that included the first complete circle flown by an airplane), and recorded his impressions. Thus, history's first published, eyewitness account of a flight appeared in 1905—and was presented in a periodical with a most unlikely title, *Gleanings in Bee Culture*.

Continuing skepticism and the Langley debacle apparently discouraged nearly everyone from seriously considering persistent rumors emanating from Kitty Hawk or Huffman Prairie, the field near Dayton where the two brothers regularly flew within sight of a well-traveled trolley line. Irregular news stories during the next few years, often highly inaccurate, were either spurned by editors or cautiously reported with elaborate disclaimers. Years later, Orville commented on this puzzling reluctance of the American public to believe in airplanes: "I think it was mainly due to the fact that human flight was generally looked upon as an impossibility, and that scarcely anyone believed in it until he actually saw it with his own eyes." Although the Wrights invited reporters to Huffman Prairie on at least two occasions, mechanical problems and poor weather prohibited those flights, and the newsmen left with their skepticism vindicated. Even the *Dayton Daily News* remained blasé. In any case, the Wrights themselves remained reticent about publicity until they felt their patent rights were secure and they had perfected certain working details.

Still, a small but influential number of Americans and some journalists believed that the Wrights had flown. By late 1906, *Scientific American* had editorially acknowledged the Wrights' success, and the brothers were beginning to assess the potential uses of their airplane. As Wilbur confided to Chanute late in 1904, "It is a question whether we are not ready to begin considering what we will do with our baby now that we have it."

With innate thoroughness, the Wrights projected a possible sequence of acceptance: Planes would first be useful for military reconnaissance, then for exploration, then for speedy transportation of passengers and freight (including mail), and finally for sport. It appears that the Wrights hoped that airplanes might actually prevent wars by creating an awesome military deterrent. Given the state of the art in 1904, however, aerial surveillance represented the earliest military potential, although commercial applications seemed promising. "It is therefore our intention to furnish machines for military use first, before entering the commercial field," Orville announced, "but we reserve the right to exploit our invention in any manner we think proper." As the Wrights expected, military officials finally made the first move toward adopting the airplane. After considerable delay and some uneasiness, as the Wrights negotiated with Great Britain, France, and Germany, the U.S. War Department concluded a contract for $25,000 on February 8, 1908, and, following successful flight demonstrations in the summer of 1909, became the owner of one flying machine.

The years 1908–9 brought wide publicity and belated acclaim. Orville's tests for the War Department at Ft. Myer, Virginia, and Wilbur's flights in Europe before enthralled crowds, including the kings of Spain and England, became convincing front-page news. Meanwhile, the flights of other pioneers, like Glenn Curtiss, stirred additional interest in aviation. Curtiss, born in Hammondsport, New York, in 1878,

Curious onlookers watch military personnel wrestle with the Wright Military Flyer during U.S. Army trials at Ft. Myer in 1908. Courtesy of U.S. Air Force.

was the first American after the Wright brothers to build and fly an airplane. Prior to his interest in aviation, Curtiss became quite successful at building engines for bicycles and set several speed records while racing motorcycles of his own design and manufacture. In 1904, he won a contract to build a power plant for the U.S. Army's first dirigible, thus beginning an important aeronautical career. Curtiss soon became interested in airplanes, and in 1907 formed the Aerial Experiment Association with Alexander Graham Bell and others. On June 20, 1908, the organization's biplane, *June Bug,* with Curtiss at the controls, made a flight of 1,266 feet. About two weeks later, on July 4, 1908, Curtiss won the *Scientific American* trophy and generated news headlines by making the first official public flight of more than one mile. With increasing publicity and public acceptance, it seemed the time had come to look forward to practical use. "I firmly believe in the future of the aeroplane for commerce, to carry mail, to carry passengers, perhaps express," Orville Wright declared. "I cannot but believe that we stand at the beginning of a new era, the Age of Flight, and that the beginnings of today will be mightily overshadowed by the complete successes of tomorrow."

But the age of powered flight was immature. The general public viewed flying as too much of a novelty to be taken seriously. When a Texas congressman introduced a bill to consider airmail operations in 1910, the *New York Telegraph* found the idea ludicrous.

"Love letters will be carried in a rose-pink aeroplane," the paper scoffed, "steered with Cupid's wings and operated by perfumed gasoline." Such public cynicism stemmed from the flamboyant character of aviation at the time. As sensational mechanical contraptions, "aeroplanes" received their first wide public exposure as star attractions in aerial exhibitions that featured various flying stunts. The attendant ballyhoo and publicity propagated the notion that airplanes were hazardous and their pilots were daredevils and fools. People were willing to pay to observe the marvel of flight and to be on hand for the deliciously chilling prospect of an airplane smash-up. Professional impresarios like Barnum and Bailey knew a box office attraction when they saw one. As early as 1907, they had contacted the Wright brothers about the possibilities of flying exhibitions, a natural act for "The Greatest Show on Earth." Although the Wrights were more inclined to deal with the U.S. Army than the circus, military purchases were not numerous and the civilian market was necessarily limited. At $5,000 to $7,500 per plane, a flying machine remained a gadget for the rich. The negligible payload continued to discourage commercial operations, which in any case would have foundered for lack of permanent landing fields, hangars, and maintenance facilities. The scarcity of alternate sources of revenue prompted the Wright brothers and Curtiss to organize their own exhibition companies and retain some of the daring fraternity of "birdmen" to fly in them.

THE BIRDMAN ERA

For their air show manager, the Wrights hired Roy Knabenshue, by now a veteran of the aerial show business of ballooning, and began to train additional pilots. During 1910, Orville Wright organized flying schools at Montgomery, Alabama, (for the winter months) and at Huffman Prairie. Early fliers trained at these locations included Walter Brookins, Arch Hoxsey, Ralph Johnstone, Phil O. Parmalee, Frank T. Coffyn, and others, among them future military fliers like Henry H. ("Hap") Arnold. There was nothing glamorous about such early flight schools. The training field at Huffman Prairie was still little more than a cow pasture. The field was marshy, and more than one errant trainee had to have his plane hauled out of the mud after trying to get airborne. A wooden shed at one end of the field served as a hangar, and at the other end stood a large thorn tree, a menacing sentinel. Under Knabenshue's direction, the Wright Exhibition Company and its newly minted pilots hit the road in 1910, playing at carnivals, circuses, county fairs, and anything else that promised a crowd and reasonable gate receipts. For the privilege of billing their flying show at such gatherings, the Wrights charged $5,000 for each plane used. Fliers got a base pay of $20 per week plus $50 per day when they flew. At these prices, pilots like Frank Coffyn earned $6,000 to $7,000 per season, and the Wright Exhibition Company annually grossed about $1 million. Aviation had become profitable.

Additional aerial exhibition teams, like the one organized by Curtiss, soon made their appearance, and specially organized flying meets became major spectator events attracting thousands of people in a single day. Most of the crowd came to be convinced.

Surrounded by a web of wires and struts, J.A.D. McCurdy, one of the well-known fliers in the fraternity of "birdmen" prior to World War I, poses in the Silver Dart, *a Curtiss biplane. Courtesy of the Smithsonian Institution.*

As Beckwith Havens, a Curtiss pilot recalled, "They thought you were a fake, you see. There wasn't anybody there who believed [an airplane] would really fly. In fact, they'd give odds. But when you flew, oh my, they'd carry you off the field. . . ." During 1910, the sky seemed to blossom with pilots and airplanes, not only at county fairs but at major flying meets across the country. One of the most publicized affairs was the worldwide meet at Belmont Park in October. Participants included the top fliers from Great Britain, France, and the United States. Perhaps the salient feature of the meet was its stature as a top society sporting event; the social elite in attendance included Harold McCormick of Chicago and Cornelius Vanderbilt, who went up for a spin with Orville Wright. The same year, ex-President Theodore Roosevelt ventured aloft as a passenger in a Wright plane piloted by Arch Hoxsey, who had to caution Roosevelt against waving too exuberantly at the crowd below. The prominent names associated with aerial events did much to increase the prestige and acceptance of aviation. After a meet at Los Angeles attracted over 30,000 people on a Sunday afternoon in 1911, Curtiss was moved to say, "I am convinced . . . that aviation is a standard and lasting thing."

One sure sign of aviation's growing popularity could be seen in its position as an

With Arch Hoxsey at the controls, Theodore Roosevelt made a memorable flight during an air show in 1910. Aerial jaunts with such well-known personalities helped to publicize aviation and dispel the fear that flying was risky. Courtesy of the U.S. Navy.

element of popular culture. Musically, the theme of flight dated back to the balloon era, although airplanes received increased attention as a vehicle for high-flying romance. The earliest aeronautical ditty, "Chanson sur le globe aerostatique," appeared in 1785, published with a cover depicting a Montgolfier balloon rising from the Tuileries in Paris. Many such early songs dealt with balloons and dirigibles, though airplanes made an increasingly strong appearance with tunes like "Come Josephine in My Flying Machine," a smash hit of 1910. Similarly, aeronautical themes became standard fare in juvenile literature. Children's books on kites and other aviation topics date from 1744, and 1900 ushered in the century with a wave of dime novels as well. Some authors, like one Capt. Wilbur Lawton, wrote at least a half-dozen titles in the Boy Aviators series (New York: Hurst & Company), including *The Boy Aviators' Treasure Quest; or, The Golden Galleon,* published in 1910. In various books, Captain Lawton's two young heroes, Frank and Harry Chester, "the Boy Aviators," ranged the world from the Sargasso Sea, to Africa, to the Antarctic, flying their redoubtable biplane, *The Golden Eagle,* "strong of wing and sound of engine."

For youngsters with a knack for mechanical devices, clubs for flying model airplanes appeared as early as 1911. "The amount of interest that is aroused is impressive to the man who is unfamiliar with the spread of the model-making movement among boys," reported *Scientific American.* Enthusiasts could buy a model kit in the stores for $2 to $5, or utilize the detailed plans and instructions in *The Second Boy's Book of Model Airplanes,* published by the Century Book Company. On the west coast, Los Angeles was experiencing the growing pains of the young movie industry. Fiction and short-story writers had already begun to exploit the romance and thrill of aviation, and almost as quickly Hollywood producers imagined the colossal emotions to be generated by bringing the airplane onto the screen. Nordisk Films presented *The Airship Fugitives* in 1913, the same year that Hermann Hemmelhaack wrote *The Air Pilot* for Globe Feature Films. Crash scenes naturally packed in a lot of suspense, and pilots who flew the action took every precaution to avoid becoming a part of the tragedy in the plot. These aerial stunt men often wore regulation football pads covered by heavy canvas coveralls over four suits of heavy silk underwear to minimize damage from wooden splinters. Typical headgear featured a motorcycle helmet armored on the outside by chunks of cork and heavily padded on the inside with inch-thick felt. The last item was a twenty-foot coil of quarter-inch rope. The pilot wound it round and round his neck and over the helmet, just loose enough to prevent choking, but tight enough to keep neck vertebra from being snapped when he crashed in front of the cameras.

The flying machine phenomenon developed several leading fliers, but few so well known as Lincoln Beachey. Beachey, often called "the Flying Fool," gave up ballooning to be trained as a pilot by Curtiss and fly for his exhibition company. In a Curtiss biplane, Beachey made one of the most sensational flights of 1911, when the audacious aeronaut careened through the mists of Niagara Falls, swooped down to within 30 feet of the thundering river below, sailed under the suspension bridge, and proceeded down the deep gorge. Anyone could perform "loops and flip-flops''; Beachey performed with more consummate skill in his 80-HP "flying kite" and specialized in flying just close enough to things to cheat death by a whisker. As he buzzed around race tracks he casually flicked up dust with the plane's wing tips on the turns. Miscalculation by an inch or two would have scattered Beachey and his plane all over the homestretch. The famous auto racer Barney Oldfield teamed with Beachey in another hair-raising stunt. According to one source, with plane and car speeding along at 60 MPH, Beachey would settle his ship just over the speeding racer and carefully plant a front wheel on Oldfield's head.

Having conquered Niagara's rocky gorges, Beachey challenged the concrete walls of Chicago's Michigan Boulevard, roaring down the street just above the heads of astonished pedestrians. At other times, Beachey felt that those watching were obligated to pay him for the privilege. At Ascot Park in Los Angeles, he noticed a group of citizens who had clustered in a tree to beat the admission charge. He banked around them with his usual alacrity, just shaving the branches. A reproving press reported that the group of harried spectators sustained three broken arms and a skull fracture while making a precipitous escape. At Hammondsport, New York, in 1913, Beachey flew too close to a hangar roof and killed one spectator while injuring three.

Such aerial tantrums only discredited aviation, yet Beachey remained a great hero. The *New York Times* quoted him as an authority on aeronautics, and, up to his retirement about 1914, he was paid as much as $1,000 a day to fly. In 1915, he came out of retirement to perform for the San Francisco Panama-Pacific Exposition for $1,500 a day. He planned to fly a special plane he had built himself, which was intended to climb vertically. While he was testing it, the wings came off in a loop. A professional to the end, Beachey intuitively shut off the engine and closed the petcock on the fuel line before he fatally crashed.

As aviation caught the public fancy, and a few companies actually produced some planes, a controversy arose over the best method of instruction for the adventurous new owners. The Wrights used a plane equipped with dual controls, whereas the technique advocated by Curtiss was highly individualistic. Under the Curtiss system, the plane's engine was throttled down and the pitch of the propeller altered to prevent its pulling the plane off the field. After ground instruction on the principles of aerodynamics and engine mechanics, the student progressed to the "grass-cutting" phase. The trainee clambered into the plane and practiced by careening over the cow pasture used as a flying field. In this way, the pilot got the "feel" of the plane—the rudiments of piloting by use of ailerons and rudder. Advanced instruction meant an increased throttle and propeller pitch, with the student making irregular hops into the air. Thus, the pilot worked into the "solo" stage. However, at San Diego, Curtiss often utilized a two-place hydroplane to give his student passengers the experience of full-fledged flight at a top rate of 50 MPH. "This speed gave men an opportunity to feel the sensation of fast and high flying," Curtiss explained, "and sometimes shakes the nerves of the amateur."

The pioneer planes were extremely frail and occasionally capricious in the air. Early ships were hard to keep on an even keel, and the pilot often had to work fast to keep them headed into the wind, like a boat, in order to keep from "capsizing" when hit broadside by a heavy gust of air. As for controls, the early Curtiss biplanes were equipped with a wheel—a push or pull controlled the elevator surfaces and turning it activated the rudder. Ailerons were controlled by the movement of the pilot's body. A yoke fit over the shoulders and, in order to balance the plane, the pilot had to shift the torso left or right to move the aileron controls attached to the yoke. There were two foot pedals: one operated the throttle, the other controlled the front-wheel brake.

Flying apparel varied. In one photograph of Arch Hoxsey, at the controls of his plane just after another record-breaking flight, he is elegantly knickered and sports a checkered cap. A stickpin secures his cravat, and his customary pince-nez are firmly in place. Being a realist, he has strapped a spare can of Havoline motor oil securely into the seat next to him. Later aeronauts appeared extensively outfitted in special flying suits, goggles, helmets, gloves, and other paraphernalia. But Orville Wright, who flew regularly until 1918, invariably took off still wearing his dark business suit and starched white collar. Occasionally, he reversed his cap and donned a pair of goggles; on cold days, he turned up his coat collar.

Those first aircraft had few instruments. The aviator often had little more than a long rag tied to the elevator, which jutted out in front of him. If the rag did not keep blowing straight back on turns, the pilot knew he was slipping too much. Other details of engineering and safety often left much to be desired. Perhaps drawing on their bicycle supplies, the Wright brothers designed their biplanes to fly with two chain-driven propellers. The chain sometimes broke, snapping guy wires and shredding wing fabric. Unfortunately, their wing-warping system sometimes flexed the already strained airframe too far, with regrettable results. There was no provision for parachutes and often no safety belt. John Moisant was killed in New Orleans when his plane was caught in a gust of air. The plane tipped at a critical angle; Moisant tumbled out and fell to his death. In 1914, an editorial in the *New York Times* commented on the similar fate of a young army aviator, Lieutenant Post, who fell from his plane when it overturned in midair. His plane "lacked the attachments necessary to make him secure," explained the *Times,* earnestly commenting further that "the unfortunate young officer's style of flight was out of date, since his life depended on keeping his machine on an approximately even keel." Even if a pilot had the chance to parachute from a stricken ship, the profusion of guy wires and spars made a clean escape from a falling plane difficult at best. Uncontrolled emergency landings had the added hazard of the heavy airplane engine and propeller situated just above and behind the pilot, who had every chance of being crushed to death in a crash.

Male and female fliers shared these dangers, as a number of women also took to the air. By 1910, two women had soloed in France. Harriet Quimby, a writer for *Leslie's Weekly,* in 1911 became the first American aviatrix and won international acclaim in 1912 as the first woman to pilot a plane across the English Channel. Particularly chic, Quimby soon had to share headlines with numerous other women fliers, like Katherine Stinson, who became a regular on the airshow circuit in 1913 and captivated crowds with loops, inverted flying, and other dazzling aerial antics. Ruth Law, another contemporary, not only performed as an exhibition flier but set a long-distance solo flight record of 512 miles in 1916. The success of female aviators brought editorial attention across the country, including minor controversies over nomenclature. In 1916, the *New York Times* dutifully directed a "glance at the difficulties to be overcome in establishing a permanent glossary of terms relating to rapidly developing science." One of the main concerns stemmed from the fact that not all pilots were men any more. Was it better to call the female flier an *airwoman* or an *aviatrix,* or was it better not to discriminate and lump all fliers under one title, *aeronaut*? The *Times* finally reported that the neuter term *aviator* was acceptable for both sexes, although *aviatrix* remained a popular term.

For men and women alike the lure of flight was potent, and increasing numbers of passengers as well as pilots took to the air. Their recollections are vivid vignettes of a unique new experience. As one anonymous writer remembered his first air trip:

> The worst part of such a journey for the novice is the waiting until everything is ready for the start. The sensation of anticipation is not unlike the feeling that one has when one is waiting for a wounded bear to break cover from the corner into which he is driven. But once the propellor starts to whirl behind you all other thoughts beyond rapid exhilaration of motion vanish.

The author even relaxed the death hold he had taken on the struts and, after the four-minute spin, it seemed "the most delightful ride that I had ever experienced. The only recollection that I have that will describe the general sensation is that of exquisite motion."

One day in 1913, Ida Tarbell told an interviewer of her faith in aviation, though admitting she had never flown. Hearing of the interview, the editor of *Flying* magazine,

Henry Woodhouse, arranged for a flight in a hydroplane. Tarbell, a worldly, muck-raking journalist, was overwhelmed by the experience. The pilot taxied into the harbor and took off. "I did not know when we came out [of the water], and looked over the side to see if I was right. The surprise of it seemed to stun me. Not that I lost consciousness, but I was literally lost in amazement at the suddenness of it," remembered Tarbell. Finally, the pilot brought the plane in for a smooth water landing and taxied up to the landing ramp to allow his passenger to disembark. But she remained seated, transfixed by her first flying experience. "I do not know how long I sat in the boat not realizing that I must get out, so overwhelmed was I at the wonder of the thing . . . so supremely superior to any other emotion that I had ever experienced."

Even observers of the new flying machines overhead found the sight to be a stunning experience. When Walter Brookins flew a Wright plane from Chicago to Springfield in 1910, a writer for the *Chicago Record Herald* reported that the plane drew out great crowds at every town along the way, as train whistles and fire sirens maintained a continuous exultant din along the entire 190-mile route. In baroque prose that captured the excitement of an era, he wrote:

> The sky-gazers looked on in astonishment as the great artificial bird bore down the heavens. . . . Wonderment, surprise, absorption were written on every visage . . . a machine of travel that combined the speed of the locomotive with the comfort of the automobile, and in addition, sped through an element until now navigated only by the feathered kind. It was, in truth, the poetry of motion, and its appeal to the imagination was evident in every upturned face.

There can be little doubt that many Americans were indeed awed by this new flying contraption and were beginning to ponder where it might lead. In a quaint rhyme of 1910, the poet wondered about the airplane's use in war, but eventually concluded on a positive note:

Brilliant, dashing, winged thing
    Moving there across the sky,
What new message do you bring
    Unto mankind as you fly?

Swift you cleave the vibrant air,
    Now you fly and now you float,
Life itself you seem to share
    Are you bird or are you boat?

What new era do you bring
    Speaking to us through the years?
Hark: Your motor seems to sing
    With the music of the spheres:

Shall mad hosts who go to war
    Look to you for deadly skill?
Will you only sing and soar
    So that men may maim and kill?

Rather may you skim the seas
   And go whirring near and far,
Far to younder Pleiades,
   Visit moon and evening star.

Waft young lovers through the air,
   Fly them straight to Heaven's own door;
Ride on sunbeams bright and fair
   Chase you cloudlets at your fore.

Go where gods in laughter sit,
   Take us where life is but kind,
Seek where elves and fairies flit
   Some new Eden for us to find.

Tiny airship, light and strong,
   Lifting upward to the sky
Like a joyous, rising song,
   You shall teach our souls to fly!

As romantic as the flying machine might have seemed to poets and enthusiastic crowds, the airplane was still a tricky thing to fly. The early products of the Wrights and others were controllable but unstable, and pilots preferred to go aloft early in the morning or late in the afternoon when the air was at its calmest. Even then, they carefully watched flags and smoke plumes and anxiously stuck wet fingers above their heads to test the breeze. This situation created problems. When a county fair listed an airplane flight at 2:30 P.M., the crowd insisted on seeing one, and if a pilot demurred, he was considered a fraud. Stalling for time one breezy day, Beckwith Havens meandered around the flying field—looking, he said, for dangerous gopher holes. The local sheriff suddenly pulled up beside him in a horse-drawn buckboard, ordered him to get in, and unceremoniously hauled him back to the grandstand to start flying as specified. Such pressures led to the deaths of many fliers. Frank Coffyn was one of four men who signed two-year flying contracts with the Wright Exhibition Company, and he was the only one who lived to fulfill it. Pilots read daily newspapers with dread, since nearly every week brought news of another friend's death in a flying accident.

The fatality rate was compounded by pressure to fly when conditions were marginal, ill-conceived stunts and maneuvers, and sheer carelessness. Ralph Johnstone, for example, was an ex-vaudeville performer who joined the Wright Exhibition Company as a pilot. He and many like him flew for thrills and money, and they literally flew until their airplane's wings dropped off. After Johnstone had superficially repaired a damaged wing, it collapsed in midair, and he plunged to his death in full view of a large crowd at Denver in 1910. These occurrences daunted neither the fliers nor the growing throngs, since the lurking element of disaster morbidly attracted more spectators. Hap Arnold, who visited the 1910 Belmont meet as a young lieutenant, recalled the ghoulish expectations of many onlookers: "The crowd . . . gaped at the wonders, the exhibits of planes from home and abroad, secure in the knowledge that nowhere on earth, between now and suppertime, was there such a good chance of seeing somebody break his neck.''

For many Americans, however, the early romantic view of aviation eventually began to turn sour, expressed in the following doggerel of the pre–World War I era:

There was an old woman who lived in a hangar
  She had many children who raised such a clangor
That some she gave poison, and some aeroplanes
  And all of them died with terrible pains.

Spectators began to tire of steep turns, spiral dives, and other gyrations. An aerial circus had been the best way to attract crowds and draw a paid attendance, but dwindling gates caused promoters to look for new means to turn a profit at a time when pressure was building for alternative ways to demonstrate airplanes. In several articles, *Scientific American* deplored the deaths of stunt fliers and asked for more competition such as long-distance flights in order to give the public a chance to "contrast the aeroplane as a practical means of transport across the country with the aeroplane as a provider of sensational amusement." This suggestion became a trend, and Calbraith P. Rodgers, who trained at the Wrights' flying school, completed the first transcontinental flight in 1911. Rodgers's plane, a modified Wright B type, was named the *Vin Fizz*, after the product of an enterprising soft drink company that helped sponsor the effort. Rodgers's main interest was the Hearst newspaper prize of $50,000 for the first coast-to-coast flight within 30 days. Heading west to California, he took off on September 17, accompanied by a special train carrying spare parts. Traveling as wind, weather, and daylight allowed, and after recuperating from injuries sustained in an intervening crack-up, Rodgers took 49 days of flying and nearly three months elapsed time to complete the trip to the coast. In spite of the time and difficulty involved, it *was* the first time anyone had flown coast to coast, at least proving that it could be done. Encouraged, *Scientific American* predicted regular transcontinental passenger and air-mail flights.

The odyssey of the *Vin Fizz* prompted similar promotional ventures, frequently disappointing and sometimes fatal. Rodgers himself died in a crash within four months of completing his transcontinental flight. One of the persistent criticisms of aeronautics voiced in the prewar era concerned the fatality rate, not only of the inexperienced novice lured by cash awards for cross-country records but also of the early birdmen and exhibition flyers. Many observers, while acknowledging some usefulness of stunting, remained concerned. Curtiss, who felt that many pilots had advanced aviation by maneuvers demonstrating the capabilities of the airplane, nevertheless castigated the "fancy flight" and "spectacular gyrations" of some flyers as unnecessary and foolhardy. Wilbur Wright said that "legitimate fancy flying" could be safe, for it gave valuable experience in handling aircraft and becoming accustomed to flying, but he was strongly opposed to extreme fancy flying merely for thrills.

Journalistic criticism continued to mount. In 1911, the *Boston Transcript* declared that "circus stunts . . . [must] be separated in the public mind from aviation proper, either as a sport or a business." During the same year, the widely read magazine *Current Literature* reprinted a cartoon by John T. McCutcheon from the *Chicago Tribune* titled "The Sirens of the Sky." It depicted an aviator being tempted by sky-sirens offering "prize money," "glory," "fame," and "applause," with Death lurking just over the pilot's shoulder. An accompanying cartoon drawn by William Allen Rogers for the *New York Herald*, featured the Grim Reaper, with scythe and hourglass, wearing a sandwich sign emblazoned, "Sensational Acrobatic Aerial Performances for the Amusement of the Public. All the Most Dangerous Feats of the Air Men." Rogers subtitled his cartoon "The Kind of Show that Kills the Sport." In 1911, after two seasons of aerial show business, the Wrights retired from flying exhibitions with distaste, although Curtiss and other troupes continued to play county fairs and other local festivals across the country.

COMMERCIAL TRENDS

During this period, the airplane was first used in a practical demonstration as a freight carrier; but compared to the excitement of assorted aerial stunts and endurance flights, the significance of transporting manufactured goods by air was almost lost. In November 1910, a department store in Columbus, Ohio, made arrangements with the Wrights to have a bolt of silk, to be cut up for souvenirs, flown up from Dayton by Phil Parmalee. In the aftermath of this first air express, the newspapers of Columbus paused to assess the significance of the event. As produced and presented by a professional airshowman like Roy Knabenshue, the flight was, not surprisingly, a drama of high adventure. An editorial in the *Columbus Journal* on the morning after still breathed the excitement of spectacle. "Our news columns will tell all about the flight," the paper proclaimed, "but the poetry of it, the romance of it, dwells in the minds like a dream, which cannot be expressed."

Yet, stirred in with awe of the episode was a dash of realism. It was recognized that Parmalee's trip was more than just a stunt, that the transfer of silk by air had added importance beyond the value of the pieces as mere souvenirs of an exciting flight. The fact remained that not only had an airplane flown from Dayton to Columbus but an airplane carrying a man and a cargo had bested the crack "Big Four" railroad express between the two cities. "The idea of a man flying up here from Dayton—where is your railroad train, your traction, your automobile now?" the *Journal* demanded. "Relegated to bygone days, along with the stage coach and the canal boat." Breaking away from a rhapsodic description of Parmalee's arrival, one reporter paused to give a nod to "the marvels of aerial navigation and aerial freight transportation" before allowing his fancy to take flight again. As the more pragmatic editors of *Scientific American* observed, the additional use of a car to whisk the silk from the landing field to the sales counter constituted a "striking demonstration of the conjunctive use of the aeroplane and the automobile in the delivery of mail and express."

Additional efforts to put the airplane to practical uses followed. There were experimental air-mail flights in 1911, and the first steps toward commercial service began soon after. Some confusion still exists as to who started the first passenger airline. One of the first was a service conducted about 1913 by Silas Christofferson, who operated a hydroplane in San Francisco Bay and offered scheduled trips between

As President Wilson and other dignitaries looked on, a Curtiss JN-4 inaugurated air mail service from the Polo Grounds, Washington, D.C., in May 1918. Bound for Philadelphia, the unfortunate pilot flew in the wrong direction before crashing. Courtesy of the Smithsonian Institution.

the harbors of San Francisco and Oakland. The operation that remains better known involved the St. Petersburg–Tampa Airboat line in Florida. Notification of its inauguration appeared on the front page of the *New York Times* in 1914, which simply announced, "Three hydro-airplanes will be put in operation by a private corporation under the direction of Tony Jannus, the air man."

The actual organizer was P. E. Fansler, a speedboat enthusiast who was attracted by press notices concerning exhibition flights of a Benoist flying boat piloted by Tony Jannus. Fansler got in touch with the plane's builder in St. Louis, Tom Benoist, and their correspondence produced a plan to try a regular commercial line to prove or disprove the possibilities of commercial aviation. An enthusiastic Benoist agreed to furnish at least two planes, while Fansler selected the route and worked out the operating and financial details.

Since this was to be a flying boat line, a location near a body of water was a prime requisite. After looking around, Fansler settled on St. Petersburg, Florida, which not only had water but was also a bustling vacation metropolis of 30,000. Fansler rounded up a group of financial backers, but his first problem was to convince them that such a contraption as a flying boat really existed. None of the potential investors had ever seen one. They finally accepted Fansler's assurances and decided to gamble on the publicity it would draw for the Sunshine City. For the privilege of having a flying boat line serve their town, they promised the support of the board of trade, gave a $1,200 guarantee, and agreed to subsidize the line at $50 per day the first month of operation, and $25 per day thereafter for each day of four regularly scheduled flights. The only stipulation was that there were to be no flights on Sundays.

The likelihood of a brisk trade from carefree vacationers was obvious, but Fansler emphasized that the line's primary goal was to provide efficient passenger transportation. After getting off the train, vacationers bound for Tampa had to make a three-hour trip across the bay by steamer or drive a 49-mile detour by land. The airplane planned to make the trip a direct 19-mile air route, taking only 30 minutes. The airline's "fleet," which consisted of one plane at first, listed two round trips daily, carrying a total of two passengers each way. According to the advertisement in the St. Petersburg paper, passengers were charged $10 for a round trip and allowed a gross weight of 200 pounds, including personal baggage. The newspaper also reported that the line would begin operation on January 1, 1914. Besides the excitement generated, there was some ridicule—many sloughed it off as a hoax.

But the airline began flights as advertised, and never had a fatality as long as it operated, although there were one or two forced landings. The service became a good draw for the tourist trade, and the St. Petersburg papers made the most of it. Before long, the rival resort of Miami picked up an itinerant pilot and his plane, and a Miami paper headed the story with the boast, "Miami Has an Airplane Too." The editor of the *St. Petersburg Independent* retaliated by reprinting the item with his own heading, which challenged: "Ours Flies; Does Yours?"

The Benoist airplane actually did quite a bit of flying. During the previous summer, it had been used to give passenger-exhibition trips at Put-in-Bay, on Lake Erie, as well as flights at Grand Rapids, Michigan, and Keokuk, Iowa. Before it became an airliner, the same plane had also performed on the Mississippi, Illinois, and Ohio rivers. During the first month of operation at St. Petersburg in 1914, the plane made 97 trips across the bay, carried 184 passengers, and logged 2,234 miles of flying. All things considered, it was an enviable record. The flights had been intended to last only three months, during the height of the tourist season, but things worked out so well that operations were planned to extend throughout the year and two more planes were readied for service. Nevertheless, Fansler's efforts were not enough to maintain a sizable air transport company, and with the coming of the war, the venture dissolved.

As far as most American entrepreneurs were concerned, neither the means nor the demand to sustain a freight or passenger business existed before the war. The Wright and Curtiss companies were involved in litigation over patent rights concerning aileron control, and their dispute tended to retard aviation because of uncertainty over the outcome. There were no real advancements in construction or design. American engine progress lagged behind that of the Europeans, and aviation also felt the dampening effects of the recession in 1913. In spite of some setbacks, aviation began to move forward in a number of ways after 1914, although the Wrights and Curtiss did not end their aileron controversy until 1917, when litigation was discontinued by mutual consent in conjunction with a cross-licensing agreement arranged by the government.

Despite these fits and starts, a market of sorts began to develop for aviation. The first company formed for the avowed purpose of manufacturing airplanes was the Aerial Experiment Association, organized in 1907. Its founders included Curtiss, Alexander

Graham Bell, and Lt. Thomas Selfridge. In March of 1909, Curtiss and others capitalized at $360,000 to form the G. H. Curtiss Manufacturing Company, selling planes for $7,500. By June, they had engaged the firm of Wyckoff, Church, and Partridge, an auto sales company in New York City, to act as the sales outlet for Curtiss planes.

From about 1907 on, dozens of airplane companies sprang up, and even aspirants like the Swivel Buggy and Wagon Company announced their entry into the new field. Other names appearing were the Morrison Automatic Flying Machine Company, the National Aerial Navigation and Equipment Company, and the Ascension Manufacturing Company. In 1910, the Scientific Aeroplane and Airship Company notified the public that it was building planes and that one could be had for the price of $5,000. The company required a down payment of one-third (cash), asked three weeks for delivery, and guaranteed the plane to fly. Unfortunately, the big aviation boom failed to materialize, and by 1915, Scientific, Morrison, National, and Ascension had disappeared along with such hopeful entrants as the Blanchard Aerial Works of America, the Aerial Yacht Company, the Edwards Common-Sense Aeroplane Company, and Scheusselburg's Aeroplane Corporation.

During the same period, however, American investors began to take an interest in airplane manufacturing. The organization of Wright interests in Britain, France, and Germany caught the eye of financiers like Cornelius Vanderbilt, Howard Gould, and August Belmont. With their backing, the Wright Company was incorporated with a capitalization of $1,000,000 on November 22, 1909. The Wrights received $100,000 outright, 40 percent of the stock, and a 10 percent royalty on every plane sold. The Wright Company at Dayton developed large and extensive financial interests in aviation during these early years, and some aviation people muttered about the creation of an aeronautical monolith. Perhaps it is to the credit of the Wright brothers that they continued to resist the demands of New York bankers to create a tight monopoly, which competitors feared eventually might rival the American Telephone and Telegraph Company.

By 1911, there were about one dozen sound firms building planes and 50 engaged in producing aircraft parts and supplies according to *Scientific American*. Among the top builders were the Burgess Company, which had reportedly sold 30 planes; the Wrights 60 to 90; and Curtiss, "a considerable number." Prices averaged $5,000 to $7,500. *Scientific American* concluded, "The business of aviation is thus growing apace, and the possibilities that are offered in the flying field for the development of industrial enterprises can probably be as little conceived of at present, as could the future of railroading one hundred years ago." Even if the numbers were overly optimistic (which appears to have been the case), *Scientific American* was certainly correct in predicting as promising a future as that of the railroad.

Even though aviation manufacturing did not become a monopolistic enterprise, corporate changes and mergers were inevitable. Saddened by the death of his brother in 1912, Orville Wright sold the Dayton factory and patents two years later. The purchasing syndicate included such men as Thomas Chadbourne and Harvey D. Gibson, of the Manufacturers Trust Company, and Albert H. Wiggin, of the Chase

In a reflective mood, Wilbur (left) and Orville Wright pose on the back porch of their home in Dayton, Ohio, prior to Wilbur's death in 1912. Although their pioneering flights had been widely acclaimed by an enthusiastic public, the Smithsonian Institution continued to endorse Langley's Aerodrome as the first successful powered plane. The dispute finally prompted Orville to send the Flyer *to the Science Museum in London in 1928, where it remained for the next 20 years. Courtesy of the Smithsonian Institution.*

National Bank. In 1916, with the promise of war contracts, this group merged with the Glenn L. Martin Company. The same syndicate also controlled the Simplex Automobile Company, and Wright-Martin was apparently one of the first to adopt the production techniques of the auto industry. By 1917, prospective war contracts had led to a consolidation of most of the smaller companies, and the early plethora of airplane firms had slimmed down to seven major groups: the Aeromarine Plane and Motor Company, the Burgess Company, the Curtiss Aeroplane and Motor Corporation, the L. W. F. Engineering Company, the Standard Aircraft Corporation, the Sturtevant Aircraft Corporation, and Wright-Martin.

As early as 1911, many thoughtful figures in American aviation circles believed that the United States had begun to fall behind in comparison to European aeronautics. France, Russia, Germany, Italy, and Great Britain all had active aeronautical laboratories with varying degrees of government support, and many of their designs were beginning to show clear evidence of superiority over American aircraft. In the United States, the National Academy of Sciences pressed the government to form a national aeronautical laboratory, but decisive action did not follow. Among the strongest aviation partisans in the National Academy was Charles D. Walcott, secretary of the Smithsonian Institution. With the support of friendly regents of the Smithsonian, like Alexander Graham Bell, Walcott decided to revive Samuel Pierpont Langley's old aeronautical laboratory. As head of the new lab the Smithsonian named Dr. Albert F. Zahm, who had chosen to write his doctoral dissertation at the Johns Hopkins University in 1898 on the physics of flight. Zahm was teaching physics at Catholic University when, in 1913, the Smithsonian dispatched him on a fact-finding tour of Europe. He was joined by Dr. Jerome C. Hunsaker, a graduate of the Massachusetts Institute of Technology, who was developing a curriculum in aeronautical engineering at the institute.

Zahm's report, issued in 1914, emphasized the galling disparity between European progress and American inertia. A mixture of national pride and a sense of urgency spurred by military aviation abroad finally led to a rider attached to the Naval Appropriations Act signed by President Woodrow Wilson on March 3, 1915. The rider created an independent group, the Advisory Committee for Aeronautics, which quickly added the prefix "National," creating the soon-to-be-famous acronym NACA. The 12 members of NACA included two representatives from each air service of the army and navy; one member each from the Smithsonian, the Weather Bureau, and the Bureau of Standards; and five people from outside the government. All members served without pay. The advisory function remained paramount, since NACA had no research laboratories of its own until 1920. This helps to explain the committee's miniscule budget of $5,000 per annum for a span of five years. NACA performed a basic task in carrying out the letter of its charter, "to supervise and direct the scientific study of the problems of flight, with a view to their practical solution." Early reports covered a variety of eminently useful topics (aerodynamics in gusts, instruments, fabrics, engines, etc.) for both civil and military aviation.

The juxtaposition of both civil and military activities within NACA's work became significant, since gathering war clouds already cast a menacing shadow over the peaceful development of aeronautics. In Europe, intense national rivalries had already flared into open hostilities in the summer of 1914, and, for the first time, significant numbers of airplanes and pilots met in combat.

WAR IN THE AIR, 1914–1918

At the start of the war, Germany mustered about 230 airplanes and 4 dirigibles; Great Britain, 110 airplanes; France, 130 airplanes. The belligerent powers used their unarmed planes primarily for reconnaissance and artillery observation, and early military pilots on both sides frequently waved to each other as they passed en route to their battle stations. As warfare grew more intense, it seemed pointless to allow the enemy unrestricted access into the airspace above one's own military operations, and single-seat "scouts" took to the air to drive off intruders with pistol and rifle fire, and eventually machine guns. The sedate sporting plane of the prewar era became a deadly fighting machine, with opposing pilots locked in bitter duels thousands of feet above the trenches below. The glamor and exhilaration of the flight gave way to the grim realities of numbing cold, the psychological strain of combat fatigue, and the fear of death in a flaming dive—there were no adequate provisions for parachutes.

The struggle for air superiority seesawed throughout the war, as first one side and the the other introduced improved military aircraft to win control of the air. More efficient engines, higher speeds, improved design, and better maneuverability were constant goals of the rival air forces. High priority also went to increased fire power, especially to a technique that would allow the pilot to aim effectively and fire his guns through the plane's propeller arc. At first, the French tried deflector plates on the propeller blades, but the Germans had more success in mounting synchronized machine guns on their planes. France and England quickly adopted the German technique. New tactics for patrol, attack, and defense were required to coordinate the masses of planes that now fought in the air. By the end of the war, the British air force alone consisted of 290,000 personnel and 22,000 aircraft.

In terms of military aviation, the United States entered the war in April 1917 as an amateur. Although American armed services had conducted several intriguing trials involving combat operations before the war, little had been accomplished in the way of systematic evaluation, training, and deployment of aviation forces. In fact, many prewar "military" experiments occurred through the activities of civilian pilots. In 1910, Glenn Curtiss first experimented with projectiles dropped by an airplane, tossing dummy bombs toward the shape of a battleship outlined by buoys anchored in a lake. Eugene Ely, one of the Curtiss exhibition fliers, that same year made the first takeoff from a ship, using a platform built over the bows of the cruiser *Birmingham*. Early the next year, Ely elaborated on this feat by landing a Curtiss biplane on a special deck constructed aboard the cruiser *Pennsylvania* in San Francisco Bay. The plane used an arrester hook, positioned to snag ropes tied to sandbags lining both sides of the deck. Following his successful landing, Ely tarried an hour to have lunch with the captain and wardroom officers, then flew off again amid an exultant clamor of cheers and steam whistles. Not to be outdone, the army got its share of attention in 1911, when Lt. Myron Crissy hand-dropped some live bombs while Phil Parmalee flew a Wright biplane during trials near San Francisco. During 1912, at College Park, Maryland, Lt. Thomas Milling piloted a Wright biplane while another officer, Charles de Forest Chandler, carried a Lewis machine gun into the air for firing trials. Finally, on July 18, 1914, army officials established the Aviation Section within the Signal Corps. The fledgling air service possessed a "paper" strength of 6 airplanes, 60 officers, and 260 enlisted personnel.

The first American use of aircraft in actual military operations involved the navy, in operations against Mexico at Vera Cruz in April 1914. Latent Mexican-American antagonisms had finally prompted President Woodrow Wilson to authorize military action, and the navy hastily loaded five Curtiss flying boats aboard the battleship *Mississippi* and the cruiser *Birmingham*. After being lowered into the Gulf waters for takeoff, the planes scoured the harbor of Vera Cruz for mines (there were none) and flew periodic reconnaissance flights over the mainland. When crew members discovered a few bullet holes in the fabric of one plane, news-hungry correspondents on the scene gleefully filed lurid news copy about the first U.S. airplanes to be fired on in anger. The army's fledgling air service finally got some operational experience in 1916, when the First Aero Squadron, commanded by Capt. Benjamin D. Foulois, accompanied Gen. John Pershing on a controversial excursion into Mexico in pursuit of Pancho Villa. Remembered as a patriot in Mexico and as a bandit in the United States, Villa launched raids across the border into New Mexico during a chaotic period of the Mexican revolution that eventually led to the Mexican border campaign under Pershing. The motley collection of planes and pilots directed by Foulois endured all sorts of problems: an inadequate and often contaminated fuel supply; an arid desert climate that dried out and cracked the laminated wooden propellers; and a lack of organizational and operational technique. Aerial scouting occasionally proved useful to Pershing's troop movements, but the exercise clearly underscored the lack of experience and preparedness of American military aviation.

The first Americans to engage in significant plane-to-plane combat did so under the flag of a foreign, albeit friendly, nation. Hoping to avoid the disastrous European conflict that had begun in 1914, President Wilson enjoined his fellow Americans to remain neutral in both thought and deed. But many Americans thought Germany to be in the wrong and often found their way to the front by joining Canadian, British, or French military units. American pilots flew in various Allied squadrons, but the best known was the Lafayette Flying Corps, an inclusive term for some 200 Yank volunteers in the French air service. The corps also included the 38 American pilots who flew as a group in the famed Lafayette Escadrille. This unique squadron, named after the French nobleman who fought in the American Revolution, got its start in the spring of 1916. Organizational support and money came from Dr. Edmond L. Gros, an important official in the American volunteer ambulance service, and wealthy Americans in Paris, like William K. Vanderbilt. The French eventually supplied a clutch of Nieuport fighters, and the Lafayette Escadrille flew into romanticized immortality. Most of the Lafayette Flying Corps joined American units following America's declaration of war in April 1917. The Lafayette Flying Corps bequeathed an important legacy of experienced aerial gladiators to lead the dozens of neophyte American aviators who began arriving in France. Ensconced in a spacious villa near Bar-le-Duc, the young fliers frolicked with a pair of lion cubs named Whiskey and Soda, engaged in youthful hijinks, and thought it grand sport to fly against the Hun. Cloaked in romance and glamor in the beginning, fliers in both the Lafayette Escadrille and the Lafayette Flying Corps soon learned the grim realities of aerial warfare; they claimed 199 air victories, but lost 63 Americans in combat.

On April 6, 1917, when the United States declared war on Germany, the Aviation Section of the Signal Corps consisted of 55 airplanes (mostly trainers), 35 pilots, and

1,987 enlisted personnel. The navy and Marine Corps accounted for an additional 54 airplanes, 1 airship, and 3 balloons, along with 48 pilots and 239 personnel. An awesome gap loomed between European capability and American preparedness. By the time America entered the war and organized the Air Service, American Expeditionary Force, to handle air operations overseas, the air weapon had already reached a new plateau of military application in Europe. Balloons provided useful, if vulnerable, observation posts, Rigid dirigibles flown by the German air force made bombing raids on Paris, London, and other cities. Under the exigencies of fierce combat, highly specialized airplane types—bombers, fighters, and scouts—had evolved, along with entirely new dimensions of tactics and strategy. The air services of nearly all the belligerents now included long-range, multiengine bombing airplanes equipped with defensive machine guns to keep attacking pursuit planes at bay. Opposing fighters fought costly duels for control of the air over the combat zones. Bombing, strafing sorties against ground troops, as well as aerial reconnaissance and aerial photography, were accepted necessities of offensive and defensive military operations.

Early in 1915, British generals at Neuve-Chapelle derived immense benefit from systematic photo reconnaissance that yielded a detailed look at German positions ranging from 700 to 1,500 yards in depth along the entire line of opposing trenches. In another instance, following up on photos of the German-held rail center of Courtrai in occupied Belgium, an Allied bombing raid caused dozens of military casualties and snarled rail traffic for three days. Similar actions were repeated dozens of times along the eastern, western, Italian, and middle eastern fronts as the combatants took advantage of air power's growing military capability. By the summer of 1918, over 8,000 military aircraft were plying the skies above the western front alone, but the Americans learned quickly. When the Allies set out to reduce the German salient at St. Mihiel in September, the combined air force of 1,500 fighters and bombers—the largest air armada yet organized—came under the command of America's rambunctious Air Service officer, Col. William ("Billy") Mitchell. Mitchell deployed his air force in a vigorous and coordinated offense, sending 500 fighters and light bombers over the front lines, strafing and bombing, while two waves of 500 planes each slammed the German rear, destroying supplies, communications centers, and transportation routes and generally throwing the German war machine into disarray.

The First World War did more than develop nascent theories of air combat; it forced military planners to include aviation in their logistical concepts. For the United States, entering a conflict three years in progress, the process of catching up to Europe in aircraft production proved to be a rude awakening. America pledged a good deal, made a concerted effort to fulfill its commitment, and settled for a good deal less than promised. In 1914, the miniscule American aviation manufacturing industry counted 168 wage-earners, who netted a total of $135,000. They worked for 16 main aircraft producers, whose value of products totaled $790,000. This was woefully below the top 56 industries, whose gross value of products was $100 million or more in 1914. It would be impossible to give the total number of planes produced after the Wrights first flew at Kitty Hawk; in addition to the planes made by "established" manufacturers, hundreds of backyard builders were busy during the prewar years. Perhaps the most reliable tabulation appeared after the war, in volume 1 of the *Air Commerce Bulletin*, published by the U.S. Civil Aeronautics Authority in 1929. According to the authority, the first reliable government records were begun in 1909, when one airplane was delivered to the army, and military deliveries were tabulated from that time on. As for civilian production, the classification "airplane" did not appear until 1912, since aircraft in the previous years had been classified with "all other cars, carriages, and vehicles." Even so, all nonmilitary planes produced up to 1917 were curiously listed as exports—according to the figures, no planes were produced for commercial or civilian use in the United States. At any rate, the total "export" production in 1912 totaled 29, plus 12 military types. By 1914, the totals had reached 49 for both types and increased to 411 planes in 1916.

When the United States entered the war in 1917, the Allies naturally expected a boost from American personnel and production potential, and America confidently expected to meet the demands and still do more than its share. On May 24, 1917, the French government requested the United States to furnish 4,500 planes for active service in time for the spring campaign of 1918. To meet this request, and to meet the needs of the United States as well, the Joint Army and Navy Technical Aircraft Board called for 8,075 training planes and 12,400 service planes. The production plan for this

During World War I, U.S. Navy pilots controlled the Atlantic approaches to French ports in aircraft like the HS-1 flying boats in the background. The photo was taken in France in 1918. Courtesy of U.S. Navy.

total of 20,475 planes was to be met in 12 months. The accompanying engine estimate called for 41,810 power plants the first year and 6,159 per month thereafter. This from an "industry" that only the year before had produced all of 411 planes.

Congress appropriated $640 million to do the job, the largest single appropriation for a specified purpose ever legislated in the United States. An awed public expected to see vast fleets of aircraft crowd the skies. Now that the funds were available, attention turned to producing results. The situation seemed anything but heartening. In the spring of 1917, the Signal Corps had contracts for 334 airplanes of 32 different designs. The orders were placed with 16 firms, only a half-dozen of which had ever produced as many as 10 planes or more. None of the companies had ever produced anything more advanced than a training plane, and there were no more than 10 designers capable of that job. Some car builders had experimented with airplane engines, but the situation in the automobile industry was little better. As for instruments, most of the dials, gauges, and other accessories required for airplanes were completely unfamiliar to American manufacturers. A new industry capable of producing such equipment had to be brought into being and learn to fabricate sensitive instruments on an assembly-line basis.

A multitude of lesser problems remained to be solved. There was spruce to be cut for wing spars and yards of fabric to be obtained for covering airframes. There was the problem of lubrication. Ordinary motor oil froze at high altitudes, and the best substitute was castor oil, even if pilots occasionally became ill from the fumes. But the United States did not have enough castor beans to produce the thousands of gallons of oil that would be required. The frustrated Signal Corps then found that seeds to grow the needed quantity of bean plants did not even exist. After scouring the globe, the Signal Corps finally obtained a shipload from India. By the end of the war, the United States was cultivating 110,000 acres of castor bean plants, one of the less useful by-products of American mobilization.

During the hectic scramble to solve these and other difficulties, the aviation industry fell prey to widespread graft and corruption as companies and managers formulated mergers, interlocking directorates, and anything else to get a share of multimillion-dollar war contracts. On the other hand, as the embryonic aircraft industry struggled to meet the demands, some extraordinary engineering feats were accomplished. Production peaked in 1918, with deliveries of 14,020 airplanes, roughly 18 times that of the previous year. By 1919, 31 companies were producing aircraft and parts valued at $14 million. All things considered, an industry worth less than a million dollars only five years before had made remarkable progress.

The country could also take pride in the remarkably rapid training of fliers and ground crews. Across the nation, hasty projects built some training centers from scratch, while other training facilities were borrowed from colleges and universities. Out of 38,000 volunteers, 23,000 potential student pilots entered flight schools, and America trained 10,000 aviators and an equal number of aviation mechanics by the armistice. Wisely, a special army commission recommended that America obtain fighter aircraft overseas (mostly French-built Spads and Nieuports) and concentrate on building its own trainers (mostly Curtiss JN-4 biplanes) while carefully selecting new combat designs. To save time, the army further decided to build foreign airplanes under contract, the best known aircraft being a two-seat British reconnaissance bomber, the de Havilland DH-4. By war's end, about 4,500 of the type had been built, and 1,213 had been shipped to Europe.

But the most notable American production record involved the Liberty engine used by the hybridized DH-4. J. G. Vincent, of the Packard Motor Car Company, and E. J. Hall, of the Hall-Scott Motor Car Company, took over a hotel room in Washington for nearly a week and designed the 8- and 12-cylinder Liberty power plants with prevalent engineering and mass-production procedures in mind. On June 4, 1916, the Aircraft Production Board immediately authorized construction; final design, manufacture, and assembly of the first 8-cylinder version was completed in the astonishingly short span of less than six weeks. By the end of August, the 12-cylinder version had passed a 50-hour test and had been ordered into mass production. Operational versions delivered up to 400 HP. As war surplus, the Liberty-powered DH-4 planes, thousands of Liberty engines, and JN-4 trainers all played a useful role in the evolution of American aviation during the postwar era.

Still, most American aerial units overseas flew into combat in French Spads and Nieuports. Even with borrowed airplanes, the American air combat record, considering its brief time span, was impressive. The first aerial victories came on April 14, 1918, but American squadrons shot down an estimated 850 planes and balloons by the November armistice. Eddie Rickenbacker, the leading American ace, piloted a Spad through 26 victories and, during the last eight weeks of the war alone, recorded 20 downed enemy aircraft. By that time, American air strength had achieved an impressive presence. At the close of hostilities, the Army Air Service (the Aviation Section had been divorced from the Signal Corps and reorganized as a distinct entity in May 1918) could muster 195,024 personnel, including 20,568 officers. The Air Service had 3,538 airplanes in Europe and 4,865 based in the United States. Naval and marine aviation represented an additional 6,998 officers and 32,882 men, with a total of 2,107 aircraft, including 1,172 flying boats.

For the United States as well as for all of the belligerent powers in 1918 the scope and significance of their air arms represented an aspect of "total war" that developed during 1914–18. Not only were the entire resources of nation-states mobilized into a concerted military effort, but in a total war, even civilians far behind the front lines were subject to attack from death-dealing aircraft. Before World War I, many nations had relied on geographical distance or natural barriers to protect their armies, population, and industry. The advent of air power changed all that.

Not all of these ramifications were clear from early attempts to assess the impact of war—as well as civil and military aviation—on society. Largely as a result of World War I, aviation matured almost overnight. Despite all of the prewar development of aviation, many people, even informed individuals, viewed its future with a jaundiced eye. In 1915, Dr. Jerome Hunsaker, who taught one of the first courses in aeronautical engineering at the Massachusetts Institute of Technology, had received an inquiry from Glenn Martin, who wanted to employ an aeroengineer for his new airplane company. Donald Douglas, one of Hunsaker's recently graduated students, happened to be looking for a job. Douglas said he was interested, but Hunsaker thought he would be

Above the Texas prairie, a group of students practice formation flying in JN-4 trainers. Ellington Field, near Houston, was one of many new training centers set up to support the American military effort in World War I. Courtesy of the Houston Metropolitan Research Center, Houston Public Library.

better off in mechanical engineering. "This airplane business will never amount to very much," Hunsaker cautioned him.

But times were changing. By 1920, Donald Douglas headed his own airplane manufacturing company, and articles on aviation were appearing with his byline in prestigious national magazines like *Scientific American*. Writing on the subject of "The Airplane as a Commercial Possibility," Douglas cautioned that planes would cost more than cars, and this expense would inhibit rosy forecasts of planes replacing the family sedan. Nevertheless, planes would be useful in transportation, although public concern about flying safety had to be satisfied. In a broader context, some aeronautical partisans saw aviation as both a practical and a humanitarian force in the world.

Even before America had entered the war, observers had marked the developments and improvements of the airplane as aerial warfare was waged in Europe. In 1916, Henry Woodhouse, editor of *Flying* magazine, predicted a great postwar future for airlines and aviation in general. He felt that the plane would prove to be a real instrument of "progress and enlightenment," as opposed to a machine adding barbaric new terrors in warfare. On the eve of America's entry into the war in 1917, Orville

Wright still saw a benign future for aviation, predicting business trips, air cargo, and an air-mail service. Following the American declaration of hostilities, he endorsed bombing German munition production sites and advocated air supremacy to keep enemy air reconnaissance suppressed. But in November 1918, as the conflict neared its end, he wrote hopefully to a friend, "The Aeroplane has made war so terrible that I do not believe any country will again care to start a war."

Thoughtful figures in American aviation looked to the postwar decade with a mixture of caution, realism, and optimism. Many probably would have agreed with a *New York Times* review of an immensely popular aeronautical salon mounted in New York in March 1919. Visitors to the show, announced the *Times,* undoubtedly realized that aviation hovered "on the threshold of a new age whose developments the most imaginative can hardly imagine."

The Aviation Business, 1918–1930

Despite the slashed budgets and bureaucratic crosscurrents of the postwar decade, military aviation managed to make significant advances. Both services accumulated useful operational experience, refined their administrative structures, and learned more of the prospects and limitations of aircraft in warfare. And, in the case of the navy, the era marked the introduction and evolution of the aircraft carrier as a major innovation in fleet operations.

Beginning with the U.S. Air Mail Service, organized by the government in 1918, air mail became an important element of American business communications. When the government established contract mail routes with commercial carriers in 1925, the airline industry began to flourish, and eventually acquired large aircraft suitable for passenger transport. During the twenties, airplanes proved useful in dozens of ways: private business and corporate flying, cotton dusting, seeding, photography, and others. There were many institutional changes, including the evolution of professional aeronautical degrees, as well as a new framework of governmental regulation. Aviation became ingrained in American culture, and generated a new dimension of international relations.

THE ARMY'S WINGS

In the immediate aftermath of the Great War, American aviation strategies attempted to plumb the lessons of the conflict and devise a realistic air doctrine for the future. Although Gen. Billy Mitchell and a handful of others ardently advocated expansion of aviation's role as an independent, offensive weapon, the novelty of the airplane made bending generations of military tradition to its new role a difficult task. An early misinterpretation occurred when the General Staff decided, on the basis of fighter, observation, and bomber aircraft on hand at the time of the armistice, to make plans on existing proportions. The General Staff thus erred in stressing aviation's observation role in cooperation with the army, as opposed to a more independent role, when, in fact, England, Italy, and France had planned to increase their bomber strength by as much as 50 percent as a result of wartime experience. Bomber partisans had to bide their time until the Army Air Corps belatedly gave due attention to bombing doctrine in the late 1930s.

The Army Reorganization Act of 1920 fell far short of granting aviators the independence they wanted, leaving tactical squadrons under the control of ground commanders. The Air Service gained some momentum with the authorization for additional flight pay, along with control of its own training, procurement, research and development, and supply. Nevertheless, the Air Service entered the postwar decade poorly prepared. Within days of the armistice, the government had canceled orders for 13,000 aircraft and 20,000 engines; no funds were available for even a small number of new aircraft. By 1921, the Air Service numbered fewer than 3,000 aircraft, a poorly balanced force that included only a dozen twin-engine bombers, but some 1,500 JN-4 trainers. All of these planes, left over from wartime orders, became increasingly unsafe to fly. During one 12-month period, 330 crashes left nearly 100 airmen dead or seriously injured. With only 900 pilots and observers on active duty, such an attrition rate of planes and pilots was clearly unacceptable. By 1924, the Air Service numbered only 754 commissioned aircraft.

In the doldrums of the postwar era, the Air Service welcomed any diversion to put planes to some useful purpose and accumulate operational experience. The so-called Far West Flying Circus, commanded by Carl Spaatz, publicized postwar victory loan drives and later flew on forest fire patrol duties. Another group flew surveillance along the U.S.-Mexican border to spot illegal crossings and miscellaneous marauders. Such activities, while marginally useful, got little attention, although several long-distance junkets aroused considerable interest and prompted national headlines. Air Service planes and pilots made group flights in formation from coast to coast, navigated 9,000 miles around the rim of the continental United States, established new records for altitude (over 33,000 feet), and made endurance flights (over 37 hours) sustained by aerial refueling. In the spring of 1923, Lts. Oakley Kelly and John Macready completed the first nonstop, coast-to-coast jaunt, piloting a single-engine Fokker transport from New York to San Diego in slightly less than 27 hours. The following year, two single-engine Douglas planes hopscotched around the globe in 175 days.

Accomplishments like these often required a mixture of heroics, inspired improvisation, and a gritty determination to succeed. In the process of setting the record altitude flight of 33,000 feet in 1920, Capt. Rudolph ("Shorty," for his lanky, six-foot four-inch frame) Schroeder passed out due to oxygen starvation and carbon monoxide fumes in the open cockpit of his biplane. At about the same time, the intense cold froze the moisture around his eyes. With Schroeder unconscious, the plane plummeted earthward in a six-mile dive; Schroeder recovered at an altitude of 2,000 feet and somehow managed to land despite blurred vision and bleeding from lacerations caused by tiny shards of ice in his eyes. On their nonstop transcontinental flight, Kelly and Macready kept the plane's leaky radiator topped off with their own drinking water and cold soup. The first two planes to circumnavigate the world started out as a quartet of aircraft, but one crashed into an Alaskan mountain and a second (though equipped with floats) sank off the Faeroe Islands in the Norwegian Sea. Crews of the remaining pair of planes soldiered on in their open-cockpit aircraft, contending with capricious weather, hordes of insects, mechanical failures, and phalanxes of bureaucrats and well-wishers who swarmed around their many stops. Taken collectively, these activities exposed a new postwar generation of Air Service personnel to the intricacies of grueling flight operations on a daily basis and sharpened skills in both long-distance navigation and logistics.

In the meantime, the issues of independent air operations and bombing doctrine reached new levels of dissension. Conventional military thinking of the era placed sea power in the forefront of American national defense. The battleship still loomed as the principal element in naval warfare. But the promise of long-distance military flights led air power advocates like Billy Mitchell to the conclusion that air power had made battleships vulnerable. Undaunted by claims that modern battleships were invulnerable to air attack, the iconoclastic Mitchell finally won approval for a historic series of bombing trials, conducted off the Virginia Capes during the summer of 1921. Although various smaller vessels succumbed to air attack, many prominent naval figures believed that Mitchell's airplanes would never sink the formidable *Ostfriesland*, a captured German battleship. Even with no crew aboard to cope with fire and explosions, the vessel's thick armor and numerous watertight compartments were designed to make it virtually impregnable. Incredulous naval officers, choked with emotion, watched as a flight of seven Martin MB-2 bombers, each with a specially designed one-ton bomb, sank the *Ostfriesland* in 22 minutes.

Mitchell and his cohorts were elated, but the wrangling with higher authorities only intensified. Mitchell churned out a succession of speeches, articles, and books that elaborated his arguments for an expanded air arm with increased autonomy and reviled his critics in the army and navy. Following a series of bureaucratic skirmishes, the belligerent and flamboyant Mitchell was court-martialed for "insubordination and conduct unbecoming to an officer" in 1926. Mitchell resigned from the Air Service but continued to expound his views. For all the fireworks he touched off, Mitchell failed to win the independent air arm he wanted. That is not to say that others were unconcerned about the role of aviation in national security, or that policy efforts stagnated. From the time of the armistice in 1918 through the mid-twenties, over a dozen investigative committees, commissions, panels, and boards examined the new phenomena of aviation and military flying. Two of the most important of these, the Lampert Committee (formed by the House of Representatives) and the Morrow Board (or the President's Aircraft Board), happened to coincide with the Mitchell trial. Although neither group insisted on a separate, unified air arm, the net result won qualified gains for army aviation. In 1926, the Air Corps Act changed the name of the Air Service to the Air Corps and upgraded it, with implications for a more autonomous role in offensive operations above and beyond reconnaissance and auxiliary duties. Moreover, the War Department now included an assistant secretary of war for air, providing military aviation with welcome visibility and bureaucratic leverage.

Gen. William (Billy) Mitchell as he appeared during the 1921 bombing trials off the Virginia Capes. The plane is a DH-4B, flown as a bomber and reconnaissance plane during World War I. Courtesy of the Smithsonian Institution.

Although the Air Corps won authorization for a five-year expansion program, persistent budget cuts and the impact of the depression of 1929 meant continuing difficulties in acquiring modern equipment in adequate amounts. Dogged by parsimonious budgets and punctuated by contentiousness, military aviation nonetheless made halting progress. During the early twenties, the Air Service's "Model Airways System" encompassed several Midwest air bases and provided invaluable seasoning for ground crews as well as pilots. Additional long-distance and endurance flights of the decade accumulated the navigational skills and practical experience for long-range missions of the future. The Air Service also took important steps in organizing its research and training to reflect the growing sophistication of military aviation. At San Antonio, Texas, the School of Aviation Medicine found a permanent home, along with the Air Corps Training Center. Chanute Field in Illinois became the principal center for technical training, and McCook Field, near Dayton, Ohio, and Kelly Field in Texas became centers of logistical operations, engineering, and flight research. Officers explored new aerial doctrines at the Air Service Tactical School at Langley Field, Virginia. By the time newer equipment began to filter into squadrons in the late twenties, the Air Corps possessed the basic bureaucratic framework to cope with better aircraft and use them more effectively.

The ranks of elderly DH-4 biplanes began to thin out toward the end of the decade and were replaced by planes like the Curtiss P-6 Hawk pursuits and Keystone twin-engine bombers, which carried a five-member crew. All these planes still featured the biplane layout, although metal wing spars and fuselage formers began to replace wood, and streamlined engine cowlings made of metal became commonplace. Annual maneuvers now included dozens of aircraft instead of a token few. In terms of modern warfare, the exercises provided a practical background in planning for sizable air operations in the fluid and chaotic region of combat zones. "We grew up," Hap Arnold said of such experiences.

If Air Corps equipment still lagged behind the potential of air warfare, the twenties nonetheless became an era of gathering useful experience. Military aviation acquired more autonomy, more bureaucratic influence, and a viable organizational framework. When advanced metal monoplanes became available in the following decade, the army air force was in a much more favorable position to take advantage of them.

AIR POWER AT SEA

In many ways, much the same pattern characterized the experience of naval and Marine Corps aviation. By the end of World War I, naval and Marine Corps aviation had trained 4,000 pilots and 30,000 enlisted personnel. During the war, pilots flew from 20 patrol bases strung from England through the Continent and south to the Azores. Although some air-to-air combat missions were flown, the naval role was primarily a matter of patrol and reconnaissance. Since many naval aviators argued for independent offensive operations, as opposed to the dominant predilection for patrol work, naval aviation found itself beset by many of the same controversies as the army. Within the ranks of aerial patrol advocates, one faction stoutly supported dirigibles rather than conventional winged aircraft. But the mammoth dirigibles, despite the desirability of their long range, were too costly to build and operate, and they never proved very successful.

In any case, naval flying boats quickly garnered some of the biggest headlines of the postwar era. In May 1919, three ungainly Curtiss flying boats, the NC-1, -3, and -4, attempted the first aerial crossing of the Atlantic. The NC-4 became the only one of the trio to succeed, between May 15 and May 31, hopscotching from Newfoundland to the Azores, to Portugal, and to England. The NC-4's achievement resoundingly proved the flying boat's potential, and the navy refined the flying boat concept into a series of successful designs that performed unglamorous but heroic service in World War II.

Marine Corps aviation, saddled with castoff aircraft from active naval units, also made its mark during the twenties in spite of the antiquated equipment. In the Caribbean region, uncertainty over the future of American investments, in addition to

The Curtiss Hawk fighter incorporated several newer features, including metal fuselage components and a streamlined, metal housing for the engine. The fuselage, wings, and tail surfaces continued to be covered with fabric. Courtesy of the Smithsonian Institution.

concerns about the security of the Panama Canal, frequently prompted U.S. military intervention, carried out by Marine Corps contingents as early as 1919. During action in Santo Domingo, Haiti, and Nicaragua, Marine Corps pilots delivered supplies to remote outposts and flew medical evacuation missions, reducing a tortuous three-day ordeal through tropical jungle to a two-hour flight. They also flew strafing missions and dropped bombs in diving attacks. During World War I, Royal Air Force (RAF) pilots had made the first dive-bombing raids, but British development of the tactic faltered in the early postwar era. In remote Nicaraguan jungles, isolated troops frequently needed a rapid and decisive assist in fire power, leading to a revival of specialized bombing techniques. Marine Corps fliers in DH-4 aircraft conducted a spectacularly successful dive-bombing mission in 1927, forcing Nicaraguan rebels to withdraw from a besieged garrison defended by U.S. Marines and Nicaraguan regulars. The dive-bombing technique perfected by naval and Marine Corps fliers during the ensuing years was to become a devastating tactic against Japanese forces and shipping in the Pacific during World War II.

Like army aviation, naval aeronautics benefited from bureaucratic evolution, especially after the organization of the Bureau of Aeronautics in 1921. The bureau's chief, Capt. William A. Moffett, a man of vision and energy, helped to lay the foundation for several significant developments. His tragic death, in the crash of the dirigible *Akron* in 1933, robbed the navy—and the aviation community—of an articulate and effective spokesman. During Moffett's administration, the navy began developing its bases at Pensacola, Florida, and at San Diego as major flight training centers. At Great Lakes, near Chicago, new recruits learned the fine points of overhaul and repair; Lakehurst, New Jersey, specialized in dirigible operations. With the bureau's encouragement, considerable attention was given to the wider use of metal components in naval aircraft and to the requisite use of anticorrosive material. Most important of all, Moffett supported the development of the radial engine and the aircraft carrier.

Even before the end of World War I, perceptive naval officers in America were watching British carrier trials with great interest. In the United States, a key event occurred in 1919, when an observer high in a plane coached main batteries of the battleship USS *Texas* to achieve startling accuracy during a firing exercise. Obviously, aerial spotters from an enemy fleet would have to be stopped; this would require agile fighters (not floatplanes) to operate from carrier decks. The dramatic bombing trials off the Virginia Capes in 1921 added weight to arguments that naval planes could also immobilize capital ships. Carrier-based aircraft would be required in both offensive and defensive roles.

In the summer of 1919, Congress authorized conversion of the fleet collier *Jupiter;* after nearly three years of alteration, the vessel was newly commissioned as the aircraft carrier *Langley*. With a complement of 34 aircraft, the *Langley* became a test-bed to refine operational procedures for future carriers within the fleet, perfect flight deck activities and arresting gear, and evolve new specifications for carrier-based aircraft. With accumulated experience, the navy confidently proceeded to acquire two modern carriers, the *Lexington* and the *Saratoga*. Originally laid down as battle cruisers, they were converted to aircraft carriers under the guidelines of the Washington Disarmament Conference of 1921–22. Commissioned in 1927, the new carriers deployed between 70 and 80 aircraft and achieved speeds of over 33 knots—equal to the best cruisers of the fleet. Visionary naval officers who had long wanted high-speed carriers with an offensive capability now had them.

The navy also provided its formidable new carriers with aircraft equipped with more powerful and reliable radial engines. Typical water-cooled engines like the venerable Liberty carried weight penalties in terms of coolant, pumps, and coolant lines. They also required additional maintenance and were more vulnerable to combat damage. Considering the stringent limitation on maintenance space and surplus parts aboard carriers, simpler power plants possessed obvious appeal for naval air operations. Interest in the development of radial engines had already surfaced in the National Advisory Committee for Aeronautics (NACA). In 1920, NACA pointed out that radial, air-cooled engines promised less weight per horsepower and offered increased reliability. By 1922, new radial engines, developed with naval support, were running for 300 hours, compared to water-cooled engines with 50-hour endurance limits. The new radials gave shipboard fighters a performance equal to land-based planes, and, more important, their inherent reliability in extended over-water operations not only raised the confidence of naval pilots but also paved the way for planning military missions that earlier strategists would have shunned.

The formidable potential of planes carried aboard fast carriers became evident during the fleet exercises of 1929, involving a simulated attack on the Panama Canal. The *Saratoga* was assigned to the attacking forces. Accepted doctrine at the time kept

Dropped from a Martin MB-2 bomber above, a phosphorus bomb strikes one of the target ships during the bombing trials carried out by Gen. Mitchell in 1921. Courtesy of the Smithsonian Institution.

carriers in a defensive, protected position behind the main force, but Rear Adm. Joseph Mason Reeves won permission to detach an escorting cruiser and carry out an independent carrier strike on the Panama Canal. A high-speed, nighttime sortie under full steam penetrated the screen of defending vessels. Using previously developed procedures for nighttime flight operations, the *Saratoga* launched a predawn air strike that caught canal defenders in complete surprise. Naval umpires judged that the *Saratoga*'s planes immobilized the canal, heavily damaged army airfields, and emerged victorious over the few army planes that got into the fray. Ships, planes, and tactics like this, developed in the twenties, would prove decisive in the Pacific theater during World War II.

AIR MAIL AND COMMERCIAL AIRLINES

Like the military services, American commercial aviation spent a good deal of time accumulating experience during most of the twenties, until modern aircraft and more sophisticated operational techniques transformed aviation toward the close of the

The USS Lexington, *commissioned in 1928, had the same design features as the* Saratoga. *The Boeing F2B-1 fighters lining the deck were the first to be used extensively in carrier operations, beginning with the older and smaller* Langley. *Courtesy of the Smithsonian Institution.*

decade. Lindbergh's solo flight across the Atlantic in 1927 implied more than a daring individual's feat; it symbolized the progress of aviation technology reflected in military and civil aviation alike. Aviation also found a utilitarian niche in the activities of American commerce. Many large businesses, continuing a trend that had begun near the close of World War I, decentralized operations during the twenties. They were able to do so for a variety of reasons, including the growth and flexibility of electrical power and advances in communications, which permitted coordination of scattered sales and production facilities. Aviation also complemented such moves, providing air-mail, passenger, and express services. Even though aviation did not radically alter business planning until the following decade, more and more firms in the twenties utilized aviation on a regular basis because it offered a more productive way of doing business.

As early as 1920–21, leaders in the aviation field felt they could foresee the specific ways that aviation would display its commercial potential. Donald Douglas stressed the advantages of speed in delivering fruit, film, repair parts, and other lightweight, low-bulk commodities with a high time-related value. The chief engineer of another wartime aircraft manufacturer predicted frequent air travel for high-salaried executives, saving them time and enhancing their working effectiveness. Even though comprehensive transport routes and regular passenger travel lay several years ahead, one existing aerial service in 1921 already provided a business advantage. The U.S. Air

Mail Service, operating under the aegis of the Post Office Department, offered American businesses an important opportunity to speed up business operations by sending correspondence by air.

During an air meet on Long Island in 1911, Earl L. Ovington flew mail on several authorized flights. Similar flights, primarily stunts, followed over the next several years, until the Post Office Department decided to offer regular air-mail service. On May 15, 1918, service began between New York City, Philadelphia, and Washington, D.C. The War Department furnished planes and pilots, hoping to stimulate the aviation industry and train more fliers, though some skeptics felt that the department was trying to counter reports of the Air Service's poor showing in Europe. The air-mail project backfired. While President Wilson looked on in Washington, D.C., the JN-4 mail plane bound for Philadelphia repeatedly failed to start (until someone remembered to fill the gas tanks); then, after flying off in the wrong direction, the pilot crash-landed just 25 miles away.

By the summer of 1920, the Post Office Department had improved reliability, took over all air-mail operations, and hired its own pilots. War-surplus DH-4 biplanes, extensively modified for air-mail service, replaced the slow, frail JN-4 trainers. By the autumn of 1920, the Post Office Department had forged a coast-to-coast air-mail route between New York and San Francisco. Restricted by unreliable instruments and a lack of adequate navigational aids, pilots of the DH-4s landed at dusk along the transcontinental run and let the railroads carry the mail until dawn, when a waiting plane took to the air again. Even with all of the transfer of mail sacks, the air-rail combination delivered coast-to-coast mail in 78 hours, nearly a full day ahead of regular mail, which required 100 hours even by crack trains.

Attracted by the time-saving advantage of air mail, the banking industry regularly began to use the service to reduce "float," or idle funds, like checks and other items in the process of collection. These idle funds represented unavailable cash and slow-moving capital that accumulated expensive interest hour by hour. The increasing value of air-mail service encouraged postal officials to plan around-the-clock flying schedules, including night flights, to eliminate the awkward railroad segments. The first attempt occurred on February 21, 1920, despite the threat of severe winter weather. The trial run succeeded but required a courageous effort by pilot Jack Knight. Leaving after sunset, Knight carried the eastbound mail across Nebraska from North Platte to Omaha. Finding the next plane unable to meet him, he climbed back into the cockpit and flew on to Chicago through a turbulent, icy night, guided by bonfires that flickered across Iowa and Illinois. Taken over by a relief pilot at Chicago, the first transcontinental mail finally reached New York in 33 hours and 20 minutes—nearly three days faster than rail service.

In order to function on a regular basis, air mail needed more than a string of bonfires. By 1924, a system of flashing beacons had gone into service on the Chicago-Cheyenne segment, expanding to a transcontinental network in two years. A series of floodlighted main terminals, plus a string of intermediate emergency fields, added to air mail's reliability. With regular long-range, all-weather, day-and-night operations becoming commonplace, American aviation developed extremely useful operational experience

An enthusiastic crowd gathered at Hazelhurst, New York, to watch a DH-4 mail plane take off on the first transcontinental run to San Francisco in 1924. Acquired from military surplus and modified to carry 400 pounds of mail, the DH-4s constituted the backbone of the U.S. government's mail service. Courtesy of the Smithsonian Institution.

that enhanced the utilization of modern passenger transports like the Boeing 247 and Douglas DC-3, which entered service in the thirties. Superiority of technique as well as equipment counted heavily for future American leadership in transport aviation.

As the sophistication of air mail improved during the twenties, the volume of banking and other financial mail began to mount. Clients included the Federal Reserve Bank and its branches as well as major banks across the country. In addition to sending drafts, securities, bonds, and stocks via air mail, a variety of companies used the service to speed delivery of manifests, bills of lading, advertising, and miscellaneous commercial correspondence. Having created a workable aircraft transportation system and attracted a permanent clientele, the government eventually acceded to popular demand and transferred its operation to private enterprise. Following the passage of the Airmail Act (the Kelly Bill) in February 1925, private lines carried air mail under contract to the POD. The Air Commerce Act of 1926 created an Aeronautics Branch within the Department of Commerce to promote aviation through additional lighted airways, navigational equipment, and other aids. Private and commercial aviation expanded at a rapid rate. The early contract mail carriers also began to consolidate, melding smaller, low-volume runs into comprehensive, viable route structures. During the latter half of the decade, now-familiar names like American, TWA, United, Delta, Eastern, and Northwest began to appear.

The availability of mail contracts not only encouraged the formation of transportation companies, it also accelerated the introduction of original aircraft designs based on the new commercial requirements of flying the mail. The young companies acquired a variety of types, ranging from single-engine biplanes like the Boeing-40 series, which could also squeeze 2 to 4 passengers into a forward compartment, to the Ford trimotor monoplane, which carried 10 passengers in relative comfort along with a load of mail. The cumbersome water-cooled Liberty engines also gave way in the mid-twenties to more efficient air-cooled radials like the Wright Whirlwind and Pratt & Whitney Wasp series, which appeared in various models from 200 to 450 HP. These power plants gave the new commercial aircraft cruising speed of up to 110 MPH. Equipped with such modern planes, the recently formed airlines took over the routes pioneered by the Post Office Department and grew into mature transportation systems. By the spring of 1929, new aviation facilities in the United States included 61 passenger lines, 47 mail lines, and 32 express lines serving trade areas that contained 90 million people; the volume of air mail ballooned from 810,555 pounds in 1926 to 7,772,014 pounds in 1929. There was also some skullduggery when the contract lines were struggling for survival. Since government payments were made by the pound, many contractors sent postcards to friends by registered mail, which required a mail sack secured by a heavy lock. "They made a hell of a lot of money on that postal card," one contemporary chuckled.

The achievement of dependable air-mail service was not without human cost, and the tenacity of the air-mail fliers was resolute. When the government began to surrender operations to private lines in 1925, only 9 men were alive out of the 40 pilots originally hired by the Post Office Department. In spite of this melancholy toll the mail was delivered with amazing regularity: 94 to 96 percent of all scheduled flights were completed in 1921 and 1922. Private operators were able to improve but little on this record by the end of the twenties. Dependability resulted from operational procedures that slowly and painstakingly evolved, as well as from the courage of the men who flew the mail. One of the U.S. Air Mail Service's most important legacies to the development of air transport concerned regular maintenance schedules that saved thousands of dollars and kept airplanes ready to fly. Along with this came the practical experience necessary for ground crews to ensure the proper care of the frequently fragile components that characterized flying equipment of the early 1920s.

Air-mail pilots, both government and contract fliers, faced other kinds of problems. Since many flights were made at night or during bad weather, the pilots were very much interested in instruments, but the panels of planes like the DH-4 were sparsely furnished with dials and gauges. The rudimentary equipment of the period usually failed. Completing a successful cross-country mail run was best accomplished by careful observation of landmarks en route. Mail pilots in the twenties had no detailed flight charts and generally relied on state road maps, although such maps normally marked only towns having post offices and gave no information at all concerning elevation, mountains, or other hazards poking into the air. Fliers in the POD compiled their own *Book of Directions* for major routes, a rather informal compendium describing railroad tracks, highway junctions, golf courses, and polo fields along the way.

Pilots also kept up their own "little black books" with vital statistics ranging from notations about threatening church steeples to locations of farms with a telephone with which to call for help after an emergency landing. Such seemingly insignificant items of information were crucial for flying in bad weather, when pilots often let down to 50 feet in order to see their way with the aid of ground references. Bryon Moore, a veteran of such adventures, recounted a foul-weather formula for landing at one field on his route: "When you come to the fork in the road, get up on the left side to miss that silo; after you cross the railroad tracks pull up into the soup [fog], count to thirty, then let down—that way you'll miss the high tension lines; when the highway angles left, take the fourth dirt road and follow it to the ravine—just across the ravine is the airport." Various hazards conspired to force down pilots en route, resulting in the classic report from Dean Smith, who sent his superiors the following dispatch: "On trip 4 west-bound. Flying low. Engine quit. Only place to land on cow. Killed cow. Wrecked plane. Scared me. Smith."

The young men who flew the mail were necessarily a hardy lot, willing to pit their flying skills against considerable odds. When Smith joined the U.S. Air Mail Service in 1920, he recalled that it was "considered pretty much a suicide club." In open-cockpit biplanes, the constant exposure to cold air was one of the worst hardships, especially in winter. Even in heavy flying suits, pilots became so bitterly chilled and benumbed that judgment was impaired. Flying became a struggle of endurance and nerves. Many times, Bryon Moore recalled, he remained in the cockpit several minutes after landing, to all appearances taking care to fill out his logbook. Actually, he was stalling for time, because he felt he could not talk evenly quite yet, and he did not want the mechanic to know he was still shaking. Whether for nerves or warmth, many pilots carried liquor when they flew, and most maintained their drinking habits when off duty. One flier of the era could remember only one pilot who did not drink—Charles Lindbergh.

It would be easy to overdraw the image of the individualistic, hard-drinking, fatalistic mail pilot. Flying the mails was not the safest job in the world, to be sure, but the pilots were not inclined to embark on patently foolhardy flights. There was a growing tendency among these fliers to view themselves as professionals. During the early years of the U.S. Air Mail Service, fliers found it necessary to take issue with bureaucrats who insisted the mail must go through, disregarding the fragile nature of the aircraft in service and the lack of truly reliable instruments for blind flying. Following the transfer of the airmail routes to private contractors after the Kelly Bill of 1925, several aviator groups sought to speak for the nation's commercial pilots. The most representative was the National Air Pilots Association (NAPA), formed in 1928. NAPA and similar groups of the 1920s represented a strong trend toward organizational unity among the airline pilots and led to the formation of the Air Line Pilots Association (ALPA) in 1931. This organization became a powerful union and culminated the growing trend toward professionalism within the ranks of pilots during the post–World War I decade.

Even before the Kelly Bill of 1925 launched successful commercial airline routes in the United States, several attempts had been made to establish private airline enter-

With special boots, two layers of heavy flying togs, fur collar, classic helmet-and-goggle head gear, and a cocky stance, pilot William C. Hopson of the U.S. Mail Service was ready for duty. He helped establish the transcontinental route in 1921, and died when his plane crashed in 1927. Courtesy of the Smithsonian Institution.

prises. American promoters were evidently encouraged by activity in Europe, where, within three years after the armistice, France boasted eight regularly scheduled lines; England, three; and Germany, two. National rivalries as well as disrupted surface transportation on the Continent immediately after the war spurred airline development and boosted airline travel, which also avoided time-consuming harassments caused by cars, buses, and trains being halted for inspection at every national frontier. Many travelers also opted for airlines to avoid delays in crossing the English Channel; in fact, the majority of passengers on cross-Channel flights during the peak summer season were U.S. citizens.

In the United States, a wave of postwar enthusiasm for airlines generated numerous plans for passenger services. One scheme was advanced by Alfred Lawson, a pioneer aeronautical publisher and promoter. But his L-4 airliner crashed during its first takeoff in 1921, and Lawson was forced to abandon his hopes for an aeronautical empire. Lawson's efforts are worth noting, however, for the L-4 and its predecessor of 1919, the C-2, were the first multiengine planes in the United States designed and built as passenger airliners. Most aviation ventures in the United States and abroad were attempting to utilize converted military aircraft, with cramped seating on passenger routes. The concept of an air transport specifically produced for passenger service anticipated a major American industry.

Still, aviation fever infected many investors, and promoters announced numerous ventures. Most of these ambitious schemes never got into the air. Travel in the United States was not impaired by inconveniences like national frontiers as in Europe, and existing train schedules for longer American distances were better than passenger air routes, which came nowhere near the comfort and dependability of overnight Pullman service. Nevertheless, a few lines were formed, representing a nascent commercial aviation at a time when the government air mail was making headlines. One of the most interesting developments of the early postwar era involved Aeromarine West Indies Airways, which inherited the experimental Foreign Air Mail contract between Key West, Florida, and Havana, Cuba, in 1920. Aeromarine used a trio of Curtiss type 75 flying boats, converted versions of the navy's F-5L patrol plane. The route appealed to the time-conscious traveler. Passengers on their way to Havana usually had to wait eight hours after getting off the train and then take an overnight boat ride; Aeromarine planned to make the 100-mile trip in 75 minutes. Many patrons were thirsty Americans anxious to escape the rigors of Prohibition in the United States. Besides attracting the parched traveler in a hurry, the line expected to draw fares from those waiting to escape sea sickness, and, as *Scientific American* explained, "The enchantment the aerial traveler experiences aloft will offer a further inducement to go via the air route."

All things considered, Aeromarine turned in quite a performance during its existence. The line operated in Florida during the winter season, December 15 to April 15, making daily departures from Key West at 10:30 A.M. and starting the return trip from Havana at 3:30 P.M. the same day. Business developed so well that the line expanded into Southern New York and Great Lakes divisions, with seasonally adjusted schedules. During the 1921–22 season, Aeromarine made 2,125 flights and flew 739,047 passenger miles without accident. Of special interest was the twice-daily service

With Curtiss flying boats, converted from a World War I design, Aeromarine West Indies Airways launched a bold experiment to carry passengers to Bahamian and Caribbean tourist spots. Bootleggers reportedly represented a large measure of the line's clientele. Courtesy of the Smithsonian Institution.

between Detroit and Cleveland. Although it did not begin until the summer of 1922, the route carried almost as many passengers (4,388) as the other two divisions combined— and this despite the high rate of $25 for the 90-minute flight from Detroit to Cleveland in comparison to fares of $9 by rail and $5 by steamer. For all of its notable activity and progress, Aeromarine finally had to suspend operations in 1924. There was no big subsidy or a liberal mail contract for its routes to make it a commercial success and its passenger volume proved to be limited. Nevertheless, by way of example, Aeromarine probably contributed more to the development of commercial air transportation than any other operation at that time, with the exception of air mail.

The failures of Aeromarine and other fledgling concerns underscored the value of adequate mail contracts from the government that paid well enough to promise a profit. The Kelly Bill offered an individual cushion for companies who still looked forward to the time when carrying passengers in airliners would become a principal reason for commercial aviation. In the mid-twenties, this period of transition created an awkward situation for potential passengers. By 1927, adventurous souls with $400 for the fares could theoretically piece together a 32-hour transcontinental flying trip in a variety of open-cockpit, two-place biplanes and enclosed-cabin, four-place Boeing 40-B4s. Since the airlines' chief revenue still came from government postal contracts that paid by the pound, passengers had to sign a proviso that allowed them to be dumped anywhere along the line if the company could pick up a more lucrative cargo of correspondence. Under such conditions, many would-be patrons remained earth-bound. Even the first regularly scheduled transcontinental passenger operations in 1929, like the early air mail, turned over their passengers to the railroads at dusk.

In the meantime, however, daytime passenger airline services on a regional basis began to develop in the mid-twenties. Numerous new airlines, organized primarily to operate the government's contract air-mail routes, acquired a variety of aircraft that could also carry two to four passengers—sometimes in enclosed cabins and sometimes not. Either way, seating accommodations were decidedly cramped. On routes that seemed likely to generate increased passenger traffic, some companies equipped themselves with larger transport planes like the Boeing, Fokker, and Ford trimotors, carrying anywhere from 8 to 16 passengers in comparatively spacious, fully enclosed cabins. At speeds of about 120 MPH, these planes boasted ranges of 600 to 800 miles. With the luxury of more available cabin space and longer flying times, airlines added passenger amenities, such as stewards who served luncheons while en route. "Comfort stops" were synchronized with passenger and/or refueling stops on the ground. The airlines lured increasing numbers of time-pressed travelers into the sky as Americans attempted to squeeze more working hours into their crowded schedules.

Business flying in those early years was not always a sedate experience. It was necessary to shout to be heard in a Ford trimotor in full flight. On cold winter days the cabin temperature hovered at a shivery 50° F. Nor would a businessman flying coast to coast consider five days fogged in at Omaha as part of a "fruitful" commercial journey. Nevertheless, a traveler who had endured all these tribulations still commended the airways. One should check the safety reputation of an airline, he cautioned, but if it seemed adequate, one certainly ought to fly. "You will save a lot of time and come down safely," he added reassuringly. Despite occasional inconveniences, an advertisement by the National City Company (New York bankers) appearing in the *Review of Reviews* in 1929 suggests that flying on business had indeed attained a reasonable degree of prestige and acceptance by the end of the decade. The illustration depicts a man with a briefcase, purposefully striding toward a waiting plane. The accompanying copy notes that people live at a fast tempo, reflected by this busy executive using a swift airliner to attend an important meeting.

The statistics indicated a growing inclination to fly on business; the number of U.S. airline passengers climbed from 5,782 in 1926 to 173,405 (including 13,654 fares to foreign nations) in 1929. Although commercial aviation was not really competitive with bus and rail transportation until after 1945, the airlines' clientele was an influential one. High-level businessmen comprised the bulk of airline fares, and the social effect of mobility, even in the twenties, was probably greater than the figures suggest, as more and more companies authorized their key personnel to patronize the airlines. A survey in 1929 shows that most American passengers were traveling for reasons of business, and that some 80 leading corporations allowed employees and executives on company business to put air fares on expense accounts. It is not surprising that most airline fares were drawn from major population centers. In the spring of 1930, one

At Boston Airport passengers board a Ford trimotor of Colonial Air Lines, a predecessor of American Airlines. The rugged, all-metal Fords performed yeoman service into the mid-thirties. Courtesy of American Airlines.

airline kept a record of every passenger's name and address; over a month-long period, about 48 percent of the airline's business originated in seven large metropolitan areas. The company also found a remarkably wide geographic range reflected in its clientele of 2,500: 42 states and 407 cities.

Another aspect of commercial aviation, cargo by air, had a prophetic, if unsuccessful, inauguration in the winter of 1919, when the American Railway Express (ARE), predecessor of the Railway Express Agency, loaded 1,100 pounds of freight into a converted Handley-Page bomber bound for Chicago. Frozen radiators delayed the start from Washington, D.C., and the plane was finally forced down on the way, making an ignominious landing on a race track in Ohio. Undaunted by this initial failure, the company followed the progress of aviation in the ensuing years and made new plans in 1927 for the commercial airlines to fly express. National Air Transport and Colonial Air Transport would offer services to the East, and Boeing Air Transport and Western Air Express would cover the West. Depending on the length of the haul, rates would vary from 25¢ to 50¢, based on each quarter pound or 50 cubic inches. Shortly after midnight on September 1, 1927, National Air Transport started the first run from Chicago to New York with a motley consignment of newsreels, machinery parts, advertising copy, trade journals, candy, and Paris garters.

The availability of air express proved to be of immediate value to American companies. A production line shutdown was averted when air express delivered magneto parts to Detroit and auto lamps to St. Louis. A printing plant on a 24-hour schedule lost only hours rather than days when a replacement part was shipped by air from Connecticut to Illinois. Paul Henderson, a former government air-mail executive who became a vice-president of National Air Transport, cited these and similar cases to underscore the usefulness of aviation in paring inventories of small-bulk, expensive parts and materials. Occasional air cargo shipments could effect substantial economies by reducing the size and carrying charges of inventory, and retail stores throughout the country soon made everyday use of air delivery to restock supplies of lightweight items. Even if a business could not use air shipments on a regular basis, it could still operate at closer tolerance without extreme risk. "Not one item out of a thousand may have to be shipped by air," acknowledged the authoritative journal *Factory and Industrial Management,* "but the fact that any one of the thousand *can be* shipped by air, if it is necessary, on a few hours' notice, makes it safe to lower the reserve a little all along the line."

Airline officials themselves were disappointed that air express did not generate anticipated revenues. On the other hand, a definite upswing in volume verified that more and more businesses found the service worthwhile. From 45,859 pounds transported in 1927, air express climbed to 257,443 pounds in two years, and topped 1,000,000 pounds per year by 1931. The routes of the ARE, moreover, did not constitute a monopoly on air express. Henry Ford's private airline, which ran mainly from Detroit to Chicago, Cleveland, and Buffalo, carried 1,000,000 pounds of company freight when it started in 1925, and averaged better than 3,000,000 pounds annually by the close of the decade. A few specialized independent companies did as well or better.

By 1929, the vague outlines of aviation's function in commercial operations had come into sharper focus, vindicating the confidence expressed years before by Donald Douglas and others. The development of the DH-4B marked a notable if not revolutionary improvement in capability and reliability, advancing the state of the art of communications by aircraft. Despite their limitations, such secondhand types provided an advantageous service and attracted a clientele. Moreover, the DH-4B could benefit from the existence of the transcontinental lighted airway. The combination of these two factors—the airplane and the "guidance system"—into a 24-hour operation that put the West Coast some three days closer to New York signaled an advance in postal communications. After 1925, a new generation of engines and transport aircraft of original design created new opportunities for businesses in air express and passenger travel as well as in correspondence. Through continued development, aviation technology increasingly attracted endorsement from the country's businesses. This commitment, plus the evolution of operational skills and a firm technological base, contributed to the emergence of aviation as an integral factor in the conduct of business and established the pattern of airline development in the United States.

THE GENESIS OF GENERAL AVIATION

As military aviation struggled to find its place in America's defense establishment and the U.S. Air Mail Service gave way to scheduled commercial airline and air

With helmets and goggles, these five insouciant fliers in Connecticut typified the growing ranks of pilots who embraced aviation as an avocation as well as a vocation during the 1920s. Courtesy of the Smithsonian Institution.

express service, another sector of the American aeronautical scene began to carve out its niche in the twenties. Sometimes called "private aviation" in the post–World War I era, it is known today as "general aviation," a sort of aeronautical catchall for a variety of aircraft performing myriad activities. Planes in this category were generally smaller than the types used on airline routes by the close of the twenties and served in a remarkable variety of capacities: training, pleasure, business, agricultural dusting, photography, air shows, cargo, and more. Specific definition of the general aviation sector, therefore, was not always easy, and general aviation aircraft, even in the twenties, included airline types like the Ford trimotor. Nevertheless, whatever the activity or size of airplane employed, the general aviation sector emerged in the postwar decade as a significant element in the framework of American aeronautics.

An airplane can do hundreds of things, and the fliers who first put planes through their paces were called the "flying gypsies" of the postwar era. The gypsy, later dubbed "barnstormer," was the offspring of postwar demobilization, born in the spring and summer of 1919 when the government released its cadres of service fliers and hundreds of serviceable military aircraft. The most popular and least expensive surplus plane was the Curtiss JN-4, better known as the "Jenny," a tandem-cockpit, two-place biplane produced as a primary trainer during World War I. Over 8,000 JN-4 types were manufactured, of which 2,600 comprised the JN-4D, the standard trainer for thousands of service pilots. The JN-4D had a 90-HP engine that gave it a cruising speed of 60 MPH. The planes cost the government $5,000 but were sold as surplus for a few hundred dollars. For ex-military airmen seeking their aerial fortunes, the Jenny quickly became the principal trademark of the flying gypsy. Working their way across the country, gypsies gave thousands of people their first ride in an airplane at a dollar a minute, making a convincing demonstration of the flying vehicle so many had read about but had never seen for themselves.

The lure of aviation captured the imagination of dozens of young Americans, and barnstorming offered a means of apprenticeship for those who had yet to learn how to fly. Charles A. Lindbergh performed in Nebraska as a parachute jumper and wing walker to scrape together the $500 to buy his own Jenny. When he took possession of the plane, Lindbergh had little instruction and had never soloed, but he did not have enough money for additional lessons. So he gamely clambered into the cockpit and the engine was started. In his first solo attempt, Lindbergh made an erratic climb to an altitude of four feet, thought better of it, and brought the Jenny down again on one wheel and the wing skid. A sympathetic pilot generously conducted a cram course of 30 minutes' instruction that launched the young flier into a new career.

The gypsy/barnstormers were popular as picturesque figures, but the sensational stunts and attendant fatalities created a misunderstanding and fear of aviation that took years to erase. By mid-decade, the penchant for sensational aerial showboating had begun to wane. Still, the gypsy/barnstormers helped make the country "air-minded," and their ranks provided a nucleus for other kinds of commercial aviators. Some barnstormers began to settle down and centralize their activities. These fliers became the local "fixed-base" operators whose hangars, repair shops, and maintained runways laid the basis for general aviation and aviation services. An organization known as the Maycock Flyers, located in Michigan, claimed that it began nonscheduled flying service all over the country as early as March 1919. Such groups as these and other ex-gypsies worked with great enthusiasm, in the belief they were helping to construct the foundation of a nascent technology. Their most important contribution to aviation was "aerial service"—imaginative application of the particular advantages of aircraft in surveying, photography, crop treatment, emergency service, and so on, entirely apart from regularly scheduled mail, freight, and passenger service.

Although Bob Johnson did not start out as a gypsy, the Johnson Flying Service in Missoula, Montana, typified the regular fixed-base operator of the mid- to late twenties. Johnson began earning money from conducting scenic tours, carrying sportsmen to remote forest areas, offering flight instruction services and commercial flights for local businesses, and conducting other operations that only airplanes made feasible. His equipment included a two-place Swallow and a four-place Travelaire. A "snowline run" into the rugged back country maintained contact with ranchers who were running low on stock feed. He delivered items like two 1,600-foot metal cables to a snowbound gold mine in the high Cascades. Landing in a ski-equipped plane, Johnson picked up a seriously injured park ranger and delivered him safely and quickly to the hospital, eliminating a four-day ordeal by pack train through wilderness in the dead of winter.

Airplanes were obviously connected with marvelously inventive minds, and this association bloomed into all sorts of intriguing operations. In the commercial age of the twenties, the thrill of aviation lent itself to techniques of sensation and ballyhoo. A

The American Eagle, with room for two passengers in the forward cockpit, was typical of the open cockpit biplanes flown by barnstormers and fixed-base operators of the 1920s. It cruised at approximately 80 MPH. Courtesy of the Smithsonian Institution.

dashing Englishman, Maj. Jack Savage, caught the eye of many advertising executives when he announced himself with a big "Hello U.S.A." scrawled in chemical smoke across New York's horizon. The major was signed to a thousand-dollar-a-day contract by the American Tobacco Company, which quickly perceived the suggestive appeal of having a smoky "Lucky Strike" inscribed in midair. In fact, the company was so delighted with this advertising medium that it planned to have Major Savage puff his way across America, lettering the message over selected cities. Politicians flew from one rally to the next during gubernatorial campaigns; newspapers relied on planes to carry reporters and photographers trying to keep up with the politicians, as well as to cover stories on floods, fairs, football, and other fast-breaking news events. Until much of their utility was displaced by Wirephoto services, airplanes turned in a colorful performance. By the late twenties, several municipal governments had purchased planes for law enforcement work, and the Coast Guard deployed Loening amphibians to scan for rum-runners during the Prohibition era. Rum-runners, in turn, acquired planes in order to elude the Coast Guard's aerial patrols.

In addition to some of the more unusual or dramatic roles of general aviation, private flying on business also became an accepted feature of American commerce. The consistent value of aviation was its speed and flexibility, although airplanes used for business flying in the early twenties frequently carried an oversized company logo on the fuselage, suggesting that many owners viewed their planes as vehicles for public relations and advertising. The evolution of business flying undoubtedly began in the early twenties, although it seems to have become widespread only in the latter half of the decade. It is likely that the growth of fixed-base operations over a period of years eventually provided an operational base for business flying, while the development of the air mail and airlines around the middle of the decade helped generate the air-minded attitude that brought attention to the possibilities of air travel by personal plane.

The industrial diversification of the post–World War I era, noted earlier, also encouraged more business flying. Despite the growth of commercial airline routes throughout the United States, dozens of sizable cities lacked scheduled air service, and many time-conscious executives discovered the expediency of a company-owned or even a personal executive aircraft. By the late twenties, the *Magazine of Business* had endorsed private business flying in a series of matter-of-fact, dollars-and-cents articles—evidence of the growing awareness and appreciation of the value of aviation in a fast-paced, modern world. In the summer of 1927, the magazine's publisher, A. W. Shaw Company, hired the services of a professional pilot and bought a six-place Stinson to be used by R. L. Putman, a company vice-president based in Chicago. In spite of some mishaps, generally inadequate ground service, and a scarcity of convenient airfields, the Stinson still proved to be a valuable business tool. "With the airplane," Putman said, "we have accomplished many things that simply could not have been accomplished by any other means of transportation." Given the rudimentary status of flying aids and services available at the time, Putman and his Stinson logged a remarkable 44,327 miles over a 12-month period, mostly in the East and Midwest, with excursions as far south as Florida, as far north as Minnesota, and as far west as Wyoming. Even more remarkable is the report that hundreds of acquaintances were made with other flying businessmen, including the ubiquitous oil executive, a steel man from Pittsburgh, an Ohio manufacturer visiting his several plants, a baker with a chain of outlets, and an advertising representative.

Taking notice of such activity, manufacturers of single-engine, two- to four-place airplanes began to beam an appeal to the potential market for business aircraft. The Waco and the American Eagle, for example, were typical biplane designs with two open cockpits; the forward cockpit accommodated two passengers. With 100-HP engines, such planes had a cruise speed of about 80 MPH, landed at 35 MPH, and cost between $2,000 and $2,500. The Stinson and Fairchild companies produced enclosed, five- to six-place cabin types. These designs were high-wing, 220-HP monoplanes that cruised at 100 MPH, had ranges of 500–700 miles, and cost $12,000. For their time-pressed, high-salaried executives the larger companies acquired multiengine, long-range airplanes. These impressively large, prestigious vehicles of the new executive fleet were replete with executive amenities. The interior of John Hay Whitney's Sikorsky amphibian was appointed with a lounge on one side of the cabin and comfortable chairs on the other, all upholstered in boldly patterned fabric. J. C. Graves, vice-president of the Richfield Oil Company, posed for a photograph in the company's trimotor, which was fitted with a buffet and berths for overnight hops along with a businesslike desk graced by a dictaphone and a vase of fresh blossoms.

Despite such refinements, the svelte, airborne executive suites were no less efficient and profitable than their junior counterparts of smaller size and spartan comforts. The Ford trimotor *Stanolind*, flagship of the Standard Oil Company of Indiana, was based in Chicago, making the company's most remote office in Minot, North Dakota, as

Conoco was among several corporations that acquired large aircraft like the Ford trimotor to carry top corporate executives in a style comparable to that of contemporary airlines. Courtesy of Conoco, Inc.

close in time by plane as Quincy, Illinois was by train— eight hours. "Even at present costs," said the editor of *Factory and Industrial Management* in 1928, "the company-owned plane is almost indispensable where fast emergency transport may at times be a matter of life and death, as in carrying relief equipment in remote mine disasters. The plane is an economy also where it can be used more or less continuously in shortening the travel time and enlarging the range of action of executives." An incomplete roster of industries that flew their own planes by 1929 runs into the dozens, from petroleum corporations, breweries, supply firms of all types, construction engineers, and mining companies to familiar trade names like A-C Spark Plug, Jell-O, Lorillard, and Walgreen Drugstores. By the decade's end, it was estimated that aircraft for business use accounted for one-third of all civil aircraft purchased in the United States.

Because official government data on air travel were not compiled until the advent of the Air Commerce Act of 1926, accurate passenger statistics on this type of business flying are difficult to establish. Even after that date, statistics are apt to be confusing, since personal business flying was included as one category of general aviation, a generic term that also embraced instructional and pleasure flights. Regardless of the paucity of precise statistics, rough figures drawn from several sources indicate a surprising degree of flight activity. Secretary of Commerce Herbert Hoover testified that general aviation accounted for 80,888 passengers in 1923, and the *Aircraft*

Yearbook (1930), official publication of the Aeronautical Chamber of Commerce, claimed an increase to 205,000 passengers two years later. In contrast, scheduled airlines carried a total of only 5,782 patrons in 1926. By 1929, general aviation accounted for 2,955,530 passengers annually, compared to 173,405 fares for scheduled passenger lines. The number of passengers in general aviation who flew for business is an open question; nevertheless, it seems reasonable to assume that the dimensions of personal business flying assumed significant proportions by the close of the decade—perhaps double that of the scheduled airlines.

In short, the full spectrum of private flying unfolded and flourished in the twenties, ranging from aviation as a spectator sport to pleasure flying to business flying. These aspects all employed aircraft in considerable array, numbering well into the hundreds. Other general aviation developments, though not requiring nearly so many planes, nevertheless had a discernible influence on American life in the twenties. Two principal trends stand out: aviation in agriculture and aerial surveying.

AIRCRAFT IN FIELD WORK

In the summer of 1921, U.S. Air Service personnel at McCook Field in Dayton, Ohio, collaborated in a striking new use of aircraft. Cooperating with representatives of the Ohio Agricultural Station, the army supplied a plane, a pilot, and rudimentary apparatus for aerial dusting. The target was a catalpa grove near Troy, Ohio, similar to other commercial stands of timber around the state. Hordes of the catalpa sphinx and catalpa midge attacked the grove, threatening destructive defoliation unless swift action could be taken. Trailing a swath of lethal white dust (arsenate of lead powder), the Air Service plane made six passes over the menaced grove, completing many hours of work in only a few minutes. Within weeks, enthusiastic reports of the aerial dusting experiment began appearing in aviation periodicals, entomological journals, newspapers, and *National Geographic*.

The successful operation at Troy emphasized the value of airplane dusting to control other insect pests, especially those that attacked cotton. In 1922, the first cotton-dusting experiments against the cotton leaf worm were conducted under the direction of Dr. B. R. Coad around the Delta Laboratory of the United States Bureau of Entomology at Tallulah, Louisiana. Again, the Air Service supported the experiments, donating pilots and all maintenance personnel for its three Curtiss JN-6 planes. One of the planes was equipped for aerial photography; the aerial maps it furnished proved valuable aids for the pilots to locate and study their target areas beforehand. Coad's coworkers perfected dusting paraphernalia, formulas, and application techniques that justified a confident prediction for the future value of airplane dusting on the basis of economy, effectiveness, and speed. Depending on the location of the various fields he had to cover, a pilot could dust from 240 to 500 acres per hour. Additional work in 1923 used DH-4B aircraft equipped with improved hoppers or larger capacity, and new trials at Tallulah made Coad's group confident that aerial dusting could also control the cotton boll weevil. To help publicize this promising new facet of farm operations, the

Department of Agriculture persuaded the Air Service to cooperate in producing a 2,000-foot motion picture, *Fighting Insects from Airplanes*.

Although aerial dusting proved valuable for applications to peaches, pecans, walnuts, wheat, alfalfa, tomatoes, and peppers, the cotton boll weevil remained the prime target for airborne operations that treated 500,000 acres of cotton in 1927. Commercial aviation groups quickly sensed the possibilities, and several airplane dusting companies began operations around the mid-twenties. Huff-Daland, one of the earliest dusting organizations, was always closely associated with Coad's experiments at Delta Laboratory. Huff-Daland's men worked with Coad as early as 1923, and with his encouragement, they developed a special airplane for the particular requirements of dusting work. In addition to Huff-Daland, the Delta Aero Dusters of Monroe, Louisiana, operated twenty planes at ten bases. Their activities included dusting orchards as well as cotton, potato, watermelon, and tobacco fields. Quick Aeroplane Dusters of Houston, Texas, which started in 1925, flew ten aircraft and dusted 75,000 acres in 1928. Depending on the size of the field, prices varied from 25¢ to 60¢ an acre for each application—usually three to five. Although most aerial dusting was carried on by companies of the South, the Morse Agricultural Service of New York ranged as far afield as Indiana in search of dusting contracts. Huff-Daland even conducted international operations. In 1927, the company crated up seven aircraft and shipped them to Peru, where they attacked a cotton boll weevil infestation.

The versatility of aircraft in dusting spurred trials in a variety of other tasks, frequently involving the cooperation of military aviation services with other government agencies. Aerially applied substances to control locusts and malarial mosquitoes proved successful. The Department of Agriculture used aerial observation to make crop estimates, and the Coast Guard spotted schools of fish for commercial fishermen. Aircraft efficiently seeded flooded rice fields and scattered grass seed across logged-over areas of the Pacific Northwest. The promising use of aircraft in forest-fire patrol, particularly in remote and mountainous regions, received strong encouragement and support from influential technical journals like the *Engineering News Record* and the *Journal of Electricity and Western Industry*. As it wrestled with the constant task of updating its charts of seaboard cities and the continuously changing coastline of an entire continent, the Coast and Geodetic Survey turned to aircraft and aerial photography.

In an age of rapid urban development, aerial surveying provided a new means for tax reassessments for the benefit of taxpayers and growing communities running short of revenue. Up-to-date tax maps were essential because equalization of tax rates was necessary to maintain confidence of present and future city taxpayers. Unfortunately, the cost of mapping at a rate that would keep up with urban sprawl was prohibitive, at least until the advent of aerial photography. The use of aircraft in tax equalization was inaugurated in 1927 by Edward A. McCarthy, an expert in municipal appraising, whose Municipal Service Company had experience with aerial maps as well as older survey methods. The first city to be reappraised with the use of air maps was Middletown, Connecticut, whose 42 square miles were mapped in 60 days at a cost of $4,000. By way of comparison, McCarthy's own home town, New Britain, was only

13 square miles, but the old survey methods used there had cost $48,000 and required four years to complete.

Aerial photography was an obvious aid in regional planning of public power and related programs. R. C. Starr, a construction engineer of the San Joaquin Light and Power Corporation, explained the role of aviation in the King's River project, a 10-year construction program for 11 plants with a combined capacity of 500,000 HP. For projects of this type, engineers needed highly accurate figures on the precipitation in a particular locale, the nature of the drainage area, and the volume of runoff in order to compute the stream flow and the potential of the hydroelectric plant. It was possible to make an estimate for a proposed project, using the available records of stream flow in similar regions. Variations occurred, however, and accurate records had to be compiled before realistic estimates could be made for a given area. Aerial survey became the only method by which mountain-summit area drainage reservoirs could be economically and accurately estimated. Ground survey crews working along the trails might miss small springs and streams beyond a ridge; planes could continue work during seasons when deep snow would bring ground operations to a complete standstill; surveying parties were spared the hazardous and time-consuming negotiations of dangerous cliffs and trekking over snow fields. After using aerial photography, Starr commented that "the results obtained . . . have proven beyond doubt the value of these [aerial] surveys for preliminary studies of large hydroelectric projects."

The rugged, remote regions of Alaska, surveyed and mapped for the first time in 1926 and again in 1929, provided one of the most imaginative and dramatic tests of the usefulness of aerial photography in natural resource inventories. The Department of the

A stable and reliable design, the FC-2 was extensively used in photographic work and was favored by many bush pilots. In Canada and Alaska, the plane was usually equipped with skis or floats, depending on the season. Courtesy of Fairchild Republic Company.

Interior asked the navy to do some aerial mapping in order to augment the knowledge of the mineral and other natural resources of Alaska, and the project eventually included the cooperation of various government bureaus: the Bureau of Fisheries and the Lighthouse Service needed various types of information; the Forest Service wanted timber estimates; the Division of Roads in the Department of Agriculture was interested; and so on. The two surveys yielded aerial photographs for maps covering 23,000 square miles of territory. It was a remarkable project. The area of operations spanned nearly 1,000 miles from Ketchikan to Anchorage over inpenetrable forests, mountain fastnesses, and vast glaciers, where a crash or even a forced landing left virtually no chance for escape or rescue. As if these hazards were not enough, the grumbling crews found it necessary to set up a night watch at one point because their planes were being endangered by errant ice floes. It was a demanding test for cameras, planes, and pilots. Photographers on the expedition produced striking portraits of a magnificent wilderness while recording the potential of its natural resources. The discovery of important watersheds for power sites, coupled with accurate timber estimates, prompted negotiations for new pulp mills in Ketchikan. With an estimated capacity for pulp production equaling one-third of the daily requirements in the United States, this promised to be an important source of indispensible newsprint. The expedition returned with a set of previously nonexistent daily weather reports for the area and presented the chart makers with new problems for revision.

The most widespread application of aerial surveying was probably in the petroleum industry. The first companies to launch such programs began in 1927, and by 1929, nearly every major company doing work in the West and Southwest was using aircraft to make survey maps. Aerial photos were used to pinpoint promising features, eliminating the considerable amount of time normally consumed by casting around on the ground. Companies also discovered the advantages of making a quiet survey of a promising area from the air without arousing the interests of speculators. Around the end of the decade, aerial photography was used to make the accurate survey required to lay down a 600-mile pipe line, and Fairchild Aerial Surveys had begun petroleum surveys out of its Dallas office that covered 30,000 square miles by 1933 and extended into Mexico. In practice, geologists learned to follow clues of vegetation and differentiation of soil types in order to trace an exposed rock ledge that continued underground. In one case where surface indications were hard to define, geologists used aerial photos to locate the contact point of two promising formations. The photos revealed a definite connection that could be seen under a stand of grain three feet high. When called for, accurate photographic features could be constructed by using stereoscopic techniques.

The advent of aerial photography also brought a new dimension to archaeology and anthropology. During the 1920s, European archaeologists elaborated some of the photographic techniques learned in World War I and effectively used oblique photography to reveal hitherto hidden details of many well-known archaeological sites. Aerial photos, properly taken and analyzed, revealed the outlines of buildings, drainage ditches, and other features where variations in soil composition caused vegetation to take on different, identifiable hues. American researchers used similar methods to record several dramatic discoveries in the United States as well as in Latin America during the 1920s and 1930s.

As aircraft flew over remote regions of the world, additional archaeological discoveries were made. In 1929, during an air route survey of the Caribbean for Pan Am, Lindbergh and a companion realized that their aerial perspective had revealed several pre-Columbian ruins in Mexico's Yucatan Peninsula. Later that year, with support from Pan Am and the Carnegie Institution, Lindbergh and professional archaeologists flew back and forth over the Yucatan for nearly a week and discovered a number of previously unknown sites. In 1931, an aerial expedition directed by the American Geographical Society led to one of the most memorable finds in South American archaeology—the discovery of the "Great Wall of Peru." Experiences such as these continued throughout the interwar years, culminating in 1938 during an anthropological expedition in remote highlands of the island of New Guinea, in the South Pacific. Undertaken by the Dutch government and the Smithsonian Institution, the expeditions flew in a twin-engine PBY Catalina flying boat over a hidden valley deep in the interior. Its discoverers named it Grand Valley and found that it stretched for 40 miles, supporting some 60,000 inhabitants, who engaged in a remarkably complex agricultural system. Although it is unlikely that anthropologists will ever uncover a totally unknown population and social complex comparable in size to the Grand Valley, aerial discoveries were still enriching archaeology into the 1960s.

Scanty and incomplete records available for the twenties give little indication of the total number of planes, civil and military, that were flown for agricultural treatment, forestry patrol, survey, photogrammetry, archeology and similar uses. It seems certain that the figure would be much lower than that of the total number of aircraft used in business flying, as is the case in the current aviation scene. Still, the airplane as an aerial implement found a place in a wide variety of utility roles. In such functions, the small numbers of available aircraft completed innumerable tasks faster and at lower cost when compared to conventional procedures of the era. Because of the speed and flexibility of the airplane, the measure of its increasing influence was not an arithmetic but a geometric progression. Within 25 years of the Wright brothers' first flight, the airplane's social impact was more complex than the Wrights had imagined, and the economic significance of utility aviation was far greater than might be indicated by the comparatively small number of aircraft involved. Aviation in agriculture became a widely accepted function within the United States and in international operations. The work in Alaska epitomized some of the most valuable assets of aerial photography: the ability to penetrate formidable geographical barriers; comprehensiveness; speed; and the bonus of aesthetic charm.

TECHNOLOGICAL TRENDS

The impressive progress of military, commercial, and general aviation owed a great deal to sundry technological advances. The extensive and planned research and development conducted by NACA is only one example. Meanwhile, new cadres of professionally trained aeronautical engineers began entering the ranks of the aviation industry, having completed specialized courses in coherently organized aero-engineering departments at colleges and universities across the country. In addition, a new foundation for aviation operations evolved, including the study of meteorology and weather forecasting as well as techniques for instrument flying. In short, the elaboration of a distinct technological infrastructure of aeronautics emerged in the post–World War I decade.

Breaking away from a purely advisory capacity, NACA had acquired by 1920 its own laboratory facilities, including a wind tunnel, at Langley Field, Virginia. NACA's growing team of engineers contended with the conventional problems of trying to correlate results from wind tunnel models with measurements from actual planes in flight, but with little success until a breakthrough came from the outstanding NACA researcher Dr. Max Munk. Munk had been an associate of Professor Ludwig Prandtl, whose original and brilliant work in theoretical aerodynamics at Göttingen, Germany, had contributed so much to German and European leadership in aeronautical science. Following his arrival in the United States in 1921, Munk probed NACA's tunnel problems and suggested compressing the air in a tunnel to 20 atmospheres. By using a one-twentieth scale model under these conditions, he argued, test results should correlate with data from a full-scale plane at normal atmospheric pressure.

NACA's first variable-density tunnel went into operation in 1923, yielding basic data for a series of pioneering NACA reports on wing improvements and setting the precedent for similar tunnels in the United States and abroad. But the modest five-foot design proved inadequate for propeller research, NACA successfully sought appropriations for a throat of 20 feet designed to accommodate a complete fuselage, engine, and propeller. Judged by the standards of 1927, this apparatus was colossal. It was also an unusually productive unit, contributing, among other things, to conclusive studies on the value of retractable landing gear and the alignment of engines on the leading edge of the wing for multiengine aircraft, factors that drastically reduced drag penalties. Finally, the big tunnel was used to develop one of the most successful aeronautical innovations of the twenties, the engine cowling.

Most American planes of the postwar decade mounted air-cooled engines, with the cylinders exposed to the airstream to maximize cooling. NACA's tunnel tests revealed that this style of installation accounted for as much as one-third of a plane's total drag. Although significant work on cowled engines proceeded elsewhere, particularly in Great Britain, NACA work provided the most dramatic success. After hundreds of tests, a NACA technical note by Fred E. Weick in November 1928 detailed a cowling design that enclosed the engine in a way that enhanced cooling while sharply reducing drag. During a transcontinental test flight, a 157-MPH plane equipped with the new cowling averaged 177 MPH. In the judgment of a contemporary expert, E. P. Warner, the introduction of the NACA cowling was "staggering." By 1932, according to one estimate, the increased operational efficiency of cowled American aircraft had saved $5 million.

Additional improvements in aircraft design and in engines during the postwar decade paved the way for economic success in air transportation. Compared to other,

NACA wind tunnel research during the 1920s contributed to many aerodynamic improvements. The first full-scale plane to be tested in NACA's propeller research tunnel, shown here in 1927, was the Sperry Messenger. The engineer standing in the tunnel opening shows the size of the facility, which was unusually large for its time. Courtesy of NASA.

earlier biplane designs, buttressed by a tangle of supporting struts and wires, streamlining became a prominent feature of later, modern aircraft. Aerodynamic research to reduce drag had begun prior to World War I, although the most successful designs were the Junkers, Fokker, and Dornier aircraft with cantilevered (internally braced) wings produced in Europe after the war. In Germany, Junkers and Dornier also pushed ahead with metal wing and fuselage structures. Early in the twenties, examples of all-metal, cantilevered European monoplanes attracted the attention of NACA, whose influence prompted further examination and exploitation of these features. One important result was the Ford trimotor of 1926, with its metal structure and cantilevered wing. Other research during the postwar era involved stressed-skin construction, using the aircraft skin itself to carry more of the load imposed on the aircraft in flight. This approach eliminated many internal trusses and braces within the wing and fuselage and contributed to a lighter, more efficient airframe design. Perfected at about the same time by designers in America and Europe, stressed-skin construction became a hallmark of modern aircraft of the 1930s.

The navy's radial engines benefited from improved cylinder heads and hollow, sodium-cooled valves devised by S. D. Heron, an English engineer employed by the Air Service at McCook Field. Heron joined the Wright Aeronautical Corporation in

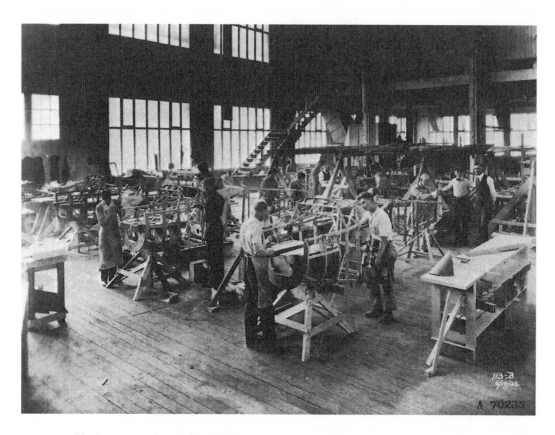

The Boeing production line of 1922 was a setting redolent of the pungent smells of hardwoods and glue. Courtesy of The Boeing Company.

1926, where he played an important role in designing the Wright J-5 engine, a classic motor that powered Lindbergh's *Spirit of St. Louis*. Together with Pratt & Whitney, Wright Aeronautical continued to introduce refined radial power plants that gave American aircraft an enviable record of efficiency and reliability, contributing to U.S. leadership in world aviation. Additional research at McCook Field concerned propellers. Increased engine power and high propeller tip speeds began to exceed the safe operational limits for conventional wooden propellers. Frank Caldwell, who presided over the propeller test section, finally devised a workable aluminum propeller in 1925. Also, by the end of the twenties, variable-pitch propellers were available on some engines for optimum settings at takeoff and at cruise. In the "thinner" air of higher altitudes, the turbo supercharger, perfected by Sanford A. Moss of General Electric, used engine exhaust gases to spin a turbine that forced additional air to the carburetor, approximating sea-level conditions. Producing high-octane, antiknock fuels for aircraft engines became feasible in the mid-twenties as a result of work by Thomas Midgely, and volume production of several grades of high-octane fuel became a reality at the decade's close.

The kaleidoscopic progress of the era embraced a number of "nonhardware" trends as well. From a handful of prewar courses dealing with aeronautical engineering, universities like the Massachusetts Institute of Technology evolved a panoply of professional course work leading to both undergraduate and graduate degrees in the subject. The Daniel Guggenheim Fund for the Promotion of Aeronautics provided money for similar programs at several other schools. In 1929, a survey by an aviation magazine reported that 1,400 aeroengineering students were enrolled in more than a dozen schools across the United States. The California Institute of Technology became a major beneficiary of the Guggenheim Fund's foresight. Although America possessed the facilities to train engineers and NACA offered superb facilities for practical research, the country lacked a nerve center for advanced studies in theoretical aerodynamics. Germany led the world in this respect, until the Guggenheim Fund lured the brilliant young scientist Theodore von Karman to the United States. Von Karman accepted a Caltech offer in 1929 and occupied his new post the following year. Within the decade, not only did Caltech's research projects enrich the field of aerodynamic theory, its graduates began to dominate the discipline in colleges and universities across the nation. The Guggenheim Fund's largesse was a tremendous stimulus to aeronautical engineering and research, as it was to the dozens of other aeronautical subjects that it supported. Between 1926 and 1930, this personal philanthropy disbursed over $3 million for a variety of fundamental research and experimental programs, including flight safety, that profoundly influenced the growth of American aviation.

Aviation medicine, generally neglected until World War I, began to win appreciation as a distinct medical specialty in the postwar era. Lt. Louis H. Bauer, graduate of the Harvard Medical School, played a prominent role in the field, first within the Air Service and later as medical director of the Aeronautics Branch of the Bureau of Air Commerce. He wrote a basic textbook, *Aviation Medicine* (1926), and helped to form the Aero Medical Association and its scholarly publication, *Journal of Aviation Medicine,* in 1929. Physiological research into issues of aptitudes for military flying helped furnish the basis for the technique of instrument flying. The transition from theory to practice owed much to Guggenheim funding, army support, ancillary research on flight instruments, and the skills of Lt. James H. ("Jimmy") Doolittle. Doolittle was already well known for several speed records and other aeronautical exploits. Behind his sensational headlines and apparent flamboyance, however, Doolittle combined sound airmanship and a thorough understanding of aeronautics, represented by his doctor of science degree in aeronautical engineering acquired in 1925 from MIT. Beginning in 1928, Doolittle helped to refine the necessary equipment for instrument flight. This included a radio homing device with visual cues; the Sperry artificial horizon and the Sperry directional gyroscope; and a more accurate barometric altimeter. The first complete "blind" instrument flying occurred on September 24, 1929. Flying the plane from a cockpit completely hooded to block his external vision (a back-up pilot occupied the front cockpit), Doolittle took off on instruments, flew a complete circle, and landed—without once seeing the ground.

Instrument flying was to be essential for commercial airlines striving to establish dependable service based on published timetables, as well as around-the-clock military operations in poor weather. Instrument flying represented another milestone in aeronautical development. In a related endeavor, the Guggenheim Fund spurred substantial progress in meteorological research and weather forecasting as an aid to flight operations. The U.S. Weather Bureau markedly expanded its activities because of the growth of flying, and the entire nation eventually benefited from regular, increasingly accurate forecasts. "Aid to American aviation," wrote Donald Whitnah, a historian of the weather bureau, "has involved more innovation, technological development, and opportunities for expansion of Weather Bureau facilities than any other field of service."

AN AVIATION COMMUNITY

The elaboration of an aeronautical infrastructure not only encompassed technical foundations, such as engineering and technical procedures for instrument flying, but also involved the rationalization of operations formerly conducted on an ad hoc basis, such as the layout and operation of airports. The maturing aeronautical infrastructure became evident in terms of recognized corporations that began to attract increasing interest from the investment community. Attendant issues such as insurance coverage, a legal framework, and the prospect of federal regulation led to concrete developments within the decade. Distinctive, aviation-related trade associations became more prominent in national activities. This institutionalization of aviation marked it as a mature technology, with increased public visibility and supported by its own interest groups. It developed its own unique financial, legal, administrative, and regulatory identity.

In 1929, Victor S. Clark's authoritative study of American industry, *History of Manufactures in the United States,* declared that the manufacture of aircraft occupied a position of "minor importance" in the hierarchy of American enterprise. Although the value of aircraft manufactures had doubled from $7 million to $14 million between 1925 and 1927 and the future held promise for growth, the study stated that the main value of aircraft manufacture to the engineering industry lay in its innovations in engineering technique.

Judging the industry by its peak of wartime achievement, as compared to its postwar record in 1929, the prospects for aircraft manufacturing may have seemed unimpressive. From April 1917 to November 1918, American producers delivered 13,984 planes valued at $113,721,043, with the capacity to turn out 21,000 aircraft per year at the time of the armistice. The war's end choked off the lucrative military contracts, and a mere 780 planes were built in 1919, all but 8 of them for the military. Until the middle of the decade the military services continued to claim most of the production, but the development of commercial aviation suggested new possibilities. The effect of the Kelly Act of 1925, which turned over mail routes to private contractors, is seen in a comparison of production figures in 1925, when 445 of 789 planes were manufactured for the armed services, and in 1926, when the military claimed only 478 out of 1,186.

In 1929, after the Air Commerce Act and Lindbergh, 779 military types were produced in contrast to 5,414 civilian aircraft; a total value of $70,334,107 for all aircraft, parts and equipment. As an increasingly accepted mode of modern transportation, aviation acquired enough momentum to weather the troubled economic climate of the 1930s. In spite of the depression, the book value of the three main sectors of the transportation industry indicates a steady increase by aviation. The automotive industry held the dominant position, but the aircraft industry did experience a significant rate of growth, even in the depressed thirties.

Aircraft exhibitions became more than an event for curious onlookers; they became showcases for buyers, suppliers, and investors. Economic activities notably increased in tempo following Lindbergh's 1927 flight. Several brokerage houses as well as banks published detailed prospectuses of aviation as an aid (and enticement) for potential investors. Between March 1928 and December 1929, $1 billion of aviation securities were traded on the New York Stock Exchange. With money available, the aviation industry experienced consolidation through acquisition and merger, a trend that could be seen in manufacturing as well as in airline companies. One of the largest and most broadly based combines was United Aircraft and Transport Corporation, organized early in 1929. By the end of the year, United's various properties included over half a dozen airlines and aircraft companies as well as engine and propeller manufacturers. The crash of 1929 did in many high-flying stocks and conglomerates, although Roosevelt's New Deal dealt the most telling blows in the antitrust fervor of the 1930s.

The burgeoning aviation industry of the late twenties created new job opportunities in many urban areas across the country, while presenting new issues of airport development and management. The Northeast, with an established network of special-ized factories and skilled workers, remained a major employer, particularly in power plants and accessories. The airframe industry spread westward to Kansas, where the prairies had few threatening topographical features, and to California, where a sal-ubrious climate encouraged year-round flying and where heating cavernous aircraft construction facilities and hangars cost less. Many cities, regardless of location, wrestled with the decision of whether to invest in airfield facilities. Under the Air Commerce Act of 1926, the government supplied funds for air navigation facilities but excluded airport development. Hence, private and municipal investment accounted for the overwhelming share of airport funding in the twenties. As air travel increased and the demands on airport facilities intensified, airport planners had to formulate guide-lines for auto traffic access, runway construction and drainage, lighting, and efficient runway layout. Such requirements prompted subspecialties in civil engineering and a new genre of technical books on the subject.

Indeed, the burgeoning of civil and commercial aviation through the decade necessarily spurred a number of ancillary specialties. Insurance agencies developed new procedures for underwriting aerial transportation, although most companies hesitated to issue any comprehensive policies until the Air Commerce Act of 1926 brought stability to aviation. This legislation helped establish a coherent body of laws and regulations for air travel. Legal and regulatory trends were complemented by a professionalized approach to air law reflected in the Air Law Institute at Northwestern

University, organized in 1929 (with substantial Guggenheim funding), and in the publication of its journal, the *Air Law Review*. Institutionalization of aviation was also apparent in groups like the National Aeronautic Association, which encouraged private flying and certified all official flying records, and the Aeronautical Chamber of Commerce, the voice of manufacturers, suppliers, and sundry commercial aviation interests.

AVIATION AND AMERICAN CULTURE

An increasing number of clues indicate that aviation was becoming a firmly fixed phenomenon in American society—an integral component of life and culture. Airplanes popped up in paintings by the popular Americanist Thomas Hart Benton and in musical scores ranging from avant-garde modernists like George Antheil ("Airplane Sonata") to the king of march music, John Philip Sousa ("The Aviators"). After Lindbergh's trans-Atlantic flights, dozens of popular aeronautical tunes appeared, and dozens of aviation books—both fictional and technical—rolled off the presses. Lindbergh's own autobiographical *We* enjoyed brisk sales, and other fliers wrote accounts of their respective record flights. In films, comic strips, and magazine advertisements, aviation emerged as an accepted token of popular culture (see chapter 8).

Women continued to be active participants in aviation development. Although a good deal of their activities tended to be promotional and sexist (the idea that if a woman could fly, anyone could), they earned increasing respect as aviation professionals. Amelia Earhart hit the headlines as the first woman to fly the Atlantic—as a passenger—in 1928. But she also began to amass her own flying records, became aviation editor for *Cosmopolitan*, and served as a consultant for several airlines. In 1927, Ruth Nicols received one of the country's first air transport licenses, and a number of women began to build careers as engineers and managers in the field of aviation.

Inevitably, the myriad social ramifications of aviation caused scholars to take note. Aviation received favorable discussion in a major contemporary study of the era, Preston W. Slosson's *The Great Crusade and After, 1914–1928* (1930), part of the distinguished multivolume series *A History of American Life*, edited by Arthur M. Schlesinger, Sr., and Dixon Ryan Fox. The effect of air transport on time and distance, bringing the nation closer together, merited notice in an article from the scholarly journal *Sociology and Social Research* appearing in 1929. Aviation won positive comment from such diverse authors as literary critic and essayist Gilbert Seldes, philosopher and historian Will Durant, and social critic and economist Stuart Chase.

In reducing aspects of time and distance, the role of aviation loomed particularly large in international relations. Even though the first airplane flights over the Atlantic occurred in 1919, most Americans read little about the impact of aviation on international relations until the U.S. Air Service completed its globe-circling tour in 1924. Lowell Thomas compared the flight to the voyage of Magellan, opening new avenues of commerce and ushering in a new era of civilization binding all nations closer together. In an address to the Franklin Institute in Philadelphia while the flight was in progress, Maj. Gen. Mason M. Patrick, chief of the Air Service, remarked that it had more than military significance. Aviation, he said, could also be a great boon to international communications, allowing people from the remote corners of the earth to come together to know one another better. Understanding was the result of knowledge, continued Patrick. Airplanes, the most destructive military weapon ever made, could well become important vehicles to promote peace throughout the world.

The odyssey of the *Spirit of St. Louis* crystallized the image of aviation as a factor in global friendship. The airplane became a symbol of a new human spirit of inter-nationalism. This was the aspect of Lindbergh's flight that stirred Harry Guggenheim most of all. "The world was impressed by the significance of the airplane not only as a lessener of this distance between the habitations of men, but between the minds of men as well," he declared. Writing for the *Review of Reviews*, J. J. Jusserand, the French ambassador to the United States from 1902 to 1925, remarked on the wild acclaim and unabashed joy with which the usually clannish Parisians welcomed Lindbergh as one of their own. On this occasion, Jusserand recalled thoughts expressed in an essay by the famous chemist Pierre Marcelin Berthelot just before his death in 1907. Of all the inventions destined to make great changes in the world, Berthelot had written, surely one of the most significant was the airplane. In the mastery of the air there would be no more frontiers to separate nations. Just so, said Jusserand; Lindbergh and the *Spirit of St. Louis* had inaugurated a new state of human relations. Myron T. Herrick, the American ambassador who received Lindbergh in Paris, emphasized the feeling of international good will generated by Lindbergh's flight over the Atlantic—a feeling so strong that it affected negotiations between the United States and France leading to the successful conclusion of the Kellogg-Briand Pact of 1928. As diplomatic historian Robert Ferrell assessed the flight later, it was "an adventitious event of breathtaking interest to the entire world, an event which tightened remarkably the hitherto loose bonds of Franco-American amity."

Subsequent flights of the *Spirit of St. Louis,* like others of the late twenties, effectively promoted international amity. The value of such flights in promoting international relations did not escape some shrewd diplomats seeking a dramatic gesture of American good will. A contemporary article by E. W. Burgess, professor of sociology at the University of Chicago, noted that the first premeditated project to promote international friendship through aviation was Lindbergh's Latin-American tour in 1928. Dwight Morrow, the American ambassador to Mexico (and Lucky Lindy's future father-in-law), had been searching for some action to improve Mexican-American relations, recently brought to a low ebb by the possibility of armed American intervention. Lindbergh embarked on an aerial excursion through Central and South America, making one triumphal entry after another in thirteen capitals. Climaxing a two-month tour, he arrived in Havana, Cuba, on February 8, 1928, to appear before a session of the Pan-American Congress. Dispatches from correspondents of the *New York Times* all along the route testified to the emotional response to his trip, which helped created a tremendous feeling of good will toward the United States and aroused a new interest in Latin America within the United States.

Taken in the mid-twenties, this photograph of Charles Lindbergh shows him as a working pilot in the days when he flew for Robertson Air Lines in open-cockpit mail planes. Courtesy of the Smithsonian Institution.

INTERNATIONAL AIRWAYS

Against a background of international amity, commercial interest, and national security, the United States forged its first significant intercontinental air routes with Latin American countries. Prior to the Lindbergh tour, business interests had made efforts to start air-mail links, but the major impetus occurred about 1924–25, when a Colombian airline operated by German interests and staffed by German nationals made strong moves toward an inter-American air-mail service. Not only did the move threaten United States hopes, but the route ran ominously close to the Panama Canal. American officials rebuffed the German-Colombian airline and in the meantime encouraged an American counterpart planning air routes between the United States and Latin America—Pan American, Incorporated.

Pan American eventually spread its wings under the guidance of Juan Trippe, an aviation enthusiast well connected with influential bureaucrats and eastern financiers. When the POD formally opened the bidding for the Key West–Havana route in the summer of 1927, Trippe adroitly outmaneuvered several rivals. The original Pan American group sold out to Trippe, who renamed it Pan American Airways, familiarly known as Pan Am. Before, during, and after these machinations, Trippe's government and business contacts won him the benediction of the POD, who saw Pan Am as the "chosen instrument" for U.S. foreign air routes and insulated Pan Am from other American competitors in order to make it a strong contender against European aeronautical interests in Latin America.

With this kind of unassailable patronage, Pan Am flourished, drawing Latin America much closer to the United States in terms of communication and travel time. The company augmented its original equipment of Fokker trimotors with Sikorsky S-38 flying boats and other long-range planes as operations expanded. The inaugural run between Key West and Miami, in 1927 reached only 110 miles, but with good management—as well as a virtual monopoly on foreign air-mail contracts—the company's route mileage soared. As of January 1, 1929, Pan Am had covered 261 miles, then grew to 5,275 miles by March, and expanded to 11,075 by August, with 4,500 more miles to be added soon. In the first six months of 1929, Pan Am flew 1,000,000 passenger miles, carried 7,000 patrons, and delivered 250,000 pounds of mail with enviable efficiency. Pan Am was not just an air-mail route. In an area of primitive land transportation and intermittent sea transport, the ability of the airline to save days and even weeks of travel made it a vital element in a new era of inter-American commerce. Panama could be reached in two days instead of twelve. Mail to Chile was delivered in nine days as compared to twenty; the three-and-a-half week ordeal by boat and burro to Tegucigalpa, Honduras, could be made in a day and a half by air. Two years after commencing operations, Pan Am had advanced from a 110-mile shuttle service to an international carrier delivering passengers, mail, newspapers, specialized cargo, and government correspondence to twelve countries over a route that tied together two continents.

The concept of an air-age world began to evolve. The first trans-Atlantic crossing by the NC-4 in 1919; the army's globe-girdling feat of 1924; Lindbergh in 1927; the intercontinental schedules of Pan Am by 1929—all prompted new thinking about global relationships. The universal sky destroyed the traditional geography of time-distances influenced by land and sea barriers and required a new type of air-age map. Two-dimensional maps had inherent limitations as to true shape, size, location, and distance. Some of these aspects were necessarily distorted in order to represent specific features, for each map had its advantages when designed for a particular need. The Mercator map was highly useful for a mariner in charting straight-line courses. For an ocean-basin geography, the Mercator map served admirably as a navigational aid, but it was not suited to the navigational capability of aircraft, whose straight-line courses were not limited to the availability of a navigable body of water.

One of the most suitable maps for the air age was the azimuthal (directional) equidistant map, in which parallels of latitude appeared as equally spaced circles along the meridians, as in the familiar polar projection. On such a map, all air routes from one city to another appeared as straight lines with their true directions preserved. To travel from Los Angeles to Moscow, a plane would not fly east (as one would suppose from a Mercator map) but north, over the Arctic. This dramatized a new element of geographical relationships brought about by the air age. People who were accustomed to thinking in geographical conceptions based on a Mercator map considered Los Angeles to be closer to Rio de Janeiro than to Moscow. But air-age geography put the Russian capital closer to Los Angeles and forced a reconsideration of the importance of Alaska and the Arctic. In 1925 the geographical implications of aircraft were recognized and treated by A. P. Berejkoff, a radio engineer for General Electric, in an article appearing in *Aviation Magazine*. Berejkoff called his map the "equidistant zenithal projection," resembling the standard azimuthal equidistant map in theory and appearance. Air transport lanes of the future, he pointed out, would follow a "great circle route"—a point worth noting, because such a route in the Northern Hemisphere usually came close to the Arctic.

A series of spectacular flights dramatized the potentialities of great circle and polar routes. In the spring of 1926, Comdr. Richard E. Byrd, veteran of a pioneering aerial exploration of North Greenland in 1925, organized the first flight over the North Pole. Accompanying pilot Floyd Bennett (another veteran of the Greenland operation) in a Fokker monoplane, the *Josephine Ford,* he departed Spitsbergen, Norway, reported reaching the pole, and returned on May 9, 1926. Just three days later, the American explorer Lincoln Ellsworth accomplished the same feat in the dirigible *Norge.* Between August 23 and October 31, 1929, in an often forgotten aerial feat, a Russian plane followed a great circle route while slicing over the Arctic fastness on a remarkable flight from Moscow to Seattle, and then proceeded to New York. Byrd capped this adventurous era of polar flights with the first trip over the South Pole on November 28–29, 1929. These ventures proved that aircraft could successfully operate within the

A Boeing 40 skims the Rockies during a mail run. Even with open-cockpit biplanes like this, aviation technology dramatically reordered time-distance relationships in the process of surmounting geographic barriers. Courtesy of The Boeing Company.

severe environment of polar regions, and, in particular, brought Arctic great circle routes of the future closer to realization.

During the late twenties, it became necessary to consider the entire globe from the standpoint of air transportation. All inhabited parts of the earth were affected by the "shrinking world" brought about by faster ocean travel and the development of the railroad and the automobile—and accelerated by the airplane. The airplane especially altered the significance of old routes of communication and prompted the need to reconsider the relationship of continents in the new light of air-age geography. Harry F. Guggenheim, who directed the Guggenheim Fund, wrote a book in 1930 titled, appropriately, *The Seven Skies*, an allusion to the universal "ocean" of the air. The book summed up extraordinary aeronautical progress—a plateau of maturity— particularly within the post–World War I decade. Although others shared this view, including influential magazines like *Aviation*, Harry Guggenheim put it most succinctly in his book. Announcing the termination of the Guggenheim Fund, he declared, "The pioneering period has come to an end."

Adventure, Airways, and Innovation, 1930–1940

In the 1930s, the accretion of the technology from the previous decade began to produce handsome dividends. Following the 1929 crash, the country's economy continued its downhill slide, but the curve of aviation progress ascended strongly. A series of innovations explored in the twenties reached a nexus of development in a new generation of modern airlines. These innovations included the concept of streamlining, variable-pitch propellers, wing flaps, engine cowlings, a miscellany of engine improvements, and other advances. Taken individually, such trends had not been particularly advantageous when adapted to the standard biplanes, trimotors, and conventional wood/fabric/wire construction of the twenties. In the thirties, advanced designs that featured cantilevered wings, retractable gear, and stressed-skin construction made these prior developments highly attractive for the first time. The confluence of technological currents was best represented in the trend-setting Boeing 247 and the Douglas DC-3. The general aviation sector, severely buffeted by adverse economic currents, eventually stabilized and began to incorporate aeronautical refinements similar to those featured on the larger aircraft entering airline service. The decade also embodied new advances in flight simulators, vertical flight, and high-altitude research. Finally, during the thirties the evolution of rocketry approached the point of practical application as a group of dedicated researchers perfected propulsion machinery, guidance systems, and structures.

THE ADVENTURERS

A succession of broken records for speed and long distance symbolized milestones along the path of aeronautical progress in the 1930s. In the years before World War II, the sheer novelty of airplanes brought more attention to these events than in later years, especially during the postwar era, when high-speed fighters, long-range bombers, and transoceanic transports became commonplace. During the thirties, record flights still attracted close attention, but the decade also marked the close of an epoch for dashing aeronautical exploits. The National Air Races, dating from the twenties, embellished by the Thompson Trophy for a closed-course circuit and the Bendix Trophy for a transcontinental dash, emphasized speed. In 1929, a speed of 194 MPH took the Thompson Trophy; Roscoe Turner posted a 283 MPH mark to win in 1938. Turner, with his powder-blue flying togs and waxed mustache, reigned as one of the flashiest pilots of the era and became renowned for aerial jaunts around the country in the company of a pet lion, trademark of his sponsor, the Gilmore Oil Company. Jimmy Doolittle and Jacqueline Cochran, among others, awed the huge crowds that made the annual National Air Races a major media event. Thousands more followed the races in papers and national journals that also gave daily coverage during the progress of sundry long-distance flights.

In 1931, Wiley Post piloted his Lockheed Vega, the *Winnie Mae,* around the world in less than nine days. Harold Gatty flew along as navigator for the flight, which began and ended in New York City. The episode dazzled public and press alike, and photos of Post, with his rakish eye patch, proliferated in tabloids across the country. Post topped his record feat in 1933, girdling the globe along a similar route in under eight days. Again in the *Winnie Mae,* its engine modified for high-altitude flight, Post completed a precedent-setting substratospheric research flight in 1935. Since the *Winnie Mae* could not be pressurized, Post solicited the aid of the B. F. Goodrich Company in developing a pressure suit. The flight demonstrated the feasibility of using the west-to-east jet stream as an assist for high-speed cruising, and Post's pressure suit set a precedent for suits used in research involving the X-15 research plane of the 1960s and for astronauts. When Post and the popular humorist Will Rogers died in a 1935 plane accident, the nation felt a profound sense of loss.

A flurry of speed, distance, altitude, and publicity flights punctuated the decade. Although several American and European women garnered headlines, Amelia Earhart held center stage. Earhart completed the first solo crossing of the Atlantic by a woman in 1932, set other records, and became an international celebrity before she mysteriously disappeared over the Pacific during a world flight attempt in 1937. Wealthy, enigmatic Howard Hughes built a remarkable single-engine monoplane, the H-1, and used it between 1935 and 1937 to set a new world's speed record of 352 MPH and a new transcontinental record of approximately seven and one-half hours. Among other things, the H-1, with a radial engine, proved that a carefully designed plane of this type, with a radial's large frontal area, could match the speed of in-line engine designs having less drag. The H-1's example carried over into many successful radial-engine aircraft of World War II. In 1938, after three years of careful planning, Hughes and a four-member crew flew a twin-engine Lockheed 14 transport around the world in 3 days, 19 hours, and 17 minutes—cutting Wiley Post's old record in half—and landed in New York before a clamorous throng of 25,000. The outbreak of European hostilities in the next year, along with deteriorating peace in Asia, closed the opportunities for similar world flights. An era ended.

Assessment of this period of record breaking poses problems. In headlines and

Amelia Earhart (fourth from the right) poses with friends during the National Air Races in 1929. The event also marked the origins of the Ninety-Nines (so named for the number of charter members), an organization of women pilots that assisted many others in entering aviation careers by offering scholarships, encouragement, and successful role models. Courtesy of the Ninety-Nines, Inc.

crowds, the men and women who flew the planes received the sort of acclaim granted to astronauts a generation later. Among other things, the air races presumably contributed to improved fuels and better engines. But racing planes were highly specialized, with designs not always suitable for agile combat planes or reliable, economical airliners. They were also dangerous. All seven models of the successful but controversial Gee Bee racers of the early thirties crashed, killing five pilots. Some knowledgeable experts looked askance at the continual tinkering with finely tuned engines that required carefully concocted formulas for fuel; skeptical engineers saw no fundamental contributions to flight technology. On the other hand, the air races brought attention to certain mechanical accessories, like retractable landing gear. The striking, long-range efforts of the decade highlighted enhanced reliability in aeronautics and focused attention on time/distance factors for national and international geography alike. Nowhere were these factors more evident than in the airline industry.

THE NEW AIRLINERS

European designers expended considerable time and money to develop aircraft to compete with the 247 and DC-3, which entered service in the 1930s. Ironically, many of the innovations that made the American planes so successful had originated in Europe. America seems to have assumed leadership because a handful of imaginative designers and engineers plied their trade in an environment less influenced by conservative bureaucrats in a nationalized or government-influenced industry than was generally the case in Europe. Moreover, intensive airline competition in the United States was a key factor in American advance to the forefront. In their authoritative study, *The Technical Development of Modern Aviation,* Ronald Miller and David Sawers concluded, ''The only basic difference between the American and European industries that provides any logical explanation for the former's greater liveliness is the existence of a competitive market for airplanes among the airlines of the United States, larger than that for military aircraft.'' The Boeing 247 and the DC-3 both owed their birthrights to this competitive environment as well as to elements of a technical legacy originating in Europe.

Early in the twentieth century, F. W. Lanchester in England and Ludwig Prandtl in Germany published studies on the greater efficiency of truly streamlined aircraft by reducing skin friction resulting from airflow. Some designers remained skeptical about the advantages of streamlining, and others found it difficult, because of construction techniques characteristic at the time, to come up with streamlined shapes like internally braced, cantilevered wings. And so, wood and fabric biplanes, carrying heavy drag penalties from their numerous struts and wires, were still favored. Eventually, Hugo Junkers, professor of mechanics at Aachen Technische Hochschule in Germany, and Anthony Fokker, a Dutch designer, began using cantilevered wing structures covered with metal and wood. With cantilevered wings, Junkers all-metal planes and Fokker plywood-wing aircraft were used successfully by several airlines in the mid-twenties. About the same period, Adolf Rohrbach designed several German planes using stressed-skin wings and fuselage, a technique he explained in a paper delivered in the United States during a meeting of the Society of Automotive Engineers in 1927. Thus, the stage was set for an important new period of development. Cantilevered wings permitted monoplane construction, a great advantage in streamlining and aerodynamic efficiency compared to biplanes, and stressed-skin construction opened the way to lighter, more efficient structures. In America, NACA reported these developments with great enthusiasm and began serious research on stressed-skin, all-metal wings, a move that influenced several new American designs of the 1930s.

Other means of reducing drag concentrated on the blunt contours of the era's radial engines. With its big, new wind tunnel, completed in 1928, NACA developed a full-engine cowling of greater efficiency that did not cause the engine to overheat. The virtues of these several developments were demonstrated by the Lockheed Vega, a design that became a household word in the wake of record flights by Earhart, Post, and others.

The Vega was designed by John K. Northrop. The original model, introduced in 1927, lacked a cowling for its engine and included no streamlined wheel covers for its fixed gear, but it featured a nicely streamlined fuselage of molded plywood topped by a cantilevered wing that incorporated Fokker's technique of stressed-skin, plywood construction. After being fitted with a NACA cowling, the Vega's cruising speed rose from 135 MPH to 155 MPH, a convincing demonstration of what aerodynamic

Ruth Nichols in front of a Lockheed Vega. Record-breaking flights as well as promotional junkets became familiar media events of the depression era. Courtesy of the Smithsonian Institution.

research could do for aircraft. The Vega was followed by the Northrop Alpha, a low-wing, all-metal plane. The Alpha incorporated two more significant construction techniques. Using research from the Guggenheim Aeronautical Laboratory of the California Institute of Technology (GALCIT), Northrop fitted the Alpha with wing fillets, or fairings, where the wing joined the fuselage, which enhanced its aerodynamic efficiency. The second technique involved wing construction, described by Miller and Sawers: "Aluminum sheet was stamped into a channel section, and the channels were riveted together to make up the . . . wing. The horizontal members formed the skin and the vertical members the wing spars, while additional sheet members were riveted in to form ribs." This significant Northrop innovation provided a rugged stressed-skin wing and a rugged aircraft. Alphas acquired by TWA did yeoman service as fast mail planes during the 1930s.

Boeing's Monomail was similar to the Alpha in its general layout. Like the Alpha, the Monomail benefited from stressed-skin metal construction, wing fillets, and full cowling. Moreover, the Boeing plane had retractable landing gear. Although the Monomail saw only limited service, its development influenced the Boeing B-9 bomber of 1931, which in turn influenced the design of the Boeing 247 airliner. Following refinement of engine cowlings, NACA investigated the optimum location of engines on the leading edge of the wing and learned to place the engine farther ahead of the leading edge in order to minimize aerodynamic interference between the propeller and the wing. The twin-engine Boeing B-9 and the Martin B-10 were designed to take advantage of this research, a vast improvement over the uncowled, underslung mountings of trimotors like the Fords and Fokkers. When the faster Martin B-10 won the bomber competition, Boeing reorganized what it had learned from the B-9 and designed the 247 airliner.

These basic advances in aircraft design were complemented by the maturation of several other key devices (including variable-pitch propellers and wing flaps) and the refinement of large aircraft engines. The idea of a variable-pitch propeller originated in Europe in the 1870s and was tried out on some airships prior to World War I. Although significant work was carried out in Britain and Canada in the twenties, it remained for F. W. Caldwell, an experienced American propeller designer, to conduct conclusive tests in 1930. Earlier mechanisms had been mechanically operated and seemed to work only in smaller engines because of the high stress placed on propellers by powerful motors. Because the newer, high-performance planes like the Monomail, Alpha, and others were definitely penalized in performance by using fixed-pitch propellers, Caldwell set aside his more complicated automatic controls and devised a two-position control for takeoff and for cruising. This propeller, supplied by Hamilton-Standard, was ready in time to be ordered for new transports under development, and enhanced their performance from airfields situated in the Rocky Mountains. Other companies soon entered the field, replacing the early hydraulically controlled designs with electrically operated, automatic controls. Several lines of research on multi-dimensional wing flaps had developed during World War I, particularly in England and Germany. In America, airplanes at the start of the 1930s used a simpler design called the split flap. The low drag of planes like the Alpha and others gave them a shallow gliding angle for landing, required longer runways, and complicated landing approaches to airfields bordered by hills or other obstructions. Split flaps not only increased the drag and permitted a steeper descent, but also increased lift at lower speeds, enhancing control during the approach and landing as well as takeoff.

Finally, there were the engines. Most of the engines available in the early twenties had evolved out of World War I combat experience. They were designed to give high power between frequent overhauls; neither combat planes not their engines were expected to have a long service life. These engines needed overhauls every 50 hours or so; of the total operating costs for an airplane, airlines had to allow about half for engine maintenance alone. By the late twenties, American engine manufacturers were receiving as much as half of their income from airline sales, and the airlines put more emphasis on durability and longer intervals between overhauls. Manufacturers responded to airline requirements. Early versions of the Pratt & Whitney Wasp needed overhaul every 150 hours; later models rose to 300-hour intervals in 1929 and reached 500 hours in 1936. By this date, maintenance costs had dropped 80 percent, and these savings represented perhaps the biggest contribution of engine manufacturers to airline development. Moreover, designers squeezed increased power from engines of basically the same size. One model of the Wright Cyclone put out 550 HP in 1930, yet was producing 1,100 HP by 1939. The power-weight radio of power plants like these improved while the fuel consumption dropped. The remarkable progress in engine durability, horsepower, and efficiency was partly due to improved designs of the cylinder heads and pistons. Equally significant were better quality fuels, identified by antiknock ratings and improved by antiknock additives. The advantages of reliability, increased horsepower, and fuel efficiency made it feasible to produce planes as big as the DC-3 in the more efficient twin-engine configuration. Not only did engine trends contribute to the phenomenal success of the DC-3, they also helped account for the successful development of later generations of four-engine transports and flying boats for transoceanic service.

These collective developments set the stage for the evolution of the DC-3. The decision-making process that brought them together began on March 31, 1931, when a Fokker trimotor of Transcontinental and Western Air (later, Trans World Airlines, or TWA) lost a wing and crashed in a Kansas farm. All eight passengers and crew died, including the idolized football coach of the University of Notre Dame, Knute Rockne. Shocked and outraged, the American public demanded stern measures. Under authority delegated to the Aeronautics Branch of the Department of Commerce, the government grounded all Fokkers in service in the United States, and then restricted them to mail service only—no passengers. When TWA began scouting for a replacement, the most attractive candidate was the Boeing 247. But Boeing was part of a large conglomerate, the United Aircraft and Transport Corporation, which also included United Air Lines. Within this cozy corporate family, United held options on all 59 Boeing 247s under production, and Boeing would not supply any of them to TWA until all of United's planes were delivered. Prodded by TWA's vice-president in charge of operation, Jack Frye, the airline's engineering staff quickly pulled together specifications for a new trimotor and started shopping for a supplier. The only interested manufacturer was the Douglas Aircraft Company.

Matching the Boeing would take some doing. The 247 used just about every advanced construction concept available at the time: an all-metal, stressed-skin mono-plane; a pair of cowled engines faired into a low-slung wing; retractable undercarriage; sound-proofed fuselage; and performance to carry 10 passengers coast to coast in approximately 20 hours. "It was the challenge of the 247 that put us into the transport business," Donald Douglas recalled. Within five days, Douglas had put together an initial proposal, with engineers ironing out the last fine points while en route to TWA's offices. Although TWA had originally asked for a trimotor, Douglas felt that the memory of the Rockne tragedy would discourage passengers from traveling in them. Douglas proposed a twin-engined model, the DC-1, that included all the benefits of the Boeing 247 plus other features to give TWA some crucial advantages over its chief rival, United.

For one thing, the DC-1 was bigger. With specifications for 12 passengers, the Douglas design promised more favorable seat-mile costs. Unlike the 247's short-chord cowling, the DC-1 had the highly efficient NACA design, and its wing flap installation made it far more efficient to operate from a variety of airports. The DC-1 also had variable-pitch propellers. As for passenger comfort, the 247's wing spar ran right through the fuselage, forcing patrons to climb over it during every entry and exit. The Douglas fuselage rested atop the wing, with a much more efficient spar structure. This last detail had an interesting background, and also helped account for the later DC-3's phenomenal longevity. Douglas had acquired the Northrup Corporation in 1932. Northrup's stressed-skin, multicellular wing construction, as used in the Alpha, was incorporated into the DC-1 wing design. Since TWA had used Alphas before and had been impressed with their ruggedness, TWA insisted that this technique be adapted to the DC-1. The Douglas transport also benefited from the Alpha's low-mounted wing design, which permitted a roomy, uncluttered passenger cabin. Moreover, the evolution of the new Douglas transport decisively benefited from the new research facilities at GALCIT. Wind tunnel tests there resulted in the addition of special fairing around the wing root to end buffeting. Several different wing and flap arrangements had to be tested before a successful scheme was struck, and GALCIT research gave the wing its distinctive swept-back appearance to solve longitudinal stability problems. Finally, GALCIT advice was instrumental in the decision to use NACA cowlings for the engines.

On July 1, 1933, eleven months after Jack Frye's first step toward a Fokker replacement, the DC-1 prototype flew. Even though it met or surpassed specifications, it never went into production. A more powerful Wright engine was available, so Douglas and TWA immediately agreed to produce the more powerful DC-2. The DC-2 went into service in 1934 on TWA's Newark-Pittsburgh-Chicago run, completing the trip in 30 minutes less than United's 247 flight over the same route. United attempted to upgrade its fleet with the 247-D, an improved model with more powerful engines and NACA cowls, controllable-pitch propellers, and other refinements. Seeking favorable publicity, Boeing decided to fly the 247-D in the MacRobertson Air Race from London to Melbourne in the autumn of 1934. Piloted by the flamboyant Roscoe Turner, the Boeing still came in third. A specially designed de Havilland Comet racer took first; second was a standard DC-2 of Royal Dutch Airlines, which flew in with six passengers and 400 pounds of mail.

Having seized an obvious advantage over rival Boeing in the production of commercial airliners, Douglas proceeded to consolidate its lead by producing the classic DC-3. The initial step came from American Airlines, who had been using Curtiss Condor biplanes with sleeping berths on the New York–Chicago run and now wanted a larger plane for coast-to-coast sleeper service. The result was a new version of the DC-2 known as the Douglas Sleeper Transport, or DST, with accommodations for 14 berths. The day-plane version, with seven additional coach seats, became known as the DC-3. The plane's 50 percent gain in payload accrued (like the DC-2) from bigger and better 1,000-HP engines, to be supplied by either Wright or by Pratt & Whitney. And once more, GALCIT's wind tunnel resolved a potentially disastrous instability problem that had resulted from stretching and widening the DC-2's fuselage and attaching larger wings.

In some ways, the DC-3 was an example of an "overdesigned" airplane. Aeronautical engineering was still emerging as a discipline, and Douglas used materials and methods more suitable for larger and more advanced aircraft. But these design approaches help explain why the DC-3 seemed nearly indestructible. Other basic features were significant as well, such as the multicellular wing, already proven on TWA's Alphas. Moreover, with its "day-coach" capacity for 21 passengers and overall aerodynamic efficiency, the DC-3 boasted seat-mile costs as much as one-third to one-half lower than its contemporaries. In the spring of 1936, American Airlines put its DC-3s into regular service on the New York–Chicago run, inaugurating the first airline operations that made money just by hauling passengers, freeing companies from total dependence on government mail contracts for a profit. United, who had started it all with its 247 fleet, swallowed hard and began to replace its Boeing planes with DC-3s later in the same year. Two years later, an estimated 80 percent of U.S. airline passengers were riding in DC-3s, and over two dozen foreign airlines were operating the Douglas transport as well.

The new airliners unquestionably set new standards of passenger comfort. In the trimotor era, the nose-mounted engine had transmitted a heavy dose of noise and vibration directly back to the passenger cabin, where the discomfort was amplified by the proximity of two additional underwing engines. For the DC-1, Douglas hired an acoustics expert, Dr. Stephen J. Zand, of the Sperry Rand Corporation, to create a cabin at least as quiet as a railroad Pullman coach. Using data from special acoustic instruments, Zand dictated a variety of changes. Soundproofing materials went into fuselage walls and into the bulkheads dividing the passenger cabin. Zand's recommendations also extended to ventilation ducts, structural joints, engine mounts, and engine exhaust systems. Carefully upholstered passenger seats were installed on rubber mounts, and the arm rests were also padded to reduce both vibration and noise. Altogether, the soundproofing effort cut the decibel level by 25 percent.

The DC-2 and DC-3, with their greater speed and higher ceilings, also permitted smoother flying for passengers. Nonetheless, pilots and passengers alike were still subjected to memorably turbulent flights. As a copilot with an early airline, E. K. Gann remembered one such run between Albany and Newark, an area notorious for its

Polished to a mirrorlike finish, the first DC-3 to fly prepares to take off on December 17, 1935, in Santa Monica, California. The original version, the DST (Douglas Sleeper Transport), shown here, was equipped with 14 berths. The airport, appropriately named Clover Field, was bordered by a golf course, seen in the background. Courtesy of McDonnell Douglas Corporation.

thunderstorms. During the flight, the amazed Gann watched the altimeter unwind at 1,500 feet per minute when the plane hit a violent downdraft. With a sickening jolt, the DC-2 hit an updraft and surged upward again, even though the captain had the nose down and the landing gear extended. Passengers invariably suffered ill effects from such experiences. Gann vividly recalled the prevailing environment of many flights during the summers. "The . . . air is annoyingly potted with a multitude of minor vertical disturbances which sicken the passengers and keep us captives of our seat belts. We sweat in the cockpit, though much of the time we fly with the side windows open. The airplanes smell of hot oil and simmering aluminum, disinfectant, feces, leather, and puke . . . the stewardesses, short-tempered and reeking of vomit, come forward as often as they can for what is a breath of comparatively fresh air." Such was the romance of air travel across the United States in the late 1930s.

Although the DC-3s dominance in airline equipment was formidable, it was not total. Lockheed produced the Electra, a smaller, 10-passenger transport whose two Pratt & Whitney radials gave it a cruising speed of 203 MPH—faster than the DC-3. Variations of the Electra were successfully used on shorter routes in the United States and abroad during the late thirties, and later versions saw service as American and English medium bombers in World War II. With planes like the DC-3 and Electra raising domestic airline service to new standards of appeal and efficiency, additional international and intercontinental routes reached out from the United States. Using a family of Sikorsky flying boats, Pan Am served more and more cities throughout the Caribbean and Latin America. American political, commercial, and trading interests in the Pacific and Asia now led to consideration of new long-distance routes and the acquisition of equipment to cope with the inherent dangers of extended flights over the vast reaches of the Pacific.

The first plane to enable such far-flung operations was the Martin M-130, the famous "China Clipper," a remarkable airplane for its day and age. An imposing, four-engine flying boat, the Martin Clipper carried 48 passengers and a crew of six for distances up to 3,200 miles. The new transoceanic operations of the Martin Clipper required extensive route surveys and the creation of a web of support facilities throughout the Pacific area. Because the planes might have to make unscheduled stops where support facilities were not easily accessible, Pan Am flying crews were expected to pass stiff exams in airframe and engine studies, navigation, and miscellaneous subjects, including history and culture of areas en route—and at least one foreign language. Capt. William M. Masland, who flew the first Clipper routes, recalled that such catholic training frequently came in handy:

> I have been at the local met [meteorology office] with my navigator making my own forecast and my own flight plan while the first officer was loading the ship, the engineers beating out some needed fittings at the local blacksmith shop and the two stewards baking their own bread in an oven they had constructed on the beach. Then we loaded fuel out of a surf boat, loaded it [into the plane] to our own specifications, and were off, often to a place none of us had ever seen before, sometimes to a place no one had ever flown before.

Beginning in 1935, the Martin Clipper delivered mail from San Francisco to Manila in the Philippines, flying via Hawaii-Midway-Wake-Guam, completing the run over six days while using 60 hours of actual flying time. Passenger service began in the autumn of 1936; the next year, Sikorsky S-42 flying boats extended the route from Manila to Hong Kong—an aerial link of 8,500 miles from America to the Asiatic mainland.

Martin Clippers in turn gave way to even larger Boeing Clippers with a range approaching 4,000 miles. Boeing delivered 9 of these mammoth planes to Pan Am, who used them on Pacific routes and began trans-Atlantic passenger flights with them in 1939. Pan Am's Atlantic service, an impressive achievement over a region legendary for its formidable winds and storms, nevertheless trailed the first passenger air routes by three years. Trans-Atlantic passenger flying to North America began with the colossal German Zeppelin, the *Hindenburg,* which made 10 round trips (at a sedate 78 MPH) between Germany and the United States in 1936. Although the dirigibles suspended operations during the Atlantic's intimidating winter weather, summer schedules remained popular. But the inaugural voyage of the 1937 season ended in catastrophic destruction at Lakehurst, New Jersey, as the hydrogen-filled airship went up in a burst of flames. Fascination with passenger-carrying dirigibles quickly waned, and airplanes like the Boeing Clipper soon dominated transoceanic routes. With an empty weight of 50,000 pounds, the Boeing Clippers mounted four 1,200-HP Wright "Double Cyclone" engines and cruised at 183 MPH. Unlike dirigibles, the big flying boats scheduled arrivals and departures throughout the calendar year. Biweekly flights to Southampton, England, (via Newfoundland) or to Marseilles, France, (via Lisbon, Portugal) took about one-and-a-half days. When World War II began in Europe, the Boeing Clippers (including three delivered to the British government) provided an invaluable communications and transportation link over hostile Atlantic waters. After America became involved in the conflict, the U.S. War Department operated the big Boeing flying boats to maintain vital wartime passenger traffic, using Pan Am crews to do the flying under contract.

The era of the big flying boats represented an alluring interlude in airline travel. Competing directly with the steamship lines for a well-heeled clientele, they offered sophisticated service, vintage wines, and *haute cuisine.* The romantic adventurer Richard Halliburton had written of a *Flying Carpet,* and aircraft with the speed, range, and creature comforts of the Clippers made it a reality. Nor were the seasoned pilots who flew the intercontinental routes immune to this unique milieu and to the ease with which they now girdled the globe. "There were days or nights of great beauty," Captain Masland wrote, "a quiet night at anchor in the remote port of Augra dos Reis (Anchorage of the Kings) on the Brasilian coast; an anchor watch on a cold clear night in the Persian Gulf with the stars hanging as low overhead as the lamps in a mosque." Summing it up from the perspective of 1964, he concluded pensively, "The pattern is gone, as well as the ships."

The end of the Clipper era came with the development of a new generation of long-range, pressurized, four-engine airliners. Their enhanced reliability negated the need for a seaworthy plane if it happened to be forced down, and the proliferation of air

The Boeing B-314 Clipper epitomized the flying-boat era of the 1930s. With a range of approximately 4,000 miles, Pan Am operated Clippers on the prestigious Atlantic route in 1939, offering passenger service from New York to Great Britain and France. Courtesy of Pan American World Airways.

travel around the world provided an adequate number of alternate airports within their range. The army had begun research on the design of pressurized cockpits in 1935, and carried out successful flight tests in 1937 with the XC-35, a completely pressurized version of Lockheed's Model 12 airliner (similar to the Electra and the Lodestar). The trend in commercial airliners began with the pressurized Boeing 307, permitting it to overfly threatening weather and dangerous storms. The 307 entered service in 1940 and was followed by the unpressurized Douglas DC-4, which first flew in 1942, and the pressurized Lockheed Constellation in 1943. Pressed into military service, the new airliners did not make their full impact felt in civilian service until 1945. With more powerful engines, higher speeds, longer range, and the ability to fly above bad weather, variations of these four-engine airlines made commercial transoceanic flying commonplace in the postwar era. In his informed study on *The Modern Airliner*, Peter Brooks categorized such hallmarks as being representative of what he called "the DC-4 generation."

The original DC-4E, funded by five airlines, rolled out of the Douglas plant in 1938 and made demonstration flights over the United system in 1939, but never entered production. Rising costs and the chance to acquire Boeing 307 Stratoliners broke up the five-company consortium. In any case, Douglas had already begun work on a smaller, refined version in 1939 that subsumed the DC-4 designation and enjoyed the support of United and American. Although all of the production DC-4s went to the army on the eve of World War II, development of the DC-4E and the DC-4 reflected significant trends in airliner development. The reworking of the DC-4, in particular, symbolized the way in which postwar projects stretched, improved, and modified a basic airframe into a series of new aircraft.

The Boeing Clippers carried up to 74 passengers, and Pan Am competed with ocean liners by offering first-class service. Liveried stewards served meals in a special dining area decorated with bowls of fresh fruit and vases of long-stemmed roses. Courtesy of The Boeing Company.

In America, a nation of superlatives (tallest buildings, deepest canyons, fastest cars), public relations writers of the day trumpeted the roll-out of "the biggest transport in the world" and noted "the gasps and awed onlookers" as they beheld the size of the DC-4E. Douglas press releases at the time stressed the tricycle landing gear, a first for planes of this size, and dubbed it the "Tri-Safety landing gear." Douglas claimed that the DC-4E fuselage was larger than a rail coach; such spaciousness, one of the plane's many features for passenger comfort, became standard. Company brochures listed complete soundproofing, deeply upholstered seats, steam heat, hot and cold running water, air conditioning, lavatories large enough for changing clothes, electrical outlets for personal appliances like shavers, and a well-equipped, electrical galley. There were additional luxurious touches, interesting in the twilight of the depression era, such as all 40 seats being convertible to berths and a private compartment and dressing room available as a "bridal suite." Such embellishments, though not standard in subsequent transports, reflected the growing concern for passenger satisfaction.

In the trimotor era, professional interior designers with qualifications in industrial design seem to have been enlisted as an afterthought. By the time of the DC-4E, however, the well-known industrial design firm of Howard Ketcham, Incorporated, was brought in early in the game as a standard operating procedure. In a related move, Douglas also enlisted the R. M. Grimstead Company, efficiency experts in hotel restaurant operations. As a result of this firm's time-and-motion studies (getting food to the galley, and from the galley to the passengers), the galley was redesigned and relocated closer to a door to facilitate food services. Another impact of these studies was United's decision to set up its own flight kitchens in 1936. Located on site at major airports, the airline kitchens eliminated the hassle of transferring prepared food from caterers miles away in city centers, a trip that indiscriminately scrambled the vegetables with the desert and also resulted in cold meals. The success of "the DC-4 generation" owed much to these collective refinements.

The DC-3, DC-4, and their counterparts of the 1930s served with distinction as military transports in World War II. They did so with little modification, and were often flown by the same civil airline crews who had learned to operate them across New Mexico, Nebraska, and the Great Lakes, rather than Iceland, North Africa, or the South Pacific. It is a tribute to technology and training that planes and pilots shifted so effectively from civil operations to wartime demands and back again after 1945. There was no counterpart to this phenomenon in the air services of the Axis powers or those of the Allies.

AIRWAYS AND AVIATORS

In the summer of 1932, the Democratic party held its national convention in Chicago. The charismatic governor of New York, Franklin Delano Roosevelt, controlled a strong block of delegates but was still short of the necessary two thirds for nomination. Al Smith, who had waged a strong campaign in a losing cause against Herbert Hoover during the presidential race of 1928, emerged as Roosevelt's strongest challenger at the convention. The Great Depression formed a backdrop to political debates as partisans of Roosevelt and Smith maneuvered for victory while, at the same time, attacking the Republicans for their failure to resolve the economic crisis. Hundreds of miles away in Albany, the state capital, Roosevelt closely followed the convention proceedings. When the Roosevelt forces eventually triumphed, the chairman of the convention interrupted a noisy demonstration with a rather startling announcement: Mr. Roosevelt had sent a wire, informing the delegates that he was flying to Chicago the next day to accept the nomination in person.

Roosevelt's decision marked a bold and unprecedented move. "In abandoning the tradition by which acceptance took place in a formal ceremony some weeks after the convention," explained FDR's political biographer, Arthur Schlesinger, Jr., "Roosevelt was responding to what he perceived as a passionate popular hope for a bolder temper in national affairs." The trimotor plane that carried the presidential nominee, Schlesinger continued, "was itself a symbol of the new spirit of decision demanded by troubled times." Early on a Saturday morning, Roosevelt and a group of friends and advisers took off for Chicago. Rain squalls en route made it a rough flight, but the nonplussed FDR kept revising his acceptance speech as the plane bobbed and weaved westward. With refueling stops at Buffalo and Cleveland, plus the poor weather, the plane was two hours late when it touched down in Chicago at about 4:30 in the afternoon. Smith had already departed for the train station.

By 6:00 in the evening, the candidate, in fine spirits, entered the convention hall to a thunderous reception. In the lapel of his blue suit, Roosevelt wore a fresh red rose. Flashing a broad, infectious grin, he acknowledged the cheers of the party faithful. The delegates knew about the long plane trip and about the stomach-churning weather, neither of which seemed to have fazed the nominee. "I regret that I am late," Roosevelt told them, "but I have no control over the winds of Heaven and could only be thankful for my Navy training." And the audience loved it.

There was little doubt that the convention crowd regarded Roosevelt's trip as something out of the ordinary. A large number of Americans in the 1930s continued to view air travel as a risky business. Some businessmen said nothing of their airline trips to wives and creditors. At the same time, they enjoyed the prestige of airline travel and boasted to business friends about flying in order to keep business appointments. The venturesome image of flying eventually gave way to improvements in aircraft, as well in administration, organization, and marketing.

Dramatic changes in flight equipment brought changes in the nature of flying. As early as 1929, nostalgia for the old days had already begun creeping through the ranks of airline pilots. One Boeing Air Transport pilot, a veteran of early army and air-mail flying, admitted that he liked flying too much to quit, although he felt that the profession had certainly changed. "A flier in the old days was in life, clear up to the hilt," he said. "It hurts, in a way, to see these Boeing pilots climbing up into heated cabins, wearing business suits and straw hats, and talking to somebody over the 'phone all the way over the run." Gone was the time when the pilot was his own commander, basing all decisions on individual judgment. He now got weather advice from a battery of meteorologists, instruction from the ground by radiophone, and flew a course by radio signal which led him safely to lighted airfields. A bit wistfully, the old pilot added, "all these improvements and safety measures have taken most of the adventure out of the business." In spite of these advances, not all airline pilots were able to take advantage of them since many companies could not afford such luxuries. Moreover, airline aviators of 1929 remained untouched by refinements like automatic pilots, deicers, retractable gear, variable-pitch propellers, airline hostesses, and additional regulations formulated by the Civil Aeronautics Administration—these things all lay in the future.

The increasing complexity of American civil aviation proved frustrating to the Aeronautics Branch of the Department of Commerce. Under the Air Commerce Act of 1926, some functions were assigned to existing units of the Department of Commerce, so that the Aeronautics Research Division was placed in the Bureau of Standards, and the Airways Division ended up in the Bureau of Lighthouses. During 1933, New Deal bureaucrats realigned the government's fragmented aviation components into a more logical structure under the Bureau of Air Commerce, still within the Department of Commerce. But this did not mean the bureau steered a course clear of controversy.

Airways and flying in the United States became subject to increasing rules and constraints, which occasionally erupted into dogfights between the government and the aviation industry. One such imbroglio flared in the aftermath of a 1935 crash involving a DC-2 flown by Transcontinental and Western Air. Among the victims was Bronson M. Cutting, a highly popular Republican senator from New Mexico. The death of a respected senator seemed hardly enough to spark a full dress review of the Bureau of Air Commerce, but Cutting's death came on the heels of the army's disastrous attempt to fly the mails in 1934, along with other incidents that had antagonized the aviation community (see chapter 4). Congressional hearings concerning the Cutting affair underscored the growing complexities of radio communications, aviation weather forecasting (both factors relating to the crash), and the relationship between the airlines and the Bureau of Air Commerce, which also had responsibility for the investigation of aviation accidents. Following an investigation, the bureau placed the blame on Transcontinental and the plane's pilots. The airline's convincing rebuttal provided ammunition for critics of the New Deal, and subsequent congressional hearings uncovered worrisome inconsistencies in the bureau's objectivity when it investigated crashes that might be related to faults in its own structure and operational procedures.

Airways modernization proceeded rapidly during Franklin D. Roosevelt's second term. It must be admitted that Eugene L. Vidal, director of the Aeronautics Branch of the Department of Commerce, had been operating on parsimonious budgets. Vidal's successor, Fred D. Fagg, Jr., an expert in aviation law from the Northwestern University Law School, effectively reorganized the bureau and won a 60 percent increase in appropriations, resulting in a budget of over $10.8 million in fiscal 1938. Safer airway and "radio range" broadcast stations were deployed, and government operation of air traffic control centers was expanded. The radio range beams were audible signals that let pilots know if they strayed to one side or another of the radio range beamed outward from a landing field. The new airway broadcast and radio range stations were designed to permit simultaneous transmission of voice and radio signals on the same frequency, so that radio range beams did not have to be interrupted for weather broadcasts (such interruptions were alleged to have been one of the factors in the Cutting crash). The proliferation of air transports and air traffic underscored the need for monitoring the flight paths of aircraft, especially around busy airports. On December 1, 1935, the first airway traffic control center went into operation at Newark, New Jersey. One of its most remarkable attributes was that the airlines themselves, not the federal government, organized and staffed the center, which provided pilots with information on the reported positions of other aircraft in the Newark area, particularly during bad weather and poor visibility. By the spring of 1936, two more centers had been established in Chicago and Cleveland. All three centers were taken over by the government and new centers were added when the Bureau of Air Comerce began to expand its aviation activities during the summer of 1936.

The Cutting crash and resultant administrative tremors hastened enactment of the Civil Aeronautics Act of 1938, which created the Civil Aeronautics Authority, an independent agency to conduct all federal civil aviation activities. A three-member Air Safety Board, although within the authority, had special prerogatives for completely independent accident investigation and recommendation of preventive measures. With this move, accident investigation remained permanently divorced from the agency that set up and operated the federal airways. In 1940, a further reorganization established the Civil Aeronautics Administration (CAA), for administrative and operational functions, and the Civil Aeronautics Board (CAB), for policy making and rate setting. Within the CAA, the Air Safety Board continued, and its effectiveness was reflected during 1938–40 when 17 months of scheduled airline flying passed without a single fatality.

The scope of bureaucratic and institutional changes generated a growing sense of fraternity among employees and employers alike. The concept of professionalism in the ranks of the pilots was not new, although the individualism of the pilots made strong organization difficult. As early as 1924, pilots in Akron, Ohio, had begun a loose organization, but one of the earliest groups to prescribe standards was the Professional Pilots Association (PPA). Members had to prove familiarity with transport equipment and needed four years of commercial experience with a good safety record. But the PPA was more like a society than a union, and stronger organization floundered until trained pilots became more important to the industry. In response to moves by some airlines to lower pay rates and infringe on sundry pilot prerogatives, dissident fliers organized the Air Line Pilots Association (ALPA) in 1931. An ALPA-sponsored strike against Century Airlines early the next year successfully demonstrated the association's viability, and ALPA became affiliated with the American Federation of Labor. Management also began to form stronger organizations. The Air Transport Association (ATA), formed by a breakaway group from the Aeronautical Chamber of Commerce, coalesced in 1936. Despite occasional friction, ATA, representing the operators, and ALPA helped stimulate increased airline efficiency, operational standards, and airline safety.

Perhaps one reason for the eventual decision by individualistic pilots to unionize stemmed from a new breed of aviator in airline service. Unless they had served in the military air services, airline pilots of the late twenties and early thirties probably began as barnstormers. The chief pilot of Pan American Airways in 1932, Basil Rowe, began as a barnstormer in Latin America. "Speed" Holman, a seasoned barnstormer and wing walker, became chief pilot for Northwest Airways, which operated air-mail services in the Detroit area. Airlines hired Holman and pilots like him because their barnstorming backgrounds provided the practical experience of keeping plane, body, and soul together in areas that were often aeronautically primitive. Ex-barnstormers were adept at executing a "pancake" landing into the muddy fields that were common to most early airlines, whereas army pilots, were used to comparatively good facilities. Ex-barnstormers had other valuable attributes, such as their long and familiar association with the locale of an airline's operations. They were already acquainted with the territory's physical hazards and the vagaries of its climate—distinct advantages at a time when aerial maps and weather charts were often sketchy. With a reservoir of technique and knowledge like this, Northwest pilots flew two and a half years without an accident, giving the pilots a sense of professional pride and lifting the operation above the level of a mere shoestring, barnstorming enterprise. Even so, complete respectability seemed almost uncomfortable, and a photo of four of the line's fliers bore a self-conscious caption: "All the Pilots Wear Uniforms."

As airline traffic increased during the early thirties and some of the older "early birds" began to retire, the supply of old barnstormers dwindled, depriving many companies of valuable, experienced fliers. Men who came out of the military air services took up some of the slack, but the drafting of ex-army and navy pilots did not solve the problem. One good reason for having barnstormers as pilots in the early days of airline flying stemmed from their acquaintance with a wide variety of airplanes. When lines consolidated or merged, they had to make do with the range of equipment that came with the transaction. Barnstormers versatility was valuable, since they often made successive flights in entirely different planes: a Pitcairn or Stinson for a short haul on Monday, a Fokker out on the regular passenger run on Tuesday, and a Boeing or Ford back the next day. Capt. William B. Lester, an old barnstormer working as chief instructor for American Airlines in 1935, recalled one week when he flew 15 different types of planes on regular routes. By the early forties, airlines were concentrating on one or two types. United Air Lines, for instance flew Douglases on main line operations and smaller, 10-passenger Boeings or Lockheeds on short-haul flights.

Even though airlines eventually standardized equipment, the difference in specialized transport design and operating procedures required ex-military pilots to have special training for airline jobs. The Boeing school used a specially converted transport in which students absorbed details of instrument flying and navigation unknown to an earlier generation, who steered by water towers and farmers' barns. The new pilots learned how to fly by an invisible radio beam and qualified as licensed radio operators. They became professionals. Recognizing the need for official endorsement of professional status, the federal government issued a special airline transport rating for pilots in 1932. Census Bureau statisticians seem to have had some trouble in classifying aviators, who were designated "aeronauts" in the census of 1910 and appeared under the heading of "showmen," along with actors and acrobats. Aviators in 1920 were listed under "other professional pursuits" and transferred to "transportation and communication" only in 1930. When the exact number of certificated airline pilots became available to census takers, classification became more distinct and figures more accurate. The number of pilots who held airline transport ratings rose from 330 in 1930 to 1,587 by 1937. Salaries also improved, even in the depression. A copilot starting with an established line in 1929 could get $300 per month and progress to $600 with experience and flying time. Copilots started at $190 per month in 1938, but were usually raised to $330 in the first year. Those who got promoted after four or five years of service averaged $600 per month, up to $1,000 for experience.

Airline pilots of the thirties flew in an increasingly regulated air space. Passenger traffic jumped to more than 475,000 passengers by 1932 and grew to over four million in 1941. This meant that air traffic expanded to handle the demand; more planes around airports meant that individual pilots had to surrender the right to pick and choose their takeoff or landing approach. With the increased use of two-way radio, control towers sprouted on airfields in imitation of Cleveland's lead in 1931. These towers controlled air traffic in, out, and around the airports, and in 1938, with the creation of the CAA, the agency assumed the responsibility for licensing the personnel engaged in traffic control. Congestion increased in the air lanes as well as around landing fields, and the freedom of the individual pilot diminished more by mid-decade, when flights became restricted to a predetermined altitude and compass course allocated by the airway traffic control centers. Pilots still made their own decisions about the question of what weather seemed flyable until 1937. Minimum weather conditions were established in that year for every route and airport, enforced by company personnel who were certified and subject in turn to the new CAA. Minimum flying conditions were determined by the local terrain, facilities, and design of the airport; the peculiarities of local weather; the radio aids available; and the type of aircraft used. These minimums were also designed to be flexible in accordance with the seasons of the year.

More and more, airline flying became a standardized and regulated vocation. Not only were pilots subjected to control in their own element, the air, they also had to tend to a number of vexing and multiplying company regulations on the ground. The company flight dispatcher carefully juggled personnel, planes, and flying time in order to program the next day's flights and pilots. This meant that pilots had to keep careful flight logs, marking pertinent information on the dispatcher's flight chart when they finished every run. Failure in this requirement incurred penalties doled out by the omnipotent dispatcher. E. K. Gann, a veteran airline pilot, remembered punishments of unattractive stewardesses and cold coffee on subsequent schedules.

The availability of coffee and other food services represented still another change in airline flying. Passenger transport became a more sophisticated business, catering to airline passengers. As improved planes allowed flights of extended duration, airlines were confronted with plane loads of hungry and disgruntled passengers at the end of the trip. To keep its patrons happy, Western Air Express hired 15 stewards in 1928, along with a railroad restauranteur to instruct them in the art of serving box lunches. Transcontinental Air Transport countered with china settings, lavender napkins, and gold-plated flatware. Not every airline hired stewards, so it was often the lot of the copilot to serve meals, dispense aspirin, and minister to airsick passengers. The latter problem seems to have been most acute up to the mid-1930s. Trimotors of the era flew at low speeds in the turbulent strata close to the ground. Fumes from gas tanks and from engine exhaust easily filtered into poorly ventilated cabins. Each passenger chair came equipped with a variety of paper bags, boxes, and basins for gastrointestinal emergencies. Most passenger planes of the day boasted sliding windows, the better to inhale pollution-free ozone, according to airline advertisements, but the principal value of open windows seemed to have been their practicality for airsick travelers. Patrons immediately behind an airsick passenger learned to keep their windows closed. It was not unusual to hose out the entire interior of a plane after completing a turbulent flight.

Hostesses appeared with the seasonal upswing of passenger traffic in the spring of 1930, when United made eight appointments. It was no longer enough for a pilot just to fly a plane load of humans from one place to another; airlines perceived the need for feminine assistance to complement the "intimate" atmosphere of travel in the confines of an airliner. "By taking our home-making instincts into the cabins of the commercial airliners," explained a United hostess, "we can lend familiar aspects to which travelers may cling." Most early hostesses were required to be registered nurses, which lent authority to their specially revised menus for the air traveler. This development had

Movement in the cabin of the Boeing 247 was somewhat restricted by the wing spar across the aisle. Passengers on domestic flights received attentive service, but had to get by with one apple, coffee from a thermos, and sandwiches served in a wax paper wrapping. Courtesy of The Boeing Company.

evolved from the nurses' compilation of "the first authentic record of airsickness from actual observation during flight," explained a 1932 *Collier's* article. United employed 164 hostesses by 1936; American employed 100. Previously, pilots themselves had handled public relations on the plane and chatted with the passengers, but these duties were taken over by more appealing hostesses, who had to meet special qualifications for age, height, weight, and—predictably—appearance. Hostesses were expected to be the "well-bred American" and not the "show-girl" type. They also had to be physically fit. Early stewardesses dusted the cabin, loaded baggage, and helped pilots shove planes into hangars. In flight, recalled one early stewardess, they also had to be discreet, intercepting errant passengers "to make sure they didn't open the exit door by mistake when they were going to the washroom."

Hostess candidates were rigidly screened on the basis of health, character, and intelligence. In an article written for the *Literary Digest* in 1936, the editors noted that similar standards applied to pilots as well, and the magazine enthusiastically conjectured that a high rate of matrimony between hostesses and pilots would "produce a race of superior Americans." Some women had already passed the demanding flight

By the late 1930s, the job of the stewardess implied qualities of glamour and sophistication. Posed under the impressive prow of a DC-3, this line-up of alluring hostesses appeared in the summer of 1939. Courtesy of United Airlines.

exams for airline pilots. At least one woman was flying as a copilot on a regularly scheduled airline by 1935. Most airline executives seemed skeptical about women pilots, however, and the airlines tended to reserve a plane's controls for men, leaving women to attend to the passengers. Nonetheless, many women, like Jacqueline Cochran and Amelia Earhart, remained active professional aviators, serving as encouraging examples for dozens of other women who began to enter the ranks of general aviation and who also established careers in public relations, engineering, and management within the aviation industry.

Executives within the airline industry began to pay more attention to marketing; the presence of hostesses represented only one facet of the trend to attract passengers and make flying a regular mode of travel. Various air-fare discount schemes appeared in the late twenties and became a widespread trend by the mid-1930s, led by the aggressive efforts of American Airlines. American issued an "Air Travel Card" for a deposit of $425. The individual or company who held it was later billed at a 15 percent discount.

This led to an intercarrier agreement about 1936, and within five years, 17 carriers had joined, permitting card holders to charge trips on each of the connecting airlines at the discounted rate. By 1941, perhaps one half of the airlines' revenues came through the Air Travel Card agreement, and the airlines had also begun an installment plan to lure new travelers into the air.

In the hunt for more passengers, advertising became more sophisticated. Although the standard format continued, listing cities on the route along with departure and arrival times, American Airlines tried several new approaches. Rather than compete directly with those of rival airlines, American's ads were aimed at the Pullman railroad clientele. But the principal aim was to encourage people to fly, especially those who had not flown before. One ad in 1939 showed a relaxed executive who had finally decided to "catch up with other modern folks!" by deciding to fly. Other formats portrayed a business traveler watching a high-flying airliner and asking himself, "I wonder if I've been missing something?" Still others stressed air travel as a family affair, eminently suitable for women and children as well as macho male executives. American Airlines also encouraged more travel during the winter, when business fell sharply, and instituted carefully orchestrated internal sales conferences.

By the late 1930s, air travel had indeed picked up. From an estimated 95 million revenue passenger miles in 1932, domestic airlines reported 270 million in 1935 and 677 million in 1939. Transcontinental flights could be made in 16 hours for a one-way fare of $150. Flying travelers also enjoyed more of the convenient arrangements that characterized rail travel. In 1938, the airlines opened their own terminal in New York City at 42 Park Avenue, across from Grand Central Station. With a sleek, Art Deco façade and fashionable address, the new Airlines Terminal provided a centralized focus for departures and arrivals, with facilities for surface transportation linking it to outlying airports in the New York metropolitan area. In many ways, the presence of the Airlines Terminal symbolized the maturity of the airline industry and the significance of airlines within the American transportation network.

The impact of technology made notable changes in the nature of piloting airliners during the thirties. A certain amount of contact flying prevailed throughout the first half of the decade, even in modern transports like the Boeing 247 and Douglas DC-3. Instrument flying did not become the general rule until 1936, following the progress of radio engineering and the development of other technical improvements. In 1930, the B. F. Goodrich Company, in cooperation with the Daniel Guggenheim Fund for the Promotion of Aeronautics, perfected deicing equipment. The automatic pilot came into use three years later in answer to the need for pilots to relax from the strain of long-distance flights. Aircraft like the Boeing 247 and Douglas DC-2 and -3 series made their appearance, featuring retractable landing gear, flaps, and variable-pitch propellers. High-powered engines burned new formulas of high-octane fuel. To monitor all these functions, cockpit instrumentation became so complex that engineers were forced to devise means of combining functions of various instruments in order to reduce the confusing profusion of dials and gauges.

In the interest of operational efficiency as well as the safety of an increasing flow of passengers, the airlines and government began to intensify their investigation of the

American Airlines played a significant role in the use of advertising that stressed the benefits and enjoyment of air travel. This layout was one of a series that appeared in national publications during the late 1930s. Courtesy of American Airlines.

Opened in 1938, the special aviation terminal erected on New York City's fashionable Park Avenue underscored the determination of airline executives to promote flying as the preferred mode of travel. Courtesy of the Smithsonian Institution.

physical and mental status of pilot candidates. In 1936, the Bureau of Air Commerce named 700 specially trained doctors to check the qualifications of pilots and copilots flying passengers and mail over authorized routes. Some companies developed candidate exams running 16 pages. Airlines acquired an annoying interest in the private lives of their fliers, checking for clues that might hint of dangerous mental strain, and sometimes even kept tabs on the proximity of a mother-in-law to a pilot's base of operations. Airlines opened their own medical and research divisions to evalute pilots, and the Mayo Clinic, Harvard, Columbia, the CAA, and military services had similar projects under way in 1939. Eventually, candidates for airline training needed such prerequisites as a special instrument rating and a commercial pilot rating from the government. The candidate had to be a "college man," preferably a graduate. He needed four to five years' flying experience and 1,000 hours of flying time. He had to weigh 150 to 200 pounds, not be younger than 23 or older than 28. If there was a vacancy and he met the requirements, he was allowed to undergo the intensive six-weeks' training course. If he passed, he became a copilot.

As the individual airlines began to mature, stabilizing their operational techniques and drawing pilots from their own training schools, certain tribal characteristics began to appear, at least in the opinion of rival airlines' personnel. According to Gann, United pilots had little color and were prone to adhere strictly to regulations. American's fliers, a varied group, were querulous and rebellious. TWA pilots were highly regarded as individuals but "pitied for the chameleon management of their company." Pilots on Pan Am's routes were envied for the long distances they flew but considered "shy and backward" when it came to operations in dirty weather. Those who flew for Eastern Airlines were "singularly determined and clever."

With the standardization of techniques, of flying procedures, of equipment, and of training methods, had all the individualism been squeezed out of the pilots of the late thirties and early forties? Airline pilots had certainly become less independent in many ways, simply because the development of airline transportation demanded more specialization for the sake of operating efficiency and safety. Pilots necessarily became more dependent on an army of meteorologists and ground personnel, plus reams of company and federal flight regulations. Still, the art of flying demanded excellent individual qualities and allowed opportunity for expression of a pilot's individuality. Modern air transports also displayed distinctive characteristics, and pilots had to make the transition from one generation of planes to another. "It is the professional pilot's bounden duty to know the idiosyncrasies of each type," wrote Gann, "for he must spend a large proportion of his active career exploiting its qualities and compensating for its faults. These secrets cannot be discovered in a ground school." The DC-3 was "an amiable cow, grazing placidly in the higher pasture lands, marvelously forgiving of the most clumsy pilot." The DC-2, its immediate predecessor, was almost its exact twin in appearance but was decidedly less docile in its landing characteristics, and pilots had to allow for this. On the other hand, pilots learned that the DC-2 was a far more able performer under heavy icing conditions.

Perhaps the peak of individual expression came when a pilot landed. Gann spoke glowingly of a pilot named Ross and his ability to handle a DC-2. Ross would choose a triangle of grass at the intersection of the runways, sideslip in, and set the wheels down at the very top of the apex so softly that it was difficult to sense the actual moment of contact. This was accomplished without the usual backfiring or excess engine noise. Disdaining the use of brakes or power, Ross would allow the plane to continue its landing roll "until the final, sensuous result was that the twelve-ton airplane eased to a halt exactly in line with the waiting ramp. . . . At these moments he was the true virtuoso, performing for his own joy, lost in it, and thus quite unaware of his audience."

Although pilots could be represented as heroes to the public, they were not always the dashing caricatures of the early birds. Magazine editors realized that the profession of airline pilot had become an occupation requiring a stable and dependable character. When seven pilots received the Air Mail Flyers' Medal of Honor for saving passengers and mail after a series of accidents in 1935, *Time* magazine described them as "extraordinary in achievements" but "ordinary in appearance," even "short, baldish"—hardly the popularized flying daredevil of film and fiction. The Harmon Trophy winner for 1935, awarded to the "world's outstanding aviator" of the year, went to the 42-year-old chief pilot for Pan Am, Edward Musick. A subsequent *Time* article again suggested that airline pilots should be regarded as anything but exotic fellows. The world's outstanding aviator had piled up two decades of accident-free flying and was never known to have stunted; he was the "supercareful, monosyllabic antithesis of a grandstand flyer."

By 1940, fliers themselves had recognized that airline pilots represented a type apart from the rest of their aeronautical cohorts. There were three acknowledged groups of pilots, wrote flier Biernie Lay—military, commercial (or airline), and private—and there were notable differences among them. The main distinction occurred between the military and the commercial pilots as the two most professional groups. Temperamentally, military pilots were the more carefree. They were given superior equipment, but accidents cost them little; if they had to bail out of one plane today, another would be waiting tomorrow. The commercial, or airline, pilots developed a more sober outlook and altitude because their employer was in the air to make money—a purely commercial enterprise. Moreover, if pilots indulged in untoward caprices, they not only risked a costly airplane and their job, they jeopardized their own life and the lives of their passengers as well.

The combination of responsibilities and regulations unquestionably altered the style of airline flying. The other branch of civil aeronautics—general aviation—operated within the same airspace. General aviation planes and pilots, if not as visible to the public at large, became increasingly subject to the same guidelines for physiological requirements and airway regulations established by federal agencies. Similar to the airline experience, these changes were also accompanied by significant transitions in flying equipment.

THE LIGHT-PLANE INDUSTRY

Stimulated by the "Lindbergh boom" of the late 1920s and early 1930s, the light-plane industry flourished. The urge to conquer oceans, set speed records, fly higher than anyone else, or maybe pilot a big airliner inherently demanded flight training and the accumulation of flying experience. This meant an interim level of training in smaller, more sensible airplanes suitable for the swarms of novice Lindberghs on their way to fame and fortune. Inevitably, hundreds of these aspiring aerial heroes never made it to the air races or the headlines but kept on flying for the love of it or because they found other ways to use aviation in a variety of business enterprises. These individuals—the private pilots—provided the market for the new light-plane industry.

Compact engines for the small craft were frequently hard to come by. In addition to motorcycle engines, an early standby was the three-cylinder Szekely, a fickle, rough-running radial engine of 30–45 HP. Imported engines always posed the problem of finding parts and qualified repair service, so manufacturers welcomed new American designs in the 1930s. In 1931, the introduction of the horizontally opposed, four-cylinder Continental A-40 engine of 37 HP provided a dependable, steady engine with a small frontal area. When similar power plants produced by Lycoming and Franklin became available, the American light-plane industry possessed a stable of reliable engines of various horsepower for airframe designers to work with.

The dreams of thousands of would-be fliers ended as the depression deepened. For hundreds more, determined to get into the air and get a plane, the high cost of flying also eventually became too discouraging. With depression era prices running $850 for a Chevrolet sedan and 25¢ a pound for steak, at $8 to $10 per hour, flying lessons were a dear commodity. The typical light plane of the era retailed between $1,500 and $2,500, and maintenance costs on a Piper Cub flown 150 hours per year amounted to $8 per hour. Between 1931 and 1936, according to one survey, about 5,000 new planes were produced and 15,000 owners were registered. Approximately one-third of the owners sold their planes within a year, and 83 percent gave up their aircraft within two and a half years. A small group of aviation enthusiasts and government officials hoped to resolve the problem by mass-producing a light, cheap plane, which could be manufactured and distributed on a scale comparable to that of the auto industry.

Chief spokesman for this coterie was Eugene L. Vidal, who became director of the Aeronautics Branch of the Department of Commerce in 1933. Vidal advocated a plane to retail somewhere between $700 and $1,000, although the idea was consistently referred to as the "$700 airplane." In the face of considerable skepticism from manufacturers, Vidal continued to lobby for the project. Despite some interesting types developed for a bureau-sponsored design contest in 1934, the $700 airplane remained an elusive dream. Vidal's campaign, however, publicized the light-plane industry and stimulated research into more economical and safer designs. One tangible result was the development of the Ercoupe, based on the work of NACA engineer Fred Weick. Designed as a spin-proof plane, the two-place Ercoupe with a fixed-tricycle landing gear went into production just before World War II. It enjoyed brief postwar popularity, and several thousand were manufactured. In the meantime, other designers and manufacturers were able to survive the depression years into the postwar era.

The goal of a truly lightweight, inexpensive sport plane was partially met by the Aeronca C-2, introduced in 1929 at a price of $1,495. The deep-bellied silhouette of the single-seat C-2 earned it the jocular but affectionate sobriquet of "the flying bathtub." Its 29-HP engine gave it minimal performance, but it did, after all, fly, and it served the important function of providing an economical plane for popular flying throughout the thirties. But the C-2 was not the complete answer for private flying. There remained a need for a basic two-seater for instruction that offered a bit more in comfort and performance. Additionally, the light-plane industry needed a spectrum of products to satisfy various requirements of utility, personal business, and corporate flying. Such planes finally evolved, setting the pattern of single-engine general aviation types that lasted into the postwar era: light singles; medium-performance singles; and high-performance singles. At the close of the decade, twin-engine corporate aircraft were also introduced. A bewildering array of manufacturers introduced aircraft in all of these categories, some more successfully than others. Three companies stand out, because they and their aircraft not only survived the depression era but also became leaders in the postwar general aviation industry: Piper, Cessna, and Beech. Two others deserve mention for the innovative trends they represented, Luscombe and Spartan.

William T. Piper was an oilman in Bradford, Pennsylvania. In 1928, he acquired a few hundred dollars' worth of stock in Taylor Brothers Aircraft, whose guiding spirit, C. G. Taylor, had designed a couple of small sport planes with tandem seating. Taylor's "Cub," model E-2, licensed in 1931, became the progenitor of a remarkable series of light aircraft. After bankruptcy of the airplane company and a split with the moody Taylor, Piper abandoned the depleted oil fields around Bradford and launched the Piper Aircraft Corporation in 1938. In the meantime, Walter C. Jamouneau, a Rutgers graduate, joined the company as one of its first engineers. Although Rutgers did not offer a degree specializing in aeronautics, Jamouneau had taken courses in aerodynamics and aircraft structure. In 1935, William Piper gave him the opportunity to modify the model E-2. Jamouneau rounded off its squared wing and tail surfaces, and the plane was awarded a different alphabetical designation, chosen for its recent stylist: the J-2. It became the bread-and-butter airplane of the Piper Company. When fire wiped out Piper's manufacturing facilities in Bradford, the company relocated in an abandoned silk mill in Lock Haven and began production of a refined version of the Cub with a modified rudder, called the J-3. The new Cub's seats were upholstered for more comfort, and, in addition to an altimeter, rudimentary engine gauges, a tail wheel, and brakes, the new plane boasted Piper's first compass and airspeed indicator. The J-3 was painted bright yellow with a black "speed-stripe" down the side. Although Piper offered fancier versions like the two-place, side-by-side Coupe and the three-place Cruiser, the classic J-3 continued as the mainstay of Piper's business until the postwar era.

Piper struggled through the lean years of the depression by a combination of enforced austerity, shrewd publicity, and an aggressive phalanx of salesman-pilots. William Piper himself kept a sharp eye on production costs and insisted that the price of Piper airplanes be kept at a minimum. In 1939, a new Cub could be had for $995 at the factory. Piper aircraft invariably showed up at air shows around the country; one act featured a Cub landing atop a platform attached to a moving automobile. The company provided eight hours of free flight training with each new plane, and dozens of aviation writers who showed up at Lock Haven were rewarded with familiarization flights and a few hours of solo time at no cost. Such policies paid off in continuing sales. In the spring of 1940, *Fortune* magazine ran an article on the company, which was building more airplanes "than any other manufacturer in the world."

By 1940, pressure from competitors like Aeronca, a rejuvenated Taylorcraft, and others had prompted Piper's designers to introduce additionally modified and embellished models. With a 75-HP Lycoming, the three-place Piper Cruiser was definitely a cut above the tandem trainer Cub and enlarged the scope of taking to the air with partners on a business trip, or flying for pleasure with a family in a plane that still listed at under $2,000. The two-place, side-by-side 65-HP Coupe represented an overt concession to increased passenger comfort and eye appeal, both inside and out. After Piper hired an industrial stylist from Detroit, the Coupe definitely displayed automotive

The Piper Cub became a hugely popular plane for utility and recreational flying as well as for instruction. Variants remained in production until 1982. Courtesy of Piper Aircraft Corporation.

refinements. And why not? As one aviation journalist approvingly reported, "There has always been too much of a let down in connection with . . . driving out to the airport in a well appointed, comfortable car and getting into a ship of twice the price of the car and less than half its appearance of quality." The Coupe's fully enclosed cowling sported decorative stainless steel striping across the air intake openings and over the crest of aerodynamically shaped wheel pants. Inside the cockpit, blue and gray naugahyde upholstery on the seats and side panels set the color scheme for the chrome-accented instrument panel. And the Coupe had a muffler. "With the ice broken," the journalist declared, "this is to be heralded as the beginning of the end of airplanes so noisy they are exhausting to fly."

While Piper continued to dominate the category of light singles, other companies began to tap the growing market for larger airplanes suitable for utility flying (photographic surveys, light cargo, agricultural applications) and for business flying. Before World War I, Clyde Cessna had built several monoplanes and had flown them in various air shows around the Midwest. In the mid-twenties, he had formed a partnership with Walter Beech and Lloyd Stearman, and their Travel Air Manufacturing Company in Wichita, Kansas, had produced several successful high-wing, enclosed-cabin aircraft. By the late twenties, Cessna's fascination with the cantilevered wing had led him to set up Cessna Aircraft Company. Several dozen versions of Cessna's A series were produced until the depression. The A series, with its high, cantilevered wing and seating for four began a trend that evolved into the successful Airmaster design of the late thirties.

Between 1931 and 1934, the effects of the depression forced Cessna to close his small factory, although a nearby shop allowed him to build several custom-designed racing planes. The configuration of their cowled radial engines later contributed to the success of the Airmaster series. The impetus for revitalization of the Cessna Company came from Clyde's nephew, Dwane Wallace, a student at Wichita University. Wallace graduated in 1933 with a degree in aeronautical engineering, but no job. In collaboration with his brother, Wallace got enough support from stockholders to convince his uncle to reopen the Cessna plant. The upshot of all this was the C-34 Airmaster, a four-place monoplane with a distinct heritage from Clyde Cessna's earlier passenger planes and racing machines but designed and built with the expertise Wallace had gained from his studies in aeronautical engineering. Introduced in 1934, it was one of the first American light planes to employ flaps, and after winning a series of contests in 1936, the C-34 was called "the world's most efficient plane." Although Clyde Cessna retired in 1936, the company continued under Wallace's guidance. By 1941, when World War II halted Airmaster production, a total of 186 Airmasters had been delivered, enough to keep the firm solvent.

The significance of the Airmaster and planes like it was the opportunity for economical business flying. Similar aircraft types, produced by Stinson, Howard, and Waco, used larger and more expensive engines to get equivalent performance, and the Cessna offered even better performance than similar planes with engines of comparable horsepower. With 145–65-HP power plants, the Cessnas cruised at over 150 MPH and carried their four passengers in relative comfort on flights of 500 to 700 miles. As one contemporary pilot acknowledged, "Here is a ship that knocks the tar out of an average business trip of a few hundred miles. There is no waste of space, but it is roomier than it looks and is comfortable." Piper's two-place Cubs provided entry into personal flying, but planes like the Airmaster made business flying feasible. The price also reflected the clientele: the Cessna of 1936 sold for $4,995, and the 1939 model retailed for $6,400.

For pilots who yearned for even flashier performance from a single-engine plane, there was the Beechcraft Model 17, the famous Staggerwing. After a stint as flight instructor during World War I, Walter Beech had barnstormed around the country before settling in Wichita, where he eventually became a partner with Clyde Cessna in the Travel Air Manufacturing Company in 1924. Following Cessna's departure in 1927, Beech ran Travel Air as president (his secretary, Olive Ann Mellor, later became Mrs. Beech), and the company continued to produce a series of medium-sized transports and racing planes. With 1,000 employees in 1929, Travel Air produced up to 25 planes per week. There were 95 commercial manufacturers that year who produced a total of 5,357 units, and Travel Air's contribution was 547 planes. In 1930, all manufacturers combined delivered on 1,582 aircraft. Heeding these warning signals, Beech sold out a year later to the Curtiss-Wright organization and netted a tidy profit.

Restless in his momentary retirement, Beech soon decided to re-enter the airplane business. His goal was the development of a high-performance personal plane with a top speed of 200 MPH or better; a range close to 1,000 miles; a cockpit to carry the pilot and four passengers in comfort equivalent to that in a high-priced sedan; gentle flying characteristics and landing speed of around 60 MPH. The design that left the drawing

The Cessna Airmaster series offered business pilots a fast, comfortable ship. The effectively cowled engine, streamlined wheel covers, and cantilevered wing reflected growing attention to drag-reducing aerodynamic treatment. Courtesy of Cessna Aircraft Company.

boards was a biplane with a "negative stagger," having the lower wing positioned ahead of the upper wing. Although this compromised downward vision somewhat, it decidedly improved the plane's stalling and landing characteristics and actually yielded some aerodynamic benefits in increasing the speed. Even with fixed gear, the 420-HP Model 17 met all the original design criteria when it took to the air in 1932. Later versions, with retractable gear and a choice of engines from 225 to over 400 HP, continued to approximate or even exceed the original specifications. Beech had a plane that competed on highly favorable terms with other contemporary designs in its class, like the Waco and Howard aircraft, and he won several competitions with it. The company sold Model 17 planes to pilots around the world, shipping the aircraft by sea, or, in the case of one plane, in the hold of a giant dirigible en route from Lakehurst, New Jersey, to Germany.

Other manufacturers were also building planes with advanced features. In the 1930s, the accepted mode of construction implied wooden components covered by fabric. Donald Luscombe's sporty, two-place Phantom of 1934 retained fabric-covered wings but marked an important step forward with other construction details for light planes. Not only did the Phantom's shoulder-mounted wings have metal spars and ribs, but the fuselage incorporated metal skin with semi-monocoque construction. Although early models had radial engines, redesigned Luscombe planes, appearing at the close of the decade, were produced with horizontally opposed, four-cylinder engines, giving them a modern look. With their high wings, fixed gear, and 50–65-HP engines, the Luscombes delivered an adequate performance, typical of the light single category. Among high-performance singles, the Spartan Executive of 1935 represented a similar benchmark. Powered by a 450-HP Pratt & Whitney radial, the Executive could carry

With a big radial engine, clean design, and retractable gear, the all-metal Spartan Executive looked like a typical fighter plane of the era. In many respects, it pointed the way toward high-performance general aviation types of the postwar period. Courtesy of Spartan School of Aeronautics.

four or five people while cruising at over 190 MPH. Cabin appointments included armrests, ashtrays, dome lights, window curtains, and soundproofing. Only its blunt-nosed radial engine and its tail-wheel marked it as a monoplane of the depression era. Otherwise, its principal features—all-metal, semimonocoque construction, low wing, retractable gear—were characteristic of the style adopted in the postwar years.

The machines produced by the light-plane manufacturers offered prospective customers a wide range of personal aircraft, from light, two-place training and pleasure aircraft through medium- and high-performance planes. Even in the depression such aircraft proved useful and advantageous for a wide range of business and utility flying. Activities such as aerial surveying and agricultural applications continued, using rugged, open-cockpit biplane designs, which were cheaper to build and gave better visibility. The pattern of such activities reflected those of the twenties, whereas the increasing comfort and efficiency of light planes with enclosed cockpits in the thirties won more converts to business flying in particular. Variants of the Piper, Cessna, Beech, Luscombe, and Spartan designs survived in the postwar period, indicative of the new standards of quality, comfort, and performance, and construction achieved during the thirties. Products like these were the archetypes of the increasingly significant general aviation industry. Still, production remained miniscule by later standards, or even in comparison with the halcyon twenties. Beech produced only 36 planes in 1935, and Cessna turned out 50 Airmasters in 1937—approximately the break-even point for the company.

Despite the slow pace of production, a slightly quickening tempo of demand in the late thirties stimulated the general aviation community to investigate other possible markets. In 1935, Beech began work on an advanced twin-engine design, which the company hoped to see as a deluxe executive aircraft or as a small transport for feeder airlines. Two years later, the Model 18 rolled out of the hangar doors. With its dual rudders, the all-metal Twin Beech bore a strong resemblance to the larger Lockheed Electra. The new plane was built with techniques characteristic of companies producing larger, more complex airliners. Its truss-type center section was fabricated from high-strength chrome molybdenum steel tubing, subjected to heat treating as a weight-saving and strengthening move. Heat-treated steel, generally used for landing gear and other components subjected to high stress, was also used in wing structures and other areas of the Twin Beech. In the process, Beech installed its own gas-fired heat furnace, for many years the largest of its kind west of the Mississippi. This kind of fabrication detail was one reason why the Twin Beech enjoyed a manufacturing run that lasted until 1969, one of the longest production runs for an aircraft of its type.

The mode of construction, and the equipment needed to produce it, also indicated the progress of general aviation designs from the days of wood and fabric. As a boy growing up in rural Tennessee, Walter Beech had developed an appreciation for farm machinery that was sturdy, reliable, and easy to repair. As a barnstormer, he had developed an appreciation for airplanes and parts that worked as well in the freezing weather of the Midwest as in the warm, humid South. As a result, not only were Beech planes designed to be rugged and easy to maintain, they also made extensive use of special protective finishes and other processes to enhance operations in a broad spectrum of climates. The planes cost more but promised economical operation for the owner over a longer period of time.

The Twin Beech proved to be a remarkably versatile machine. With pilot, copilot, and 6 passengers, the plane's pair of 350-HP Wright engines gave a cruising speed of over 190 MPH and a range in excess of 1,000 miles. Beech's hopes of selling the plane as a feeder transport in the United States floundered, since American airlines were intent on developing major trunk routes, which required larger aircraft. Outside the United States, the plane was more successful in remote areas, where smaller air transports could operate efficiently on routes of lower passenger density; Prairie Airways, in Canada, operated a fleet of Model 18 Beechcrafts. Even at a cost of $33,000 per plane, a few businesses in the depression era found it worthwhile. One insurance executive from Oregon used a Model 18 to visit his widely dispersed branch offices all over the Pacific Northwest, including Alaska. A Puerto Rican sugar growing and refining firm flew a Twin Beech around its various island operations in the Caribbean. Another early Twin Beech owner was the Canadian department store chain run by John D. Eaton.

Export sales comprised an important share of Beech production; by the end of 1939, owners flew Beech aircraft in 23 countries around the world, and company sales amounted to $1.3 million. Most planes went to civilian owners, although military deliveries also increased. In the late thirties, a number of specially modified Model 18 planes were sold to foreign air forces for photographic work and for evacuation of wounded troops. Twin Beeches were also selected by the U.S. Army Air Corps for the transport of high-ranking personnel (the Staggerwing was already in service for use by military attachés stationed in London, Rome, Paris, and Mexico City). The Air Corps also ordered a few Twin Beeches for high-altitude aerial photography, modifying the planes to allow them to fly several hours at altitudes of 25,000 feet. This special group of aircraft did yeoman service in mapping strategic areas of the entire North American continent, including the route of the Alaska Highway, a vitally important transportation and communications link during World War II.

Cessna also sensed the market for a light twin and flew its T-50 for the first time in 1939. For owners seeking a plane at less cost and performance just under the larger Twin Beech, Cessna designed the T-50 to sell for under $30,000 and carry five passengers at a cruising speed of about 170 MPH. Only 43 civil versions of the T-50 had been delivered by 1942. The outbreak of war in Europe caused the U.S. Air Corps and the Royal Canadian Air Force to pre-empt civilian deliveries and order modified T-50 planes (respectively, the AT-17 and the Crane) as advanced twin-engine trainers and as bomber-crew trainers. Continuing military requirements committed all of Cessna's energies, in addition to those of Beech, Piper, and others, to production for the armed services.

The impact of military flight training in the late 1930s somewhat muddles the figures pertaining to passengers carried in general aviation operations. However, a leading aviation text of the period, John H. Frederick's *Commercial Air Transportation*, published in 1942, reported specific data for 1937, before the military influence had become pronounced. Instructional flights accounted for 515,056 passengers that year, or 32.6 percent of the total. Flights made on a "for-hire" basis accounted for 799,215 (50.6 percent); business flights, not for hire, 103,359 (6.5 percent); and pleasure flights, 162,783 (10.3 percent). Since private fliers were not required to report passengers carried as guests, the figures for the latter were probably higher. In any case, assuming that the for-hire flights were probably made for business purposes, the total business and pleasure passengers for 1937 approximately totaled 1.2 million and roughly equaled the total number of business and pleasure passengers who flew on regularly scheduled airlines.

Like its counterparts in the airline transport industry, general aviation became a vital part of the national defense effort at the decade's end. Even before the impact of military contracts, the sophistication of general aviation designs sparked promising sales levels, even for the depression era. With their comfort and versatility, the improved designs of late-1930s aircraft established a quantity of orders that led to volume production—a true light-plane industry. Accelerated by wartime demands, general aviation manufacturing reached industrial maturity.

OTHER AVENUES OF RESEARCH

The remarkable progress of civil aviation and the pattern of military aviation prior to World War II sometimes obscures ancillary lines of aeronautical development during the thirties. Depression-era trends worth noting in particular include such diverse threads as flight simulators, rotary-wing aircraft, stratospheric research, and rocketry.

Electrical-mechanical simulators appeared in 1929, the handiwork of Edwin A. Link. He wanted to find a less expensive and safer way to begin flight instruction, and built the prototype models of his Link Trainer from air bellows, mechanical linkages, and miscellaneous components found around his father's organ factory. Link ran into stubborn resistance from traditional aviators who felt the best method to initiate flight students began in the air. Initially, his only steady customers were amusement parks. The navy finally evaluated the device in 1931, was impressed by the results, and placed the first large orders for Link Trainers. The U.S. Army followed in 1934, continuing a trend that was to give Link a worldwide market by the end of the decade.

Rotary-wing vehicles, like the helicopter, date from the early nineteenth century, although early devices remained unstable and difficult to control because of the extreme torque of their large rotor blades. For a time, the answer seemed to be the autogyro, which used a conventional engine for forward motion, relying on a free-wheeling rotor blade mounted above the fuselage to generate lift. Although lacking capacity for sharp vertical ascent and descent or for hovering, the autogyro's short-field landing and takeoff, plus "low-and-slow" speed qualities, generated keen interest. In Spain, Juan de la Cierva made the first successful autogyro flight in 1923, followed by extensive development in England. In the United States, Harold Pitcairn began autogyro work during the late twenties and sold an early model to the navy in 1931. Other designers entered the field, and the army, navy, and marines seriously evaluated autogyros for battlefield liaison, evacuation, and similar tasks. Some enthusiasts predicted extensive use in civil aviation and foresaw door-to-door aerial commuting, but the autogyro never caught on. The start of World War II upset further work on the autogyro, and the development of a practical helicopter convincingly displaced the autogyro fad.

The world's first successful helicopter, flown in Germany in 1936, overcame the torque problem with a twin-rotor design in which outrigger struts supported each rotor on either side of the vehicle's fuselage. This layout required careful synchronization of the rotor blades and presented control problems. In the United States, the versatile Russian emigré Igor Sikorsky designed and built the VS-300, which used a single main rotor for lift and a small vertical rotor mounted on the tail to counteract torque. The main-end tail rotor arrangement achieved encouraging controllability and set the pattern for most helicopter designs of succeeding decades. After the VS-300 made its first tethered flight in 1939, free flights and hovering trials soon followed. On the strength of eyewitness reports and movie clips, the army gave Sikorsky a contract in 1940, setting the stage for the world's first production helicopter. Deliveries of the final production version, the R-4B, began early in World War II, and a few models eventually reached combat units overseas.

Other experiments probed the extent of the physiological aspects of human flight at extreme altitudes. Since aircraft of the thirties did not yet have upper-stratospheric capability (between 34,000 and 180,000 feet), an era of remarkable balloon flights ensued, with pressurized gondolas and oxygen supplies for the crew. In Europe, Swiss physicist Auguste Piccard pioneered the way, ascending to 51,777 feet in 1931 and reaching 53,152 feet a year later. In America, intense interest in stratospheric flight within the Air Corps led to an outstanding project in cooperation with the National Geographic Society during 1934–35. During the latter year, Capts. Orvil Anderson and

In hat and overcoat, Igor Sikorsky takes the controls of the VS-300, a design that established the pattern of wartime development of the helicopter. Early test flights like this one utilized restraining cables as a safety measure. Courtesy of United Technologies/Sikorsky aircraft.

Albert Stevens soared to 72,395 feet in the helium-filled *Explorer II*, a record that held for two decades. The data and experience gathered in the process suggested practical procedures and equipment for high-altitude combat operations in the skies above Europe and Asia during World War II.

The high-flying missions in futuristic-appearing gondolas also seemed one step closer to realizing journeys portrayed in the era's equally futuristic, popular comic strip, ''Buck Rogers.'' The reach into the stratosphere via manned flight in balloons came at a time when other researchers began realistic thinking about using rockets to probe the upper levels of the earth's atmosphere and beyond. The first steps required instrumented launches, and manned missions would involve special passenger capsules and large, chemically propelled vehicles. By the late thirties, the technology that allowed serious consideration of such an adventure had begun to emerge. It was possible to consider the gondola of a stratospheric balloon as a prototypical passenger capsule. Powerful rocket boosters also seemed feasible for the first time.

Stories of rockets date back to ancient China, and authentic records tell of black-powder rockets used in both China and Europe in the thirteenth century. By the late eighteenth and early nineteenth centuries, field armies were deploying batteries of rockets as a regular tactic. In the War of 1812, British troops used powder rockets against American forces at Fort McHenry, near Baltimore, in 1814. Watching the battle, Francis Scott Key scribbled a poem, ''Star-Spangled Banner,'' with reference to "rockets' red glare . . . bombs bursting in air," later set to music as America's national anthem. Rockets also found more peaceable uses. In 1826 the British began using rockets to carry life lines to stricken ships, and a handful of rocket life-line stations along the British coast eventually saved thousands of lives.

Dreams of using rockets for space travel seem as old as the rocket itself, but there were serious drawbacks. One common denominator characterized the military, whaling, and lifesaving rockets from antiquity through World War I: they were powder-burning, or "solid," rockets. A solid rocket, although simplicity itself, had several shortcomings. There was no way to control the rate of thrust after ignition of the rocket and there was no guidance after the launch. Powder technology at the turn of the century seemed to dictate a missile with an optimum weight of about 150 pounds (most were in the 30- to 50-pound category), and the range rarely exceeded 3,000 yards. Advances in artillery in the late nineteenth century displaced the rocket as an effective weapon. For space exploration, solid-fuel rockets seemed to lack the thrust potential for reaching high altitudes. Visionaries thinking of using rockets for space exploration had to consider fuels and guidance as well as the problem of human survival in the space environment.

At the same time that powder rockets began to fall from favor in the late nineteenth century, a realistic theory and development of space flight was beginning to evolve, with strong interest in new types of liquid chemicals as propellants. Pivotal figures in the new era of rocket technology included Konstantin Tsiolkovsky (1857–1935), Robert H. Goddard (1882–1945), and Herman Oberth (b. 1894). They were imaginative men who drew their theories and experiments from the growing store of science and technology that had developed by the turn of the century. For one thing, the successful liquefaction of gases meant that sufficient fuel and oxidizer could be carried aboard a rocket for space missions. Research into heat physics helped lay the foundations for better engine designs, and advances in metallurgy stimulated new standards for tanks, plumbing, and machining to withstand high pressures, heat, and the extremely cold temperatures of liquefied gases. Progress in mathematics, navigational theory, and control mechanisms led to successful guidance systems. Reflecting early aeronautics, early rocketry owed much to an imaginative European heritage.

Although Tsiolkovsky did not construct any working rockets, his numerous essays and books helped point the way to practical and successful space travel. Tsiolkovsky spent most of his life as an unknown mathematics teacher in the Russian provinces, where he made some pioneering studies in liquid chemical rocket concepts and recommended liquid oxygen and liquid hydrogen as the optimum propellants. In the 1920s, Tsiolkovsky analyzed and mathematically formulated the technique of staging vehicles to reach escape velocities from earth.

In contrast to the theoretical work of Tsiolkovsky, Robert Goddard made basic contributions to rocketry in terms of flight hardware. Following graduation from Worcester Polytechnical Institute, Goddard completed a doctorate in physics at Clark University in 1911, and became a member of the faculty. While lecturing on conventional themes of physics, he occasionally commented, to politely skeptical students, on ways of reaching the moon. Beginning in 1916, Goddard managed to get a trickle of funds from the Smithsonian Institution to fabricate a series of test rockets. The Smithsonian published his pioneering 1920 monograph, *A Method of Attaining Extreme Altitude,* but caustic press reaction to his proposals for sending a small rocket to the moon caused him to adopt a low profile thereafter.

Goddard continued to experiment with liquid-fuel rockets, testing them on his Aunt Effie's farm, where loud, strange noises occasionally elicited irate complaints from the neighbors. A milestone occurred on March 16, 1926, when a Goddard assistant used a blowtorch to light the professor's latest rocket design. While Goddard crouched behind a wooden lean-to and watched, Aunt Effie's orchard became the site for the world's first flight of a liquid propellant rocket. In the rocket's two-and-a-half-second journey, it climbed to 41 feet and landed 184 feet distant—an abbreviated but portentous step toward space exploration.

In 1929, Charles Lindbergh learned of Goddard's work. Lindberg thought a reaction motor like a rocket might be useful to aviation, providing momentary emergency thrust if an airplane engine failed just after takeoff, and became enthusiastic about rocketry in general after contacting Goddard. Through Lindbergh, Goddard received an initial $100,000 grant from the Guggenheim family, followed by additional funds over the next several years. With this adventursome private support, Goddard began developing larger rockets at a new test site located in a remote desert area outside Roswell, New Mexico. Lindbergh frequently swung through the Southwest to visit Goddard and his small crew of loyal technicians. From 1930 to 1941, Goddard made substantial progress in the development of progressively sophisticated rockets, which attained altitudes of 7,500 feet, and refined his equipment for guidance and control as well as techniques of welding, insulation, and pumps. In many respects, Goddard laid the foundations of practical rocket technology by way of his Smithsonian research paper—a primer in theory, calculations, and methods—and his numerous patents, which comprised a broad catalog of functional rocket hardware. In spite of the basic contributions of Tsiolkovsky in theory, and of Goddard in workable hardware, both men went largely unheralded for years. Tsiolkovsky's work remained submerged due to the political conditions in Russia and the low priority given to rocket research prior to World War II. Goddard preferred to work quietly, absorbed in the immediate problems of hardware development, leary of the extreme sensationalism the public seemed to attach to suggestions of rocketry and space travel.

Although Hermann Oberth's work was original in many respects, he was more significant as advocate and catalyst, since he published widely and was active in popularizing the concepts of space travel and rocketry. Born in Transylvania of German parentage, Oberth later became a German citizen. He became interested in space through the fiction of H. G. Wells and Jules Verne, and left medical school to take up a teaching post where he could pursue his study and experimenting in rocketry. Oberth worked independent of Tsiolkovsky, though he heard of Goddard's brief paper of 1919 just as his own book, *The Rocket into Planetary Space* was going to press in 1923. *The Rocket into Planetary Space* was read widely, translated into English, and was followed by many other books, articles, and lectures by the energetic author. Oberth analyzed the problems of rocket technology as well as the physiological

By 1940 Robert Goddard (left) and his assistants, working at a site in the desert near Roswell, New Mexico, were developing liquid-propellant rockets of increasing size and complexity. Rocketry had reached the threshold of the space age. Through the efforts of Charles Lindbergh, Goddard's research received substantial funding from the Guggenheim family. During World War II, Goddard went to work for the U.S. Navy. Courtesy of the Smithsonian Institution.

problems of space travel, and his writings encouraged many other enthusiasts and researchers. During the ensuing years, Oberth continued to teach while writing and lecturing on space flight, and served as president of the Society for Space Travel, or *Verein für Raumschiffahrt (VfR)*, formed in 1927. In 1928, Oberth and others were consultants for a German film about space travel called *The Girl in the Moon*. The script included the now-famous countdown before ignition and lift-off. As part of the publicity for the movie, Oberth and his staff planned to build a small rocket and launch it. Although the rocket was only static-fired and never launched, the experience was a stimulating one for the work crew, which included an 18-year-old student named Wernher von Braun.

The existence of organized groups like the *VfR* signaled the increasing fascination with modern rocketry in the 1930s, and there was frequent exchange of information among the *VfR* and other groups like the British Interplanetary Society and the American Interplanetary Society. Even Goddard occasionally had correspondence in the American Interplanetary Society's *Bulletin*, but he remained aloof from other American researchers, cautious about his results and concerned about patent infringements. Because of Goddard's reticence, in contrast to the more visible personalities in the *VfR*, and because of the publicity given the German V-2 of the Second World War, the work of British, American, and other groups during the 1930s has been overshadowed. Their work, if not as spectacular as the V-2 project, nevertheless contributed to the growth of rocket technology in the prewar era and to the successful use of a variety of Allied rocket weapons in the Second World War. Although groups like the American Interplanetary Society (which later became the American Rocket Society) succeeded in building and launching several small chemical rockets, much of their significance lay in their role as the source of a growing number of technical papers on rocket technologies.

But rocket development was complex and expensive. The cost and the difficulties of planning and organization meant that, sooner or later, the major work in rocket development would have to occur under the aegis of permanent government agencies and government-funded research bodies. In America, significant team research began in 1936 at GALCIT. In 1939, this group received the first federal funding for rocket research, achieving special success in rockets to assist aircraft takeoff. The project was known as JATO, for jet-assisted takeoff, since the word *rocket* still carried negative overtones in many bureaucratic circles. JATO research led to substantial progress in a variety of rocket techniques, including both liquid and solid propellants. Work in solid propellants proved especially fortuitous for the United States; during the Second World War, American armed forces made wide use of the bazooka (an antitank rocket) as well as barrage rockets (launched from ground batteries or from ships) and high velocity air-to-surface missiles.

The most striking rocket advance, however, came from Germany. In the early 1930s, the *VfR* attracted the attention of the German army, since the Treaty of Versailles, which restricted some types of armaments, left the door open to rocket development. A military team began rocket research as a variation of long-range artillery. One of the chief assistants was a 22-year-old enthusiast from the *VfR*, Wernher von Braun, who joined the organization in October 1932. By December, the army rocket group had static-fired a liquid-propellant rocket engine at the army's proving grounds near Kummersdorf, south of Berlin. During the next year, it became evident that the test and research facilities at Kummersdorf would not be adequate for the scale of the hardware under development. A new location, shared jointly by the German army and air force, was developed at Peenemuende, a coastal area on the Baltic Sea. Starting with 80 researchers in 1936, there were nearly 5,000 personnel at work by the time of the first launch of the awesome, long-range V-2 in 1942. Later in the war, with production in full swing, the work force swelled to about 18,000.

Having completed his doctorate in 1934 (on rocket combustion), von Braun became the leader of a formidable research and development team in rocket technology at Peenemuende. Like so many of his cohorts in original *VfR* projects, von Braun still harbored an intense interest in rocket development for manned space travel. Early in the V-2 development agenda, he began looking at the rocket in terms of its promise for space research as well as its military role, but found it prudent to adhere rigidly to the latter. Paradoxically, German success in the wartime V-2 program became a crucial legacy for postwar American space efforts.

Air Power at War, 1930–1945

D uring the lean years of the 1930s, the military air services began to acquire modern planes with aerodynamic construction, retractable gear, and other features that characterized trend-setting transports like the DC-3. But the depression era meant closely trimmed budgets, and neither the army nor the navy acquired funds to purchase airplanes in the quantities they desired. Only the drift toward war in Europe led to large procurement orders—from England and France, as well as the U.S.—beginning a hesitant build-up of production capacity that reached full stride after American entry into the conflict. Early combat sorties underscored the comparative inexperience of U.S. aircrews; aerial warfare on a global basis strained logistics operations to their limits. American military aviation, in terms of numbers, quality, and doctrine, resolutely overcame deficiencies and eventually met the enemy on even terms, finally emerging as the world's leading air power.

MILITARY AVIATION BETWEEN THE WARS

The role of NACA became singularly important for military aviation in these years. In the early thirties, the establishment at Langley had been the world's leading aeronautical laboratory, and NACA had developed a close beneficial relationship with commercial aviation. Some critics felt that NACA was too cozy with the big manufacturers and transport corporations and charged that its research only duplicated work performed by other agencies, like the Bureau of Standards. NACA employees successfully lobbied on their own behalf in Congress and began to stress aeronautical research for the armed services, tactics that enabled the agency to survive bureaucratic storms as the depression deepened. Still, NACA was forced to admit in 1937 that it was no longer one of the leaders in research. This change occurred largely because of

The Boeing P-26 was the standard Army pursuit plane of the early to mid-1930s. Although it represented a departure from the biplane designs, the gear remained fixed and the cockpit open. Courtesy of The Boeing Company.

advanced military research in Europe, where international tensions were already boiling over into active conflicts in Abyssinia (Ethiopia) and in Spain. Under Hitler, Germany's drive for leadership in air power resulted in five new German aeronautical laboratories in the late thirties.

Reflecting the tenor of the period, Gen. Oscar Westover was named in 1938 as chairman of the Special Committee on the Relation of the NACA to National Defense in Time of War, and other studies recommended a new NACA lab for California, where the principal airframe manufacturers were located. Not long after authorization of the new NACA facility at Moffett Field, a former naval air station, Hitler's invasion of Poland in 1939 and a growing sense of an American deficiency in engines gave added impetus to the establishment of a third research lab, for power plants, located near Cleveland, Ohio. A rising sense of national emergency brought NACA increasingly under the direction of the armed services. The crisis atmosphere of aerial rearmament in the late thirties meant that NACA spent less time on basic research and more time on applied research and problem solving for military airplanes in service and entering production. The state of the American air arm on the eve of World War II required a lot of catching up. NACA work in areas such as low-drag wings, high-speed cowls, and tail-group arrangement paid early dividends for both army and navy aircraft.

Strategic bombardment, a dominant theme of Army Air Corps doctrine in the thirties, required the development of the long-range bomber, and the Martin B-10, which had entered service early in the 1930s, fit the bill. With twin engines and retractable gear, the all-metal B-10 had a ceiling of 28,000 feet and could fly over 200

MPH, a performance that surpassed many pursuit ships of the era. Arguing that fast, respectably armed, high-flying bombers could hold their own against opposing pursuit planes, many Air Corps leaders pressed for larger bombers with longer ranges. Early in the decade, army and navy agreements gave the army responsibility for land-based air defense of the American coastline and overseas possessions, including reconnaissance and operations up to the limit of the radius of action of Air Corps planes, like bombers. Following promising demonstration maneuvers on the West Coast in 1933, the Air Corps advocated formation of a general headquarters (GHQ) air force, a central striking unit of reconnaissance and bomber aircraft for defense against seaborne attack, which would act directly under the command of the army. Although not the "independent air force" championed by Billy Mitchell and others, the GHQ air force idea gained impetus from this concept and from the mission of coastal defense.

However, the growing momentum for such a plan hit a snag in 1934, when the Air Corps became ensnarled in an abortive attempt to carry the U.S. mail. Acting on information divulged by newsman Fulton Lewis, Jr., Sen. Hugo Black of Alabama launched an investigation of irregularities in air-mail contracts awarded during the Hoover administration. At a time when Pres. Franklin Delano Roosevelt's new administration had begun prying into other assorted corporate improprieties, Senator Black's exposure of alleged favoritism in parceling out Post Office air-mail contract awards to big companies, at the expense of smaller operators, was a sure way to attract headlines and public attention. In the midst of all the publicity, FDR canceled existing contracts on February 9, 1934, and called on the Air Corps to fly the mail until new arrangements could be worked out. Gen. Benjamin D. Foulois, chief of the Army Air Corps, had only a few hours to digest the implications of the emergency but felt sure the Air Corps could follow through. His confidence was to be badly shaken.

The Air Corps hurriedly modified a motley collection of bombers and pursuit ships—mainly open-cockpit biplanes—by tearing out armament and extra seats and by converting bomb bays and rear cockpits into makeshift mail compartments. Most of the pilots were inexperienced in bad-weather and night flying, and many of the planes lacked landing lights, as well as illuminated cockpit instruments for nighttime navigation. In hasty training before formal operation began on February 19, three pilots were killed; during the week that actual flights began, vicious February storms swept across the country and added to the toll: two more fliers dead, six injured, and eight planes destroyed. Eddie Rickenbacker, now a vice-president of Eastern Airlines, castigated the whole affair as "legalized murder."

Amid the rising outcry of an appalled public and consternation within the administration, the Air Corps desperately tried to regroup and salvage what it could on its commitment to keep the mail flying. The harsh weather of the first week had been unusual, but the criticism of glaring deficiencies in Air Corps operations and equipment that had been cruelly revealed was justified. And it did not help to know that an Eastern Airlines DC-2 had taken off from Los Angeles on February 18, the day before the president's cancellation order, and had made a record transcontinental run, climbing to 20,000 feet to outrace a storm in the Alleghenies, and descending at Newark in good visibility. After a week's respite for more training, the Air Corps returned to the task with better flying aids and better planes, like the B-10. But by the time operations ceased on June 1, when the airlines took over again under new contracts, a grim total of 12 dead pilots and 66 crashes or forced landings had been recorded.

Following the debacle of the mail flights, the Air Corps began to equip its planes with adequate night-flying aids and navigational equipment, and investigating boards assessed new moves to revitalize American military aviation. The report of a War Department board, chaired by former Secretary of War Newton Baker, hedged on the issue of an independent air force but finally endorsed the GHQ air force (an opening wedge for independent air force adherents), which was created in 1935. This event gave an impetus to the research and development of long-range strategic bombers like the B-17 and B-24 and, ultimately, the B-29. Prototypes of the B-17 were flying by the summer of 1935. But the role of this new air weapon remained unclear. In a demonstration of the big bomber's capacity for coastal defense in 1938, three B-17s successfully "intercepted" the Italian ocean liner *Rex* 725 miles out in the Atlantic as the ship steamed toward New York. The navy, instead of being delighted, became alarmed that its traditional prerogatives were being threatened and engineered new guidelines limiting GHQ air force operations to 100 miles offshore. The practicality of using long-range bombers for 700-mile missions against lone ships in open seas during wartime was in itself open to question. In any case, the restrictions certainly compromised the vigor of the GHQ air force just as it was gathering momentum.

Not until late 1938 and early 1939, after the Czechoslovakian crisis, did the Air Corps begin to win the budgetary and bureaucratic support it needed. Concerned by the European situation, President Roosevelt made a firm commitment to the development of American air power. "Our existing [air] forces are so utterly inadequate that they must be immediately strengthened," Roosevelt warned early in 1939, and the Congress responded with a $300 million Air Corps appropriation. A new procurement program called for 5,500 planes, with 3,251 to be delivered within the next two years.

But the immediate problem involved catching up to other air forces, who were far ahead of the United States in trained personnel and first-line equipment. Gen. Frank M. Andrews caustically described army aviation as a "fifth-rate air force." In 1939, the Air Corps numbered 26,000 personnel. The British Royal Air Force (RAF) numbered 100,000 and the German Luftwaffe totaled some 500,000. The Air Corps had approximately 800 modern aircraft, compared to the 1,900 of the RAF and the 4,100 of the Luftwaffe. The numerical inferiority of American planes was further aggravated by qualitative inferiority. The P-36, a standard first-line Air Corps fighter in 1939, was outclassed by British Hurricanes and Spitfires and the German Messerschmitt Bf-109; German attack bombers were far superior to contemporary American counterparts. The one encouraging spot was in heavy bomber aircraft, where the B-17 surpassed comparable types. As 1939 drew to a close, the Air Corps mission was expanded to the extent of defending the entire Western Hemisphere, providing an opportunity to implement long-range bombing concepts of big-bomber adherents.

Beguiled by unwarranted confidence in long-range aerial bombardment, American pre-eminence in bomber design had been secured at the expense of fighter equipment. "There can be no doubt that this attitude hindered the development of fighter planes," stated one postwar air force historian. "This belief that big bombers could be made to go as fast as pursuits and could defend themselves against fighters contributed much to the failure to provide escort fighters for the heavy bombers in the early years of World War II—one of the biggest mistakes of the air war." American fighters like the P-40 and P-38, designed in 1936–37, were still not available in quantity when the United States entered combat in World War II, and American equipment generally trailed that of the British and Germans until the availability of the P-47 and P-51 in 1943. For all its weaknesses in 1939, American air power proved to be an invaluable weapon in the Second World War, and the story of its evolution is indeed remarkable. Administrative improvements helped, particularly when they provided influence for individuals with the right balance of experience, foresight, and resourcefulness. This fortunate circumstance occurred in June 1941, when the U.S. Army Air Force (AAF) became a reality, commanded by Maj. Gen. H. H. Arnold.

The U.S. Navy had experienced its own share of troubles in the prewar era. Despite the sobering loss of the dirigible *Shenandoah* in 1925, partisans of the mammoth airships in the navy wangled funds for two more of them at $6.5 million each. The USS *Akron*, half again the size of the *Shenandoah*, was commissioned in 1931. In order to reduce the airship's vulnerability to air attack, the *Akron* carried five biplane fighters within its helium-filled hull. To launch and retrieve the planes, the *Akron* extended a rigid, trapezelike device to release the little fighters one at a time and to pull them back inside the hull at the end of their missions. Even with this fighter escort, the *Akron* frequently proved vulnerable during maneuvers. In any case, the airship's weakness in heavy storm conditions was driven home again in the *Akron*'s tragic breakup off the Delaware coast in the spring of 1933. Only three crew members survived. Among the 55 men and 18 officers lost was Rear Adm. W. A. Moffett, chief of the navy's Bureau of Aeronautics. Commissioned not long after the *Akron* went down, the fighter-equipped *Macon* was lost during maneuvers off the California coast two years later. Stubbornly, the navy then recommissioned the *Los Angeles,* a German-built airship acquired in 1924 as part of German reparations for World War I. Retired at Lakehurst in 1932, the refurbished *Los Angeles* made occasional forays from 1935 until 1940, when it was scrapped. For all the tribulations of the giant airships, the navy learned much about the operation of lighter-than-air vessels and began development of smaller, nonrigid types for their obvious advantages in sustained reconnaissance over expanses of ocean. A direct result was the K-class blimp, respectably successful in convoy escort and submarine patrol missions during World War II.

In the meantime, the navy operated a series of flying boats and perfected their role as patrol bombers. The Consolidated P2Y, a twin-engine, modified biplane design, set the pattern for these long-range, seagoing aircraft before it was superseded in the mid-thirties by the Consolidated PBY Catalina. This flying boat's remarkable versatility for flying cargo and for long-range patrol, rescue, mine-laying, or bombing made it an invaluable workhorse in naval missions throughout World War II. But the most valuable naval contributions to military aviation in the thirties were the navy's continued development of the aircraft carrier, a complement of carrier-based aircraft, and a coherent doctrine for carrier warfare. The actions of the *Lexington* and the

PBY-5 Catalina flying boats compiled a remarkable service record in every theater of the war. They rescued downed pilots, conducted grueling reconnaissance missions, sank enemy submarines, and performed a multitude of other duties. Courtesy of General Dynamics.

Saratoga during the maneuvers of 1929 had begun the process of winning more converts to carrier operations within the navy. Even those who continued to regard the battleship as the principal actor in fleet operations began to recognize the significant offensive and defensive fire power of aircraft carriers. After the 1930 exercises, in which the *Lexington*'s planes "scored" heavily against defending battleships, the idea of a new tactical unit, the "carrier group," consisting of a carrier, four cruisers, and two destroyer squadrons, gained currency. During war games in 1931, the effectiveness of dive bombers against surface ships and the success of air patrols against submarine attacks spurred further refinement of these airborne tactics and the acquisition of additional aircraft carriers. The problem for the navy's Bureau of Aeronautics, like the Army Air Corps, hinged on the miserly appropriations available in the early thirties, not only for ships and planes but also for effective refinement of a host of ancillary equipment ranging from magnetos and spark plugs to gun sights, bomb racks, and arresting gear for carrier flight decks.

From 1934 onward the navy profited from President Roosevelt's support for allocations through the Public Works Administration (PWA). Funds from the PWA, for example, provided the money to lay the keels of the carriers *Yorktown* and *Enterprise* in the spring and summer of 1934. Such financial bootstrapping evolved from the president's warm regard for Rear Adm. Ernest J. King, who began a three-year tour as chief of the Bureau of Aeronautics in 1933. This period also included development of several new types of naval aircraft that performed valiantly after 1941. The biplanes of an earlier era finally gave way to preliminary designs for the Douglas SBD Dauntless dive bomber (the scourge of so much Japanese shipping in World War II), the Grumman TBF Avenger torpedo bomber, and the Grumman F4F Wildcat fighter (originally designed as a biplane). These metal monoplanes, with retractable landing gear and powerful radial engines, turned in stalwart performances in the opening phases of the Far East conflict and teamed with more advanced designs to win unquestionable control of the air in the later stages of the Pacific campaigns.

Continuing trial operations yielded invaluable experiences for military operations in treacherous weather and troublesome climates. During the winter of 1931, the *Langley* conducted exercises off the New England coast to prove that naval aviation could carry on in frigid seas, a type of training that continued for the next three years in the Aleutians. Weather investigation flights near Shanghai and Guam probed the dangers of typhoons and added to the knowledge and flying procedures for tropical military missions. On a regular basis, massed flights of standard service aircraft to remote air stations (unlike the unique long-distance flights of special planes, as had been the practice in the twenties) demonstrated a potential mobility of air power that was not lost on war planners in the Navy Department. The war games of 1936 included exhaustive trials of aerial scouting for extended duration and led to the installation of the automatic pilot on all patrol planes, beginning with the PBY Catalina. Exercises in 1939 led to plans for small attack carriers with armored decks for fast sorties from the battle line. Thus, naval aviation began to accumulate the equipment and procedures for modern combat.

Although the various chiefs of the Bureau of Aeronautics during the thirties had not always begun their careers in naval aviation, they invariably seemed to sense the significance of naval air power and, as they moved on to different commands, leavened other areas of the fleet with an appreciation of aviation in the navy. This was an advantage for Rear Adm. John Towers, who became chief in 1939 and who guided naval aviation through the crucial early years of World War II. Moreover, Towers came to the bureau with considerable flying experience. At the time he took over direction of naval aviation, Towers was the navy's most experienced flier (the oldest living graduate from Glenn Curtiss's 1911 flying school) and an officer who had lived through all the struggles and progress of fleet aviation. He was an eminently qualified leader for the crisis about to break. His principal problem, again similar to that of the Army Air Corps, was still that of inadequate quantity of personnel and equipment. By 1941, the navy had amassed only 5,260 aircraft of all types and only 6,750 pilots, including Coast Guard and Marine Corps fliers, along with 1,874 ground officers and a total of 21,678 enlisted men. Including the aging *Saratoga* and *Lexington*, the navy had only eight operational carriers—far too few for the worldwide commitments portended by the deteriorating international situation. "Actually," naval historians Archibald Turnbull and Clifford Lord wrote later, "what had been got together represented by comparison

Its dive brakes extended, a Douglas SBD Dauntless dive bomber demonstrates its mission as one of the U.S. Navy's most destructive weapons in the Pacific. Courtesy of McDonnell Douglas Corporation.

with what would be needed hardly more than a 'sergeant's guard' with half-filled bandoliers and haversacks.''

The ''sergeant's guard'' received a sudden call to action. Japanese aggression in Manchuria and China threatened to draw in the United States, who saw its national interest in the Far East increasingly jeopardized. Diplomatic negotiations between America and Japan to avert war in the Far East reached a stalemate. Japanese militants felt it imperative to strike at Pearl Harbor in Hawaii and knock the U.S. Navy's Pacific Fleet out of action in order to fulfill Japan's scheme for conquest of the Philippines and other strategic objectives in Asia and the Pacific. On December 7, 1941, planes from six Japanese fleet carriers completely surprised Pearl Harbor. Out of a total strength of 432 planes, the Japanese launched 354 in two waves, lost only 29, and achieved staggering results. Only a handful of U.S. aircraft escaped destruction and challenged the attackers. The strike disabled all eight of the navy's battleships at Pearl Harbor, along with three cruisers and three destroyers. Over 2,000 personnel died. The functioning of the vital naval and military base remained chaotic for days.

The Japanese made mistakes. First, the attack resolved an agonizing dilemma in the United States, splintering organized isolationist opposition to the idea of collective security and galvanizing overwhelming public opinion behind President Roosevelt's call for a declaration of war against Japan. Second, the Japenese attack failed to destroy key dockyard cranes and other machinery, permitting rapid restoration of maritime support services, including repair of many vessels severely damaged during the aerial assault. Many of the huge fuel storage tanks dotting land around the anchorage likewise survived. Finally, U.S. carriers based at Pearl Harbor happened to be at sea the day of the attack on sundry patrol and transport duties, including aircraft deliveries to other American bases in the Pacific. The carriers' survival meant the navy retained a vital weapon with the potential to disrupt the Japanese timetable and ultimately restore American momentum.

AIR WAR OVER EUROPE

On December 11, 1941, Germany and Italy announced a state of war with the United States; America declared against them the same day. Even though outraged public sentiment after Pearl Harbor supported major action against the Japanese at once, American strategists emphasized treating the European crisis first. Hitler dominated Western Europe. Although the RAF had fought the German air force to a stand-off during the Battle of Britain in the summer of 1940, Great Britain remained in peril. Without the United Kingdom, the Allies would have no European industrial base, staging area, or airfields from which to launch a conclusive offensive against Germany's strong position on the European continent. The distant Pacific theater of operations, involving far-flung sea and air movements over thousands of miles of open ocean and scattered islands, would have to accept second priority for the time being.

In fact, working associations with counterparts in the British armed services had already begun. Alarmed at Hitler's conquest of Austria and Czechoslovakia by 1938, British and French purchasing missions arrived in the U.S. with orders for military equipment, including aircraft. In view of the European emergency, amendments to America's Neutrality Laws permitted U.S. manufacturers to sign contracts, and a preliminary tooling up of the aircraft industry got underway. Expanding American military budgets in 1939, following Roosevelt's rearmament program, cleared the way for aircraft production in volume. After the fall of France in 1940, the country's orders were channeled into British and American services. The Lend-Lease Program, arranged in the spring of 1941 to skirt remaining Neutrality Law constraints and to build additional planes and other military hardware for ''loan'' to Great Britain (and later to

Russia), accelerated production plans. Observers overseas and Allied personnel aided U.S. designers and manufacturers in acquiring information to modify American equipment based on European combat experiences. These were important early steps—a prelude to the long strides needed in the years ahead.

During 1942, American air forces struggled to get organized, hammer out channels of effective coordination with the RAF, and begin air strikes against German forces in Europe. None of these goals was easy to achieve. The RAF had been waging an active air war for more than two years and felt that its program of night bombing of enemy targets was the surest way to minimize losses to its bomber forces and inflict the greatest damage on the enemy. American strategists argued for daylight bombing, which they claimed would provide greater accuracy and take the best advantage of the Norden bombsight. The Norden equipment, American airmen argued, allowed a bombardier to put a bomb into a pickle barrel. For several months such arguments were academic, since the air force first had to accumulate planes in England in numbers sufficient to justify the start of combat missions.

By the winter of 1942/43, before bad flying weather set in, the air force had landed nearly 900 planes in the United Kingdom. Moreover, several sorties during the previous months had provided invaluable operational experience. In February 1942, Brig. Gen. Ira C. Eaker went to England to organize the American aerial effort, and Maj. Gen. Carl A. Spaatz followed in June in take command of the Eighth Air Force. Early American efforts were carried out in direct partnership with the RAF or put together almost as an afterthought. The first missions, using only a few aircraft, proved only marginally successful and gave little indication of the formidable war machine that the "Mighty Eighth" became as it gathered strength.

In July, six American crews in training with the RAF took off in a half-dozen British Bostons (the export version of the Douglas A-20) for an attack on German airfields in Holland. The loss of two bombers from intense flak that hindered accurate bombing did little to raise morale. Compared to the dozens of bombers deployed by both the RAF and Luftwaffe, it was obvious that American squadrons had much building to do, and even more to learn. The first American bomber raid from the United Kingdom (with RAF Spitfires flying fighter cover) was launched on August 17, 1942, when one dozen B-17 Flying Fortresses took off for a strike against enemy-held Rouen, France. The lead bomber, christened the *Yankee Doodle,* carried General Eaker himself, flying as an observer. After a successful bomb run at 23,000 feet, all 12 Flying Fortresses returned to a jubilant welcome at their base near London. In the following weeks, more short-range missions were flown against targets in occupied France as American fliers and ground crews increased their operational experience, culminating in a raid of over 100 B-17s and B-24s on October 9, 1942, on steel mills and locomotive works at Lille. Fierce opposition by German fighters threw the attacking bombers into confusion. Nevertheless, the Lille operation was an important precursor to the kind of mass raids the Eighth planned to carry out. Further, these raids were flown in daylight, to justify the American contention for continuing precision daylight raids. Up to this point, however, such missions had been comparatively short range and flown with fighter cover. Long-range attacks against industrial targets, beyond the range of accompany-

In a photo taken during July 1944, when the U.S. Army Air Force had reached new levels of destructive power, Lt. Gen. James H. Doolittle (left) confers with Gen. Carl Spaatz on the control tower catwalk of the 351st Bomb Group, based in Great Britain. Courtesy of U.S. Air Force.

ing fighters, required the bombers to fight their way through enemy fighter opposition en route to the target, drop their bomb loads, and fight their way out again. Several months passed before the Eighth Air Force could attempt this kind of mission, since priorities in the meantime had shifted to the campaign in North Africa.

Operation Torch, the combined British-American invasion of North Africa, was launched on November 8, 1942, to satisfy a combination of political and military considerations. For one thing, the United States and Great Britain felt a need to respond to Josef Stalin's insistent demands for a ''second front'' agaist Germany in order to ease intense pressures on Russian armed forces along the 1,500-mile-long combat zone in the Soviet Union. For another thing, Gen. Erwin Rommel's Afrika Korps was giving the British armies in Africa a bad time and threatening the "Suez Lifeline," used for oil shipments and for communications with the Far East. In order to win control of the skies and provide tactical air support for the North African campaign, the Eighth Air Force was shorn of a good deal of its air strength. "We were torn down and shipped away," General Eaker grumbled.

American air operations in North Africa had actually begun months earlier, in almost impromptu fashion. A detachment of B-24s en route to China had gotten as far as Khartoum, where the British persuasively argued to keep them to assist desperate British forces in the Mediterranean theater. On June 12, 1942, 13 Liberators took off on a 2,600-mile round-trip strike against oil fields around Ploesti, Romania. Only eight planes got back (one crash-landed and four were diverted to Turkey, where they were interned), and damage to the oil fields was minimal. This was actually the first American air strike against a European target, a sobering experience in the exigencies of long-range strategic bombing. During the summer of 1942, available air force elements in Africa joined with the RAF in harassing Rommel by attacking his supply ports around the Mediterranean coast and strong points inland. Following Rommel's defeat at El Alamein and his retreat into Tunisia, Operation Torch was designed to eliminate the German position in North Africa. Under the command of Maj. Gen. Lewis H. Brereton, the Ninth Air Force added more B-17s, light bombers, and fighters in a bitter contest for control of the air, which combined Allied forces won in the spring of 1943.

In northwest Africa, the AAF began to learn how to use tactical air units, especially the deployment of fighter-bombers. The task was enhanced by the creation of a combined Anglo-American air command, Northwestern African Air Forces (NAAF) under Gen. Carl Spaatz; the Ninth Air Force remained a separate entity. Two primary groups operated under the NAAF umbrella: the Strategic Air Force, under Gen. Doolittle, and the Tactical Air Force, under Air Vice Marshal Sir Arthur Coningham. Coningham, who had led RAF forces against the Germans during 1941, brought invaluable knowledge of the desert air war and of tactical operations. During the early phases of Operation Torch, heavy rains had immobilized Allied fighters, although German fighters and dive bombers continued to fly from well-prepared fields with hard surfaces. Allied ground troops, continually harassed by German air attacks, complained that the only planes they ever saw were the enemy's. As the weather improved, the British and Americans took to the air in greater numbers. The Americans also learned to cope with demanding maintenance requirements in the harsh desert environment; to perfect aerial reconnaissance photos; and to carry air attacks down to ground level against troops, trucks, and tanks. The Germans, in a desperate gamble at aerial resupply, dispatched several waves of obsolete JU-52 trimotor transports from Italy to Africa. The air force intercepted them over the Sicilian straits and shot more than 50 of the heavily laden planes out of the sky in a mission called the "Palm Sunday Massacre." Within weeks, Rommel fled back to Germany, and 270,000 German and Italian troops, their backs to the Mediterranean, surrendered in September 1943.

The Sicilian and Italian invasions, conducted in the summer of 1943, received highly effective aerial support and neutralized serious German air force resistance in the autumn campaigns. First in Sicily, then in Italy, Allied airborne troops made their first large-scale parachute drops of the war. The biggest U.S. jump occurred in September, when 1,200 paratroopers of the 82d Airborne Division dropped from AAF transports onto the beachhead at Salerno. In the course of extensive air operations, the AAF also perfected fighter-bomber tactics. Carrying eight heavy machine guns and extensive capacity for bombs or rockets beneath its wings, the Republic P-47 Thunderbolt evolved into a versatile fighter-bomber type. Other planes also served effectively. Pilots struck at trains and truck convoys, disrupting schedules and jamming up traffic behind, then realized they had even more productive objectives to attack. The elongated Italian peninsula, with innumerable communication and transportation bottlenecks along its mountainous spine, offered unusually vulnerable highway and railroad bridges and tunnels as targets. Allied fighter-bombers first wrecked these

An armament of eight .50 caliber machine guns gave the Republic P-47 Thunderbolt extra-heavy fire power; bombs and rockets carried under the wings made it an outstanding fighter-bomber. Later models had the bubble-type canopy. Courtesy of U.S. Air Force.

targets, halting freight trains and truck convoys en route to the front and creating massive backups in railroad marshaling yards, supply depots, and factories behind enemy lines. Then the medium and heavy bombers blasted the congested rail yards and storage areas as well as seaports, coastal vessels, and industrial targets. Consistently short of supplies, the retreating German forces nonetheless gave ground only grudgingly.

In the course of these operations, organizational lessons learned earlier in the Mediterranean theater were more effectively practiced, particularly in the case of tactical operations. The headquarters of the U.S. Army was located close to the newly formed XII Air Support Command, so that air and ground officers remained aware of each other's problems. Liaison officers from the army went directly to airfields to explain their situation. Air liaison officers, nicknamed "Rovers," traveled around forward areas in a Jeep. The Rovers, equipped with radios, discussed ground targets with air force planes flying directly overhead. The fighter-bombers and other planes committed to ground-support operations thus achieved far more satisfactory results.

During the Italian campaign, black combat pilots began receiving overdue recognition. Trained in segregated flight schools in the United States, bunched in segregated flying units, and often equipped with battle-weary aircraft, the black fliers struggled to overcome persistent bureaucratic obstacles. Col. Benjamin O. Davis, Jr. became commander of the 33d Fighter Group, whose aerial campaigns ranged over the Mediterranean theater, central Europe, France, and Germany, where action during one escort mission for a raid on Berlin merited a Distinguished Unit Citation.

Allied air strategists in the meantime had planned a heavy blow at German petroleum production as an adjunct to the Italian invasion and to hamper the German offensives in Russia. Again, the target was Ploesti. Unlike the small, 13-plane raid the year before, a massive force of 177 B-24 Liberators of the Eighth and Ninth air forces roared away from airfields at Benghazi in Africa at dawn on August 1, 1943. But the Germans were well prepared, and the vital oil production facilities at Ploesti were protected by dozens of carefully sited antiaircraft batteries, hundreds of machine guns, and veteran fighter pilots flying late-model Messerschmitt Bf-109s, making the target one of the best defended in Europe.

As the Liberators droned across the Mediterranean, the two lead navigational planes were lost, and heavy clouds over the Balkans separated some elements of the attacking force so that many planes arrived over the target behind schedule. Through intercepted radio messages, the Germans at Ploesti had been put on full alert for the impending attack. To compound the problems of the increasingly hapless mission, navigators mistakenly identified a checkpoint and sent dozens of planes wheeling off in the wrong direction. When the Liberators finally arrived over the target, the raid became an uncoordinated melee of bombers making runs from the wrong vectors across the paths of other attacking aircraft. The belated arrival of lagging B-24s added to the muddle. The alerted German fighter pilots and ground defenses took a heavy toll of the flustered attackers, while flames, dust, and dark, oily masses of smoke obscured the target and caused even more confusion. Only the fact that the Americans benefited from the initial surprise of bomb runs at treetop level saved many planes. As it was, the raid cost 53

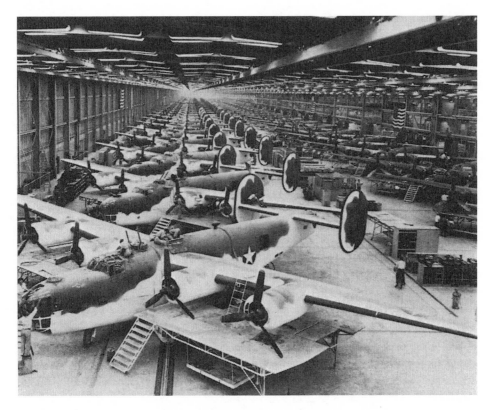

Production line for the Consolidated B-24 Liberator heavy bomber. In cavernous structures like this one, the American style of mass production represented one of the most significant contributions to the Allied war effort. Courtesy of General Dynamics.

downed bombers and over 50 more badly damaged planes. More than 400 airmen were dead or missing, with 70 more interned after they made emergency landings in Turkey. The results of the Ploesti raid continue to be debated, since there were no follow-up strikes, and the refining facilities at Ploesti were soon back in operation. Still, the Ploesti complex had been operating at 60 percent of its capacity before the air strike, and the damage inflicted continued to impede full production, depriving Germany of increased refined petroleum output at a critical time.

For AAF planners, who pinned so much of their hope on the effectiveness of long-range strategic bombing in daylight, the bitter experience of Ploesti was a grim portent. Bomber groups all over England were poised for massive strikes into German heartland, but unless losses could be diminished, the cost of daylight bombing over Europe promised to be monumental—perhaps prohibitive. The American scheme for successful daylight bombing without fighter escort against targets deep in Europe remained to be proven. At a high-level conference in Casablanca, Morocco, in January 1943, Winston Churchill had urged the Americans to give up on daylight bombing and join the RAF in nighttime missions. General Eaker, called in to argue the American case, pointed out that American bombers had been designed and their crews trained for daylight operations. Modifications and retraining would eat up precious time and energy. Precision bombing by day, he maintained, could still be significant in taking out crucial sectors of German industrial capability. Further, splitting the RAF and AAF bomber offensives into day and night operations would cut down congestion at airfields and in the crowded air space above English bases. Clinching his arguments, Eaker reminded Churchill that bombing Germany around the clock—day and night—would strain German air defenses to the utmost. With an ear for a well-turned phrase, Churchill seized on "bombing around the clock," conceded the American argument, and pressed it himself in addresses to Parliament on his return to London.

With the eighth Air Force bomber squadrons and equipment brought up to sufficient numerical strength by the summer of 1943, air force strategists devised a major aerial thrust into Germany. The air force planned to hit ball-bearing plants at Schweinfurt and the Messerschmitt factory at Regensburg within minutes of each other. On the morning of August 17, 1943, 376 Flying Fortresses lifted off from their English air bases. American fighters planned to escort the bombers as far as possible. After that, everyone hoped, the defensive fire power of the massed B-17 formations could see them through the German interceptors. The American strategy was to draw off the main German fighter defenses to Regensburg, giving the Schweinfurt bombers an excellent chance for reduced opposition on their bombing runs. In the meantime, the Regensburg formations would confuse the enemy by proceeding south to airfields in North Africa, rather than fly back over German fighter bases en route to England. The bombers attacking Regensburg indeed drew savage German fighter opposition: of 24 Flying Fortresses lost, most were shot down while flying to the objective. Fortunately, the escape route to North Africa took the Luftwaffe by surprise, and the Regensburg force successfully withdrew after a remarkably accurate bomb run over the target. The Schweinfurt mission was not so lucky. Bad weather delayed takeoff; the bombers headed across the Channel some three and a half hours after the Regensburg strike, instead of making the 15-minute interval originally planned. The lapse gave German fighter squadrons enough time to land, rearm and refuel, and get airborne again. Enemy fighters subjected the Schweinfurt strike force to continuous attack all the way to the target and hounded them as far as the Channel on the way home. Thirty-six bombers out of 230 went down, and dozens more limped home badly damaged.

Despite the heavy losses, the eighth Air Force was encouraged by the general accuracy of their bombing. Nevertheless, Schweinfurt's production of ball bearings persisted, requiring a second raid. Bad weather during most of September and early October caused mission planners to seek targets elsewhere, but heavy costs in bombers and crews continued. Still hoping for a knockout blow at the ball-bearing industry, the AAF launched its second Schweinfurt raid on October 14, 1943. Although 60 Liberators turned back after failing to make the proper rendezvous and 32 Flying Fortresses aborted the mission because of mechanical problems, 291 B-17s droned eastward toward Germany. The Germans put up virtually everything they had, including obsolete Stuka dive bombers. In a grueling aerial duel that continued for 3 hours to the target and back again, the German defenders shot down 60 bombers (5 more crashed in landing; 17 more had to be scrapped), a loss of about 35 German aircraft. Including the

Its bomb-bay doors open, a Boeing B-17 flies through heavy flak during a raid over Germany in World War II. Courtesy of U.S. Air Force.

Schweinfurt raid, over 150 bombers had been lost in just seven days. At this rate of attrition, the Eighth Air Force would soon be decimated. The Luftwaffe remained in control of the skies over Europe, and the battered Eighth Air Force called a momentary halt to unescorted bombing missions to targets deep in Germany.

It had been a tenet of faith that the defensive armament of American heavy bombers would see them through long-range attacks without the protection of friendly fighters. Mission planners hoped bombers could protect themselves by forming massive "combat boxes," spaced so that the field of fire from machine guns in each plane would overlap that of others, as would the combined fire power of adjacent formations. The German fighter pilots evaded this massed fire power in several ways. In one ploy, a lone fighter screened itself in a bomber's contrails, waiting to get close enough for a telling burst of gunfire or lethal, air-to-air rockets. The majority of fighters attacked bomber formations head on, since many early models of the B-17 lacked effective nose armament. Tail gunners eventually learned to keep a sharp eye for fighters hidden in the contrails, and additional guns inside blisters and chin turrets augmented the forward fire power of B-17 squadrons; the design of the B-24 facilitated the installation of a power turret in the nose. Improved tactics and enhanced fire power helped on subsequent combat missions but did not entirely solve the crisis for American bomber squadrons.

The real answer to the frightful loss of bombers was the provision of fighter escorts. The lessons of 1943 spurred the development of external fuel tanks that could be dropped when the escorting fighters engaged enemy aircraft. By the end of the summer, such tanks had increased the P-47's combat range from 230 miles up to 375 miles, which would reach into portions of northwestern Germany but fall short of Frankfurt and other industrial targets beyond. By December, similarly equipped P-38 Lightnings could range 520 miles from their bases into central Germany, still short of Berlin and remote objectives in the southern and eastern sections of the country. Moreover, despite the P-47's rugged construction and fire power and the P-38's speed, both planes experienced difficulty in handling the more agile Messerschmitt 109 and the powerful Focke-Wolf 190. The balance tipped more favorably for the AAF after February 1944, when the P-51 Mustang became available in large numbers. With drop tanks, the Mustang had a range of 850 miles. The penetration of American fighters deep into German-occupied Europe directly contributed to accelerating losses among German fighter squadrons. Increasing numbers of bombers arrived on target, and the rise of Allied bomb tonnages dropped into the Third Reich increased the disruption of German industrial capacity. Cumulative shortages of enemy war material contributed to the successful Allied invasion of Normandy and the eventual defeat of Germany.

The P-51 developed a reputation as one of the best fighters of World War II. Ironically, its introduction into the air force occurred almost as an afterthought. The design had originated in the dark days of 1940, when the RAF placed an emergency order with North American Aviation in California. The company had already purchased data from Curtiss-Wright Corporation that related to an aerodynamically advanced design featuring a ventrally located radiator. Under the press of time, North American cut corners and utilized certain components from their own AT-6 North American trainer: wheels, brakes, hydraulics, and electrical systems. In a series of around-the-clock design conferences, North American's engineers finalized a configuration and hand-built the first airplane in just 102 days. With careful attention to streamlining and use of a radical new NACA airfoil that reduced wing drag by half, the Mustang emerged from the drawing boards as a lean, lithe airplane. After flying an early export version powered by an Allison engine, a canny test pilot from Rolls Royce (Ronald W. Harker) realized that the more powerful Rolls Royce Merlin engine might give the Mustang a stunning increase in performance. He was right. With a top speed surpassing 440 MPH, the Mustang could outspeed and outmaneuver any comparable German fighter. Rolls Royce licensed the Merlin engine for manufacture in the United States, and the hybrid P-51B Mustang went into production for the AAF in 1943.

By early 1944, American bomber losses had been made up through increased production, and long-range escort fighter tactics had been effectively orchestrated. Beginning February 10, 1944, combined RAF and AAF missions during a seven-day period severely jolted German military production and wrested control of the air over Germany from the Luftwaffe. During the "Big Week," RAF bombers joined over 1,000 AAF bombers (many from new bases in Italy) in devastating raids against enemy

The Lockheed P-38 Lightning served capably in the European and Pacific theaters. Its nose-mounted machine guns represented a lethal concentration of fire; other versions with nose-mounted cameras were particularly successful as high-speed photo-reconnaissance aircraft. Courtesy of Lockheed Corporation.

industrial plants and air bases. The air force flew over 3,500 fighter sorties and downed some 450 German fighters. Within the next few weeks, additional bomb raids and aggressive American fighter tactics crushed the Luftwaffe. During February and March, the German air force lost 800 day fighters over the western front—a force that was irreplaceable. The Luftwaffe began to run out of competent pilots at a time when both fuel and time to train new fliers became severely pinched. Just when the Germans were girding to meet the anticipated Allied invasion, they lost control of the air. Free to range over the invasion sector, raids by fighter bombers, followed by heavy bomber strikes, made a shambles of enemy communications and supply systems. Allied fighter sweeps virtually eliminated the German air force. "If you see fighting aircraft over you," a confident General Eisenhower told troops on the eve of D-Day, "they will be ours."

Controversy over the choice of bombing objectives persists. Many strategists urged wider attacks against the comparatively few oil production targets that were hit before D-Day. Eisenhower and others seemed to feel at the time that destruction of other

Fresh from the factory, a North American P-51D Mustang with a bubble-type canopy takes to the air. Sleek lines and a series of increasingly powerful engines made it one of the best performers among piston-engine designs of World War II. Courtesy of Rockwell International.

industrial targets and tactical strikes against transportation and communications systems would do more than anything else to negate opposition in the vicinity of the invasion zones during the crucial period of establishing a beachhead. The 12 months that followed the successful Allied landings in Normandy on June 6, 1944, which included glider and parachute drops behind enemy lines, saw a shift in Allied bombing priorities from transportation and communications targets to oil refineries. In any case, once the beachhead was secure, bomber strikes on oil refineries accelerated the decline of the German army and air force.

Nevertheless, tactical strikes before and after D-Day did much to assure successful landings, followed by important breakouts from the beachhead itself. Prior to the invasion, both heavy and medium bombers decimated rail centers, even though this inevitably meant numerous casualties among French civilians in the target areas. Next, medium bombers and fighter-bombers went after bridges, thereby cutting off truck convoys. Following D-Day, rail and highway routes were so snarled that German reinforcements had to hike long distances into the Normandy region. Attacks on enemy airfields and coastal defenses also cleared the way for the massive amphibious assault mounted by the Allies. In the landing zone, engineers cleared an emergency landing strip for fighters on the first day. Within one week, several rough strips were complete, and Allied fighter squadrons were operating from them. Late in July, some 2,500 heavy bombers, medium bombers, and fighters attacked the front near St.-Lô, blasting away German defenses opposing the tanks and infantry from Lt. Gen. George Patton's Third Army. As Patton's columns fanned out, fighter-bombers flew interference ahead of them; air controllers riding in Patton's fast-moving tanks radioed to overhead planes, asking about the terrain, locating enemy strong points, and calling in air strikes. Air and ground units had attained a formidable partnership.

Although many German officers and personnel sensed defeat, Allied forces during the last 12 months still encountered stiff German resistance. Bad weather slowed air attacks on German refineries, and the Luftwaffe, hoarding available aviation fuel, continued to make determined flights against air force bomber squadrons, which cost many American planes and crews. The appearance of the twin-jet Messerschmitt Me-262 in late 1944 caused additional alarm. The Me-262 could outclimb and outrun the best Allied fighters, and the specter of squadrons of these jets attacking bomber formations was alarming. Bomb raids against jet airfields were increased, but the Me-262's real failure stemmed from Hitler's misguided earlier decision, in 1943, to use it as an attack bomber instead of the interceptor role for which it had been designed. By the time this mistake was rectified, dwindling fuel reserves and poorly trained pilots had spelled the end of the Me-262 and other German jets.

American raids in 1945 began to take on the dimension of area bombing rather than precision bombing. Attacks on sprawling railroad yards, in particular, meant that many stray bombloads hit nearby public buildings and homes. There seemed to be an increasing tendency to strike at civilian population centers as political or psychological targets to hasten the collapse of Hitler's Germany. Such reasoning led to the controversial saturation bombing of Dresden in collaboration with the RAF. The city had few significant military targets, and it was packed with refugees from the Russian offensive along the eastern front. Over the course of two days, February 13–14, 1945, British and American bombers razed the city from end to end; massive fire storms created an inferno that blazed for a week. The death toll of 135,000 people represented nearly twice the casualties that would be inflicted by the first atomic bomb dropped on Hiroshima six months later. The Allied campaign against Hitler ended in May 1945. Major industrial targets in Germany had been reduced to random heaps of rubble by heavy bombing. During the last few weeks of war, most Allied and AAF planes flew totally unopposed through German skies. In the Far East, air combat continued to be a factor in bitter battles still ahead.

AIR WAR IN THE PACIFIC

The devastating Japanese strike at Pearl Harbor unquestionably demonstrated the potential of carrier-based air power. Through remarkable luck, the Pacific Fleet's three major carriers, plus their escorts of cruisers and destroyers, were all at sea, away from

Pearl Harbor. The *Saratoga* was steaming off San Diego, California; the *Lexington* had departed the day before to deliver fighter planes to Midway; the *Enterprise* was sailing back from Wake after completing a similar mission. The *Enterprise* and its escorts sailed into Pearl Harbor on December 8, and crews lined the rails in grim silence at the destruction they saw as their ship steamed up the channel. After a hectic night of provisioning and refueling, the *Enterprise* steamed back to sea the next day, this time armed for battle.

Despite the destructive raid at Pearl Harbor, the American military position in the Far East showed some bright spots. Crucial fuel supplies, docks, and repair facilities at Pearl Harbor remained capable of serving the remaining units of the Pacific Fleet. The aircraft carriers functioned as key vessels among these surviving remnants, numbering 7 large fleet carriers and 1 small escort carrier. For the time being, carrier task forces comprised the only major means of carrying the war to the enemy or frustrating additional offensives that could permanently reduce the U.S. Navy's ability to turn the tide in the Far East. The Imperial Japanese Navy still held a slight numerical edge over the Americans, with 6 fleet carriers and 4 light carriers—a total of 10 Japanese carriers to 8 American. Since the Allied strategy envisioned the European theater as the area to receive the most support in the early days of the war, and because time was needed to rebuild the Pacific Fleet after Pearl Harbor, the role of naval carriers in carrying the brunt of military responsibility in the Pacific made their operations even more crucial. For these reasons, two historic carrier engagements in the spring of 1942 had a decisive effect on the outcome of combat operations in the Pacific theater. In addition, American carriers participated in a sensational bomb raid against Tokyo itself.

Compared to a few successful American hit-and-run raids against enemy outposts in the central Pacific, the Japanese advance during the late winter and spring of 1942 seemed inexorable. Resistance to the Japanese offensive in the Dutch East Indies finally crumbled. The British lost Singapore and Burma. General MacArthur finally abandoned the Philippines, and last-ditch American resistance at Corregidor concluded a valiant, but futile, gesture. Searching for some sort of move to boost faltering morale in the United States, the navy struck on the idea of bombing the Japanese homeland. Carriers could not get close enough to the Japanese islands for short-range navy planes to get off a strike, but twin-engine army bombers could take off from a point where the carriers would probably be safe from retaliatory attack. The bombers could proceed on a one-way mission, landing in friendly territory on the Chinese mainland.

After some experimental flights to see if bomb-laden B-25 bombers could indeed make it off a carrier deck, Lt. Col. James C. Doolittle assumed command of 16 B-25s loaded aboard the *Hornet*. With the *Enterprise* in company to provide emergency air cover, the *Hornet* headed toward Japan early in April 1942. The plan originally called for takeoff at a range of 500 miles from the target, but an unexpected sighting by an enemy patrol on April 18, 1942, prompted a hasty takeoff from 650 miles out. One by one, the B-25s roared down the *Hornet*'s deck and wallowed into the sky. Although the attack on Tokyo and nearby cities caught the Japanese by surprise and the bombers hit their targets, the early takeoff left fliers without enough fuel to make Chinese airfields. One plane limped into Vladivostok, Russia, but the other fifteen had to be ditched by

The stoutly built Grumman F6F Hellcat outfought the Japanese Zero in the Pacific. Courtesy of Grumman Aerospace Corporation.

their crews. Most of the airmen, including Doolittle, were eventually able to get back to the United States (though four of the eight airmen captured by the Japanese died). Admittedly, the damage caused by the Doolittle raid was minimal. Far more important was its effect on the morale of American troops and civilians, and the fact that some Japanese fighter squadrons were now recalled from forward combat areas to protect Japanese industrial centers. And, though less devastating than the Pearl Harbor attack, the raid once more demonstrated the mobility of the carrier weapon in World War II.

Left smarting by what had obviously been a strike launched from American carriers, Japanese strategists remained constantly on the alert to draw American carrier forces into a decisive battle to eliminate their influence in the Pacific theater. By cutting off Allied transportation and communication to Australia (the obvious staging area for Allied counteroffensives in the Pacific), the Japanese hoped strategically to neutralize Australian bases and tactically to lure American carriers into a trap. The stage was set for the Battle of the Coral Sea, when American and Japanese carrier-based squadrons came head to head for the first time.

There were many similarities as well as contrasts in the opposing carrier fleets. In the early phase of the war, both Japanese and American carriers were equipped with wooden flight decks; combat experience eventually dictated construction of new carriers with reinforced steel flight decks, which offered better protection for the crowded and vulnerable hangar decks immediately below. Carriers were particularly vulnerable because of the quantity of highly volatile aviation gasoline they carried, along with magazines full of aerial bombs, torpedoes, and machine gun ammunition for the planes. Because of this vulnerability, carriers depended primarily on their own fighter aircraft for protection, in addition to speed (30 knots or more) and an unusually heavy complement of antiaircraft guns. A typical American carrier had a crew of 2,900 seamen and carried between 70 to 90 aircraft. The planes were normally divided into three types of squadron: fighter, torpedo bomber, and scout-dive bomber. This division of aircraft by type was similar for most Japanese carriers also, although Japanese vessels generally carried fewer planes and had smaller crews.

The standard Japanese carrier fighter was the Mitsubishi A6M Type O—the famous "Zero." First flown in April 1939, the Zero was superior to most of the fighter opposition thrown against it during Japanese campaigns in the late thirties and during the early years of World War II. It had a top speed of over 340 MPH, and its light construction gave it a significant edge in speed, rate of climb, and maneuverability. But these advantages were purchased at the expense of adequate cockpit armor for good protection of the pilot, and a lack of self-sealing fuel tanks made it especially vulnerable against heavily armed opponents. Because of a confusing system of numerical designations for Japanese aircraft, Americans habitually identified most with male or female nicknames. In the early years of the Pacific campaign, American pilots learned to identify the "Val" dive bomber, an outdated design with fixed landing gear, and the "Kate" torpedo bomber, better than early American counterparts but weak in defensive armament.

Against the Zero and other combat planes during the opening phases of the Pacific war the U.S. Navy relied on Grumman's F4F Wildcat fighter. With a top speed of 330 MPH, the Wildcat was easily outrun and outmaneuvered by the more agile Zero, but the F4F was a rugged airplane that could absorb considerable punishment. And the Wildcat carried four .50-caliber machine guns with a high rate of fire that could tear the more fragile Zero to pieces. For many months after Pearl Harbor, the navy relied on the obsolete Douglas TBD Devastator as its basic torpedo bomber and frequently suffered horrendous losses when this lumbering plane attacked well-defended Japanese ships. The Douglas SBD Dauntless dive bomber, on the other hand, became a highly regarded weapon and continued as a front-line combat type and scout plane until the last months of the war.

Early in May 1942, the *Lexington* and the *Yorktown*, screened by destroyers and cruisers, steamed into the Coral Sea to retaliate against Japanese thrusts toward Australian life lines. Major action began on May 7, when planes from both American carriers sank the Japanese light carrier *Shoho*. But two of Japan's newest carriers, the *Zuikaku* and the *Shokaku*, got into position to launch air strikes against the American task force. During the hectic morning of May 8, Japanese and American planes grappled in swirling dogfights and struck furiously at each other's carriers. When the forces finally disengaged, the *Shokaku* was heavily damaged, but the crew of the crippled *Lexington* finally had to abandon ship. Losing the *Lexington* was serious, but much had been gained. The Japanese retired with only 39 planes out of an original force of 125, a tribute to accurate shooting by American naval pilots and gun crews. The Japanese invasion fleet, headed for Port Moresby, Australia, had to turn back. And the Battle of the Coral Sea marked the first naval engagement in history fought entirely by carrier-borne aircraft. The opposing ships never once sighted each other.

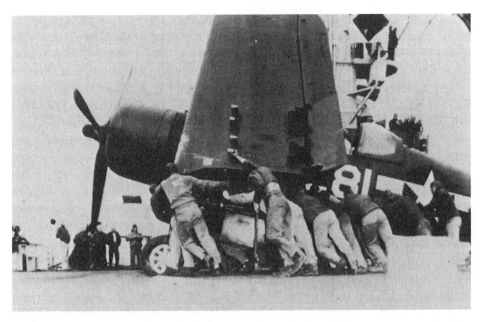

Combat operations aboard the confining dimensions of an aircraft carrier's deck required a high level of teamwork. Here, deck personnel position a Vought F4U Corsair. The folding wings of naval aircraft permitted carriers to handle many more planes. Courtesy of U.S. Navy.

Having failed to win decisively in the Coral Sea, the commander-in-chief of the Japanese navy, Adm. Isoroku Yamamoto, next proposed to entice Adm. Chester Nimitz's Pacific Fleet into battle at Midway. Yamamoto organized a fleet of 6 aircraft carriers, 11 battleships, 22 cruisers, 67 destroyers, 21 submarines, and a host of supporting vessels. The seagoing phalanx represented the most powerful ocean fleet ever assembled and included 411 airplanes, although Yamamoto diluted his strength by sending a task force of two carriers to land north of Midway, in the Aleutians. He hoped this might decoy Nimitz into an unfavorable position and continued to rely on overwhelming surprise to achieve victory. But American intelligence teams had broken the Japanese code, and the American carrier force of the *Hornet,* the *Enterprise,* and the hastily repaired *Yorktown* eluded Japanese submarines that tried to shadow them and report their position to Yamamoto. The Americans not only shrugged off the Aleutian feint, they also knew that Yamamoto had further divided his strength by sailing his carrier striking force some 300 miles ahead of the main body. The carriers, under the cautious Admiral Nagumo (who had lead the Pearl Harbor attack), launched a heavy air strike at Midway Island on the morning of June 4, 1942. The American carrier force, its exact location still unknown to the Japanese, now waited with some 233 planes for the opportune moment to attack.

The Japanese inflicted heavy damage on Midway and successfully beat off counter-attacks from army, naval, and marine planes based on the island. Rear Adm. Raymond A. Spruance now launched his carrier planes, grimly realizing that many would run out of fuel before they could return to his ships. But Spruance also calculated that he would hit the Japanese carriers at their most exposed moment, when flight decks were crowded and hectic with planes returning from Midway, and the ships would be highly vulnerable as the crews refueled and rearmed the returning aircraft. The lumbering Douglas Devastators, armed with antiquated torpedoes, were torn apart by antiaircraft fire and a protective Japanese fighter screen. On the other hand, the Douglas Dauntless dive bombers and their F4F Wildcat fighter escorts arrived over the target virtually undetected. As the Wildcats swooped down on defending Japanese planes, the Dauntless pilots nosed their planes into screaming dives. In the space of a few minutes, three of Nagumo's carriers were reduced to flaming hulks, as bombs set off furious secondary explosions and fires from Japanese ordnance and fuel lines exposed above decks. The dive bombers and fighters got away with acceptable losses, though many pilots low on gas had to ditch their planes near Spruance's escort vessels.

With the position of the American carriers finally revealed, the remaining Japanese flattop, the *Hiryu,* mounted a belated but determined counterattack. After a preliminary wave of dive bombers scored three hits on the *Yorktown,* a second wave of torpedo bombers put two projectiles into their target, ripping open the carrier's hull and flooding the boiler rooms. Spewing smoke and flames, the *Yorktown* went dead in the water and developed a list so alarming that officers issued orders to abandon ship. But the stricken *Yorktown* remained afloat, and was actually under tow toward Pearl Harbor when torpedoes from a Japanese sub finally ended the carrier's gallant career. In the meantime, the *Enterprise* sent a final strike against the *Hiryu,* catching it in the act of rearming and refueling its planes. Dauntless dive bombers scored four quick, lethal hits, putting the *Hiryu* out of action. All four blazing Japanese carriers eventually broke up and went to the bottom of the Pacific. American planes also attacked and sank the heavy cruiser *Mikuma* on June 6. Yamamoto finally withdrew after staggering losses: 1 cruiser, 4 fleet carriers, over 200 planes, and an estimated 3,000 sailors. Many of the planes and pilots were lost aboard the carriers when the ships were struck and went down. The fliers were skilled veterans, and their loss shadowed Japanese naval aviation for the duration of the war.

In spite of these setbacks, Japanese pressures on the life line to Australia continued. Discovering an enemy airfield under construction on the strategic island of Guadalcanal, the Americans set a counteroffensive in motion. On August 2, 1942, U.S. Marines hit the beaches at Guadalcanal, supported by air cover from four naval carriers (the *Saratoga,* the *Enterprise,* the *Wasp,* and the *Hornet*). In a grueling campaign that dragged into February 1943, the battle for Guadalcanal included numerous naval skirmishes and carrier engagements that cost the navy 24 vessels, including the *Wasp* and the *Hornet.* But in the process, the long campaign back toward the Philippines began, and the roster of combatwise American fliers expanded in the army and marines as well as in the navy.

More important, American industrial capacity by 1943 had begun to affect the conduct of the air war in the Pacific. During 1943, seven big new *Essex*-class carriers entered service—900-foot-long vessels of 33,000 tons that literally bristled with antiaircraft weapons and carried 100 planes. Nine light carriers and no fewer than 35 small escort carriers were also commissioned in 1943, in addition to scores of other vessels, from minesweepers to battleships. The navy won an even stronger edge against comparable Japanese aircraft with a new generation of combat planes. The Grumman F6F Hellcat matched the Zero in speed and agility and surpassed it in range, fire power, and survivability. The Vought F4U Corsair, with its distinctive, inverted gull wings, first saw action with land-based marine squadrons. One of the fastest fighters in the Pacific, at over 400 MPH, it proved to be rugged and reliable. The Dauntless dive bomber was joined by the bigger, faster Curtiss SB2C Helldiver, and the Grumman TBF Avenger gave the navy a dependable, carrier-based torpedo plane and bomber of considerable range and payload.

Equipped with better ships and better planes and using new tactics gleaned from combat experience, naval aviation relentlessly seized the initiative. A series of coordinated land, sea, and air offensives neutralized the Japanese strong point at Rabaul, in the Bismarck Islands, late in 1943. This prepared the way for landings in the Gilbert and the Marshall islands in late 1943 and early 1944. American strategy called for a dual advance against Japan: While General MacArthur swung westward through New Guinea and the Philippines, Admiral Nimitz would strike directly toward the Japanese home islands through the central Pacific. American planners intended to divide Japanese defenses and to secure island bastions in the central Pacific from which land-based bombers could strike targets in Japan. But the key islands to be attacked along Nimitz's route were well outside the range of Allied land-based planes. In this kind of seaborne offensive, the role of carriers for control of the air became crucial.

The growing superiority of American equipment and combat techniques was demonstrated in the Battle of the Philippine Sea on June 19–20, 1944. Starting with a force of 9 carriers and 430 planes, the Japanese lost 2 carriers and over 300 planes in the first 24 hours. In one of the most one-sided air battles of the war, American losses came to only 29 aircraft. With 891 planes assigned to 15 U.S. Navy carriers in this engagement, the Americans enjoyed a clear-cut numerical superiority, but there were additional reasons for the decisive edge in the ratio of American to Japanese losses. Although Japanese carriers deployed refined models of the Zero, as well as new designs in dive bombers and torpedo planes, fliers in the Imperial Japanese Navy frequently received little more than six months of flight training. American pilots got approximately two years of intensive training before combat duty, and new naval fighters like the Hellcat and Corsair proved more than a match for contemporary Japanese planes. Still, the naval aerial offensive against Japan took a toll in American ships, planes, and pilots. By the time the two-day Battle of the Philippine Sea ended, the Japanese had lost nearly 400 carrier planes and a third carrier, at a cost of 50 American planes downed in combat. The last naval air strike, against the carriers *Hiyo* and *Zuikaku*, took American planes beyond their combat radius, and many had to ditch into the sea as they ran short of fuel on return to their ships. All told, 209 American airmen were downed in combat or forced to ditch, although well-organized search and rescue efforts saved all but 49 lives. The losses bought a highly significant position, since the Battle of the Philippine Sea helped secure the Marianas island group and put American bombers like the B-29 within range of targets in Japan itself.

Larger carriers carried several types of aircraft. In the foreground are a pair of Grumman TBF Avenger torpedo bombers (TBF on the left shows its single-gunned ball turret); followed by four rows of F6F Hellcat fighters; followed by five rows of Curtiss SB2C Helldiver dive-bombers. Courtesy of Grumman Aerospace Corporation.

For American carriers forces, the Philippine Sea engagement in many ways represented the culmination of carrier aviation's role, forged during Coral Sea, Midway, and Guadalcanal. Although American control of the air over the Pacific was only occasionally contested in air-to-air combat after the spring of 1944, one of the most feared Japanese aerial weapons appeared in a different guise—the Kamikaze Corps. Named after the typhoon, or "divine wind," that had destroyed a Mongol invasion fleet in the thirteenth century, the Kamikaze Corps recruited young pilots who earnestly sought immortality through deliberate sacrifice of their lives for their country and the emperor. In a variety of bomb-laden planes, the Kamikaze Corps made fanatic diving attacks directly into Allied ships, carriers being favored targets. During the Okinawa landings in the spring of 1945, the Kamikaze Corps were primarily responsible for American losses of 36 ships sunk, 1,000 ships heavily damaged, and nearly 5,000 sailors dead. Such suicidal and destructive attacks were barely kept in check through longer-range radar detection, improved antiaircraft defenses (including proximity fuses), and vigilant aerial patrols.

At the war's end, air power originating from American carriers had shot down an estimated 6,484 Japanese planes and destroyed nearly 6,000 more on the ground at a cost of 451 U.S. Navy aircraft. Carrier aircraft sank over 700,000 tons of enemy combat vessels, destroying more military tonnage than any other naval surface or undersea weapon. Carrier planes also sank 1.3 million tons of Japanese merchant shipping, second only to the score accumulated by American submarines.

Meanwhile, out of the shambles of Pearl Harbor, the AAF in the Pacific also struggled to reorganize and regroup into an effective fighting force. In the Philippines, the most advanced outpost of the AAF, the few dozen B-17 bombers, P-40 fighters, and P-35 fighters proved inadequate to stop the Japanese advance. Even though defending American aircraft frequently shot down more Japanese fighters than they themselves lost, their continually declining numbers inevitably gave the Japanese total control of the air. Leaving a handful of battered air force fighters to fly reconnaissance missions only, since there were too few for air combat, a handful of the remaining B-17 bombers flew south to Australia, to where Gen. Douglas MacArthur had withdrawn on March 17, 1942. In the Orient, the only American forces that seemed to be making headway against the Japanese belonged to a colorful fighter unit in Burma known as the Flying Tigers.

The Flying Tigers, or American Volunteer Group (AVG), were led by Claire L. Chennault, a retired Air Corps captain. Recruited to help Chinese forces protect the Burma Road, a major Chinese life line after Japanese occupation of China's principal coastal seaports, the AVG was officially attached to the Chinese government, even though it flew Curtiss P-40 fighters and received clandestine American support. From December 20, 1941, until July 4, 1942, when the AVG merged into the AAF, the 70 or so pilots of the Flying Tigers shot down enemy planes at a ratio of nine to one. The AVG fared better than U.S. Navy fliers of the same period because the dreaded Zero served primarily as a carrier-based fighter, leaving the Flying Tigers opportunities to tangle with less advanced land-based planes. In their occasional encounters with Zeros, AVG pilots effectively followed tactics that worked against other Japanese planes: a hard-diving attack from above, arching back upward to try a second attack from high altitude. Against Zeros in particular, prudent pilots made one diving attack, then scooted for home, since the Zero enjoyed the edge in high-altitude dogfights.

The steady Japanese advance through Burma finally cut the Burma Road in June 1942, forcing the Allies to set up an aerial supply route between India and China over the towering Himalayas—a route that American pilots drily nicknamed the "Hump."

The aerial link to China was important, since it represented the basic means of keeping China in the war against Japan. In this theater, air transport missions assumed a significance on a par with bombing missions, and the flying conditions were equally hazardous. The air routes into China lay across uncharted sections of the Himalayas, where twin-engine C-46 and C-47 transports threaded their way through 14,000-foot mountain passes and over 16,500-foot mountain peaks. Four-engined B-24 bombers, converted to transports, joined the operation later. The monsoon season brought swirling storm clouds and 100-MPH gales. Pilots flew over a malign terrain, a contrast of foreboding ice fields and dense jungles inhabited by headhunters. Against these grim obstacles and persistent aircraft losses, Hump fliers increased monthly tonnages from 7,000 tons in October 1943 to 34,000 tons a year later. The rate reached 71,000 tons per month in 1945, twice as much as had been carried via the Burma Road.

The tenacious AVG and the success of fliers over the Hump denoted bright spots in a China-Burma-India campaign hobbled by bureaucratic confusion and a low priority. Late in the war, AAF units, in cooperation with the RAF, finally straightened out bureaucratic tangles and waged an outstanding air offensive against the Japanese, including remarkable airborne troop offensives far behind enemy lines. In March 1944, a glider force landed deep inside enemy-held Burma and cleared a rough airfield for dozens of C-47 sorties the next day. Overnight, 10,000 troops and thousands of pounds of supplies were flown in to create a stronghold in the midst of a tangled jungle, and the Allied forces proceeded to attack many vital Japanese supply routes.

While some air force units cooperated with naval and marine outfits during the protracted battle for Guadalcanal, the Fifth Air Force under George C. Kenney went to work in the skies over New Guinea. Complete air superiority was not achieved until December 1942, when increasing numbers of Lockheed P-38 Lightnings reached Kenney's fighter pilots. During the intervening months, B-17 heavy bombers and B-25 medium bombers made effective attacks on enemy airfields, and A-20 attack bombers perfected new tactics and weapons. In low-level strikes against parked enemy planes, the A-20s dropped fragmentation bombs attached to parachutes, slowing the descending bombs sufficiently to allow the attackers to scuttle safely away. Inventive ground crews also field-equipped the A-20 with up to eight machine guns in the nose, giving the plane a devastating concentration of fire on strafing runs. The same techniques and gun installations were utilized by the B-25s, some of which even mounted a 75-mm cannon in the nose, giving them enough punch to sink small cargo vessels. Kenney's fliers also perfected night-bombing techniques to reduce some of the losses in daylight raids and worked out procedures for skip bombing, one of the surest ways for conventional twin-engine bombers to attack enemy shipping. Rather than attempt a pinpoint bomb drop on an evasive target several thousand feet below, skip bombing took A-20s and B-25s down to a few feet above the water. They roared in toward a lightly defended Japanese cargo ship from the beam position, so that the vessel's broadside silhouette presented the largest possible target area. With practice, bombardiers learned to release a bomb so it would literally skip across the waves and explode against the ship's hull just above the water line. These tactics worked with devastating effect against an attempt by the Japanese to reinforce their position in New

The ubiquitous C-47 transport (military version of the DC-3) played a key role in medical evacuation service, saving thousands of lives and boosting the morale of combat troops. The portable steel matting shown in the photo permitted early air support for fighters as well as transports in forward combat zones. Courtesy of McDonnell Douglas Corporation.

Guinea in March 1943. In the Battle of the Bismarck Sea, land-based planes of Australian squadrons joined units of the Fifth Air Force in attacking the Japanese ships, leaving only four known surviving vessels out of a force estimated at 16 to 22 ships.

In an unusual action at about the same time, the air force plotted an aerial assassination of Admiral Yamamoto. Intercepted radio messages revealed details of Yamamoto's inspection tour through the Solomon Islands. In April 1943, a group of P-38 Lightnings from Henderson Field made a long, risky flight, intercepted the admiral's plane, and shot it down. Yamamoto's death came as a considerable blow to Japanese morale at a time when the tide seemed to be turning inexorably against them. Allied troops slowly extended their control of New Guinea, frequently by-passing Japanese strong points and leaving them cut off from supplies. With wing tanks added, speedy P-38 Lightnings escorted B-24, B-25, and A-20 bombers against major Japanese bases in raids that not only destroyed enemy supplies and equipment but chewed up the Japanese air force as well. When MacArthur's forces invaded the western tip of New Guinea in July 1944, thousands of Japanese troops were left isolated in positions to the rear on New Guinea and in the Solomons, effectively neutralized because Allied sea and air power forestalled all attempts to resupply them.

During the campaigns for the Philippine and Ryukyu islands in late 1944 and early 1945, naval aviation supplied initial air cover until army and marine air bases could be established inland. The air force Lightnings were joined by swelling numbers of P-51 and P-47 fighters; some marine units flew B-25s in addition to their single-engine Corsairs. Air combat and ground attacks on Japanese airfields had a telling effect. Whereas stiff air fighting developed over Leyte, where the Fifth Air Force shot down 314 planes while losing only 16, almost no air opposition materialized during the assault on Luzon. Although aerial opposition began to fade, Japanese ground troops continued fierce resistance, and increasing numbers of American fighters went into action, characterized by the assaults on Okinawa and Iwo Jima. Since each of them was only 400 miles or so from Japan, American possession enabled the establishment of advanced bomber bases and the opportunity to maximize the most formidable bomber of World War II: the Boeing B-29 Superfortress.

Plans for a bomber like the B-29 had begun as early as 1939. Even before Pearl Harbor, American strategists had viewed the plane as a long-range bombardment aircraft against Japan in the event of a war in the Pacific. Four 2,200-HP engines gave the B-29 a cruising speed of 225 MPH, a range of 5,800 miles, and a bombload of 16,000 pounds. The plane was unusually advanced for its time. Pilots and crewmen amidships were provided with pressurized sections for maximum comfort during long missions at extreme altitudes. The gun turrets (one in the tail and two each in the dorsal and ventral positions) were power-operated and remotely controlled by gunners amidships who used advanced electro-optical gun sights. The pressurized crew compartments and remotely controlled, powered gun turrets were the first such installations in a heavy bomber. The prototype made its first flight in the autumn of 1942, and production deliveries began in the summer of 1943. Although the Superfortress did not enter combat until the spring of 1944, its military career was one of the primary factors in the defeat of Japan.

When the B-29 began to reach air force squadrons in 1943, the United States had no bases from which it could bomb Japan. Surmounting immense logistical problems, engineers established airfields for the Superfortress in India early in 1944. Next, air force officers made plans to stage B-29 bombing raids on Japanese steel plants from forward bases in China. In the process, the B-29s had to transport almost all of their own fuel and supplies to China, and pilots of the newly formed XX Bomber Command grumbled about flying more transport runs than bombing missions. Logistical problems and primitive support facilities at the advanced Chinese bases continually hindered active operations. Although some strikes against industrial targets in the home islands were carried out, most of the operations remained directed against targets in occupied Burma and China.

During the summer of 1944, the invasion of the Marianas offered promising new bomber bases on Guam, Saipan, and Tinian. After an accelerated program to build the necessary airfields, the XX Bomber Command began to receive its B-29s in October. Following a series of shakedown raids in the central Pacific, over 100 Superfortresses took off to bomb Tokyo on November 24. This raid, like several other daylight precision missions after it, proved discouraging. The round-trip mission meant navigating over several thousand miles of open ocean. Too many planes aborted en route due to mechanical failures and other problems; the weather was bad; bombing accuracy was abominable; there were no escort fighters over the target; many planes had to be

The Boeing B-29 Superfortress, the most formidable bomber of World War II, operated exclusively in the Pacific against Japan and dropped atomic bombs on Hiroshima and Nagasaki. Courtesy of Jay Miller.

ditched into the ocean on the way home. Seeking solutions to this discouraging situation, General Arnold ordered Gen. Curtis E. LeMay, an experienced B-29 officer with the XX Bomber Command, to take over the bombers based on Guam. Meanwhile, American objectives in the Pacific gave high priority to the invasion of Iwo Jima, whose position so near to Japan would permit fighter escorts over the home islands and provide an emergency landing spot for crippled bombers returning to bases in the Marianas. Ships and planes blasted Japanese defenses for 19 days before the first marines landed, but the capitulation of Iwo Jima came after a particularly brutal campaign that lasted four weeks instead of the expected three to four days. American troops suffered nearly 4,900 casualties in the battle, but before the end of the war, Superfortresses had made 2,400 emergency landings on the island, saving the lives of approximately 25,000 crewmen. Moreover, Iwo Jima became the home for P-51s that escorted the Superfortresses as well as the operational base for additional fighters and fighter-bombers making innumerable forays against enemy airfields, ships, trucks, trains, and other targets.

To improve the destructive efficiency of the big Superfortresses, air force planners decided to try incendiary bomb attacks against Japanese cities. General LeMay not only accepted that premise but enlarged on it by ordering his planes to make incendiary attacks at night and at low level. There were immediate military advantages to American operations in these decisions. The Japanese defenses against night raids were weak, promising reduced bomber losses. By flying at low level, thereby saving the weight of additional fuel required for long climbs to high bombing altitudes the B-29s could also carry greater tonnages of incendiaries to the target. Although the accuracy of pinpoint bombing was compromised by night raids, the choice of incendiaries meant that industrial targets alone were being discarded in favor of indiscriminate fire bomb attacks against private dwellings and the civilian population in general. For the Japanese, the results were frightful. On the night of March 9, 1945, over 300 Superfortresses fire bombed Tokyo, killing an estimated 80,000 people, burning out close to 16 square miles in the city's center, and destroying 25 percent of the city's buildings. The nighttime incendiary attacks were feebly contested by Japanese fighters, and B-29s ranged virtually unopposed over all the country's major cities, even dropping leaflets ahead of an attack warning citizens to flee.

By the summer of 1945, close to a thousand Superfortresses were based in the Marianas, and phalanxes of 600 planes on night raids were not unusual. Dozens of smaller cities in Japan were literally reduced to ashes and rubble, and more than 100 square miles in the centers of Japan's six major metropolitan areas were left in smoldering ruins. The air force began to run short of reasonable bombing targets, so the B-29s turned to other missions, such as planting 12,000 mines in strategic sea approaches to major ports. Bombed, burnt out, and blockaded, many Japenese war industries had lost over 75 percent of their productive capacity by August 1945. While some realistic Japanese leaders advised surrender, influential reactionaries continued to argue in favor of military resistance. Controversy still persists whether continued blockade and bombardment alone might have brought Japan's capitulation. For Japan, the issue came closer to resolution on August 6 and August 9, 1945, when B-29s dropped atomic bombs on Hiroshima and Nagasaki, respectively. Russia also declared war on Japan on August 9, and the next day the Japanese government announced defeat. As Allied war planes droned overhead, the final surrender documents were signed on September 2, 1945, aboard the U.S. battleship *Missouri,* at anchor in Tokyo Bay.

The effectiveness of strategic bombing in World War II continues to be debated, but there can be little doubt that air superiority was immensely important in the Allied offensives. Air superiority helped to assure the capture of Pacific island strongholds as well as the successful Normandy invasion in 1944. The value of close tactical air support and the airborne deployment of soldiers and supplies, in addition to paratroops, was demonstrated many times in all theaters of the conflict. And air power delivered the most destructive weapon of the conflict, ushering in the reality of nuclear warfare and attendant horrors of an unprecedented scale.

COMPONENTS FOR A JUGGERNAUT

The ingredients of American air power success in World War II encompassed far more than fighters and bombers exchanging machine gun bursts with enemy aircraft over hostile territory. Waged on a global basis, the war effort required a vast array of miscellaneous supporting operations, equipment, organization, and personnel. Newspaper headlines of 1942–43 highlighted the bitter island campaign in the Pacific and hard-fought air wars in Europe and Africa. The role of aviation in the desperate Battle of the Atlantic seemed less glamorous by comparison, but the struggle there played a key role in the mounting Allied assault on enemy-held Europe.

In the early weeks of the war, German submarines nearly stymied the flow of war materials across the Atlantic. During 1942, ship losses to German subs frequently approached 2 vessels per day and reached a peak in May when German subs sent 47 ships to the bottom of the ocean. Heavy losses in American coastal waters, frequently in sight of the mainland itself, were particularly galling. As quickly as possible, the army and navy allocated whatever planes were available for coastal patrol and antisubmarine duty. The motley air units flew everything from single-engine light planes to B-17s. By August 1942, no merchant ships had sunk in the Eastern Sea Frontier (around U.S. coastal waters), and beginning in 1943 no merchant ships sank within 600 miles of U.S. coastal bases. But the German subs moved their operations to the mid-Atlantic, the crowded sea approaches around Britain, and Gibraltar, at the entrance to the Mediterranean. In turn, American antisub units established new bases in Newfoundland, Iceland, England, and North Africa, buttressing operations of the RAF Coastal Command. In long, arduous flights, PBY Catalinas, modified B-24s, and planes from carriers gave additional support. Aircrews often begrudged the monotonous missions over vast, empty stretches of open ocean, but these sorties helped to keep German subs submerged, reducing their effectiveness and keeping the feared U-boat "Wolf Packs" off balance. Evolution of effective antisub warfare techniques, along with improved radar, depth charges, and torpedos, continued to inflict prohibitive losses on German submarine operations. Armed with depth charges, naval blimps that accompanied convoys across the Atlantic proved surprisingly effective. In fact, enthusiastic "gasbag" crews claimed that convoys with blimp protection over the full route never lost a ship to enemy subs.

In many ways, Allied victory in World War II hinged on the prodigious American industrial contribution. The American effort also required sophisticated personnel planning and logistical organization. The production of aircraft achieved by American industry reached astonishing numbers. Beginning in the spring of 1942, factories ran 24 hours a day, six to seven days a week. Women joined the work force by the tens of thousands and helped to swell the aircraft industry labor force to a high of 2,100,000 workers by the end of 1943. Most of them were assigned to simplified tasks under the supervision of experienced cadres of employees. At first, turnover was high, but production efficiency improved as workers gained experience and manufacturing procedures were ironed out. The automobile industry was sure its mass-production techniques would work miracles but found that aircraft were far more complex to produce and required far more exacting tolerances than cars. After much trial and tribulation, the enormous Ford plant at Willow Run did reach remarkable levels of production, turning out 5,476 B-24 bombers in 1944–45. In 1944, Willow Run alone produced 92 million pounds of airframe weight—more than half of Germany's annual production and nearly equal to Japan's total for 12 months. The growing efficiency of American aircraft in production was reflected in the drop from 55,000 individual working hours required to turn out a B-17 in 1941 to 19,000 hours in 1944. All

Wartime mobilization created serious labor shortages in defense industries. Although many production managers were skeptical about women learning to master the demanding techniques of aircraft production, "Rosie the Riveter" and her sisters proved to be very adept. Courtesy of McDonnell Douglas Corporation.

together, the American aircraft industry manufactured 300,000 military planes for its own armed services and the Allies. In the spring of 1944, American industry had a production capacity of 110,000 planes per year.

The siren call of mass production created problems in combat operations. In some instances, the drive to meet production goals meant that obsolete types like the P-39 and

P-40 were kept in production when newer models would have been more successful against enemy fighters. In addition, inadequate attention was given to the production of spare parts and repair kits, which forced deliberate "cannibalization" of many planes just to acquire necessary parts to repair combat damage. The drive for mass production also enforced standardization, despite the need for constant modification of military aircraft based on combat experience. The air force finally set up no less than 28 modification centers to refit production planes in light of the latest intelligence from the front. Even so, additional changes were still often made when planes reached front-line bases, especially in bomber squadrons.

The general aviation industry made a substantial contribution in the production of primary flight trainers (like the Piper Cub) and of small, twin-engine planes as transition trainers for pilots and crews of larger transports and bombers. Painted in a new uniform of olive drab, various types of light planes were also pressed into military service for special transportation of officers and important wartime personnel. Known as "Grasshoppers," versatile and maneuverable light planes like the Piper Cub J-3 performed vital front-line service in liaison work, observation, artillery spotting, and evacuation of the wounded. Nearly 6,000 Piper Cubs served in the wartime Grasshopper fleet. Light planes also equipped volunteer units of the Civil Air Patrol (CAP), organized in 1941. Dozens of American private pilots contributed their time and aircraft to the CAP for antisub patrols along the Atlantic and Gulf coastlines. They not only spotted many enemy submarines but also located crippled ships and survivors of sub attacks.

The thousands of planes rolling out of factory doors required pilots and crews, and military training programs strained to meet the challenge. During the hasty build-up of American military forces after 1939, General Arnold persuaded Congress to support private training schools for primary flight instruction, leaving the air force its existing flight training facilities for basic and advanced training. The Civilian Pilot Training Progam (CPTP) started with an objective of 300 trainees per year in 1939, increased to 7,000 per year in the spring of 1940, and went to 30,000 per year in 1941. By 1944, CPTP had turned out an estimated 400,000 trainees. Not all of them became active military pilots, since basic and advanced training still lay ahead: between 1939 and 1945, 193,444 pilots completed all phases of flight training. Early in the war, primary, basic, and advanced flight training in the air force consumed a year, but refined instructional techniques eventually trimmed the time to nine weeks each. Fliers who passed three phases still had to complete transition training in their assigned combat planes and spend additional time in specialized formation flying and unit operations. The navy demanded still more time of fliers who wanted to qualify as carrier pilots. Aspiring fliers who washed out frequently retrained as bombardiers or navigators, although the air force's growing fleet of bombers led to increasing reliance on aptitude tests administered at the beginning of flight training. Assignments to schools for pilots, bombardiers, and navigators were made accordingly.

At the height of its strength in 1944, the air force numbered 2,411,294 people, including 306,889 officers and 2,104,405 noncommissioned personnel. The majority belonged to operational Air Corps flying units, with the rest in the Signal Corps,

Quartermaster Corps, the Engineers, and other bodies. By the end of hostilities, the air force had more than 100,000 in its Engineers, a group frequently overlooked but one that performed heroic service in the construction of airfields in every theater of the war, usually in the most adverse situations and living conditions and frequently under enemy attack. The navy's Seabees did the same.

The process of transferring thousands of planes from American factories to new airfields in combat zones all over the world depended heavily on the Air Transport Command (ATC). The principal task of the ATC involved flying military aircraft across the Atlantic and Pacific for delivery to combat units abroad, a process that required it to build a global network of airfields, weather stations, and communications stations. The ATC also developed a worldwide air transport service that not only carried key personnel but also delivered thousands of tons of high-priority military supplies and mail. By June 1942, the ATC network had touched all six continents; during the peak of wartime operations, an ATC cargo or combat plane crossed the Atlantic at an average rate of one every 13 minutes and traversed the wider Pacific Ocean once every 90 minutes. The personnel and equipment of civil airlines in the United States contributed significantly to the ATC's success. Under military control, former airline personnel shared administrative skills and operational experience in flying long-distance air routes and sustaining complex schedules. From the airlines, the

The Beech staggerwing, originally selected in 1938 to serve armed forces attachés overseas, continued in military liaison duties during the war. It and thousands of other general aviation aircraft were used in training, observation, transportation, and other roles. Courtesy of Beech Aircraft.

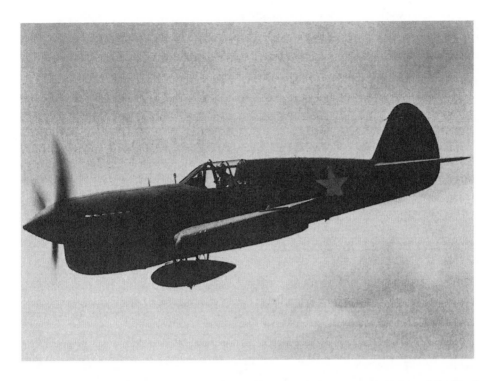

The Curtiss P-40 was flown in every theater of the war by American, British, and Soviet air forces. Delivery of thousands of combat planes like it was facilitated by WASPs. Courtesy of U.S. Air Force.

ATC also impressed dozens of DC-3 transports into wartime service, which flew under the military designation of C-47. The ATC also acquired Douglas's four-engine transport, the DC-4, redesignated the C-54 during the war. Both of these planes became workhorses on far-flung ATC routes.

The profusion of ferry flights, transport missions, and flight training put a severe strain on crew allocations and siphoned off many fliers trained for combat. The Women Airforce Service Pilots (WASPs) began to take on more and more of the burden. The idea evolved after General Arnold asked Jacqueline Cochran to find the additional pilots needed to ferry planes to England late in 1940, when the United States, still a noncombatant, was preparing to stop German aggression through the Lend-Lease Program. Missions to England meant flying in combat-zones, which discouraged the original plan, even though Cochran recruited 25 women to ferry RAF planes within Great Britain. After America entered the war, the WASPs were organized in 1942, eventually numbering more than 1,000 women. They flew demanding ferry routes all over the United States and also supported the war effort by towing aerial targets for gunnery practice using live ammunition and insr ucting hundreds of cadets. "[The WASPs] were marvelous instructors," Cochran said. "Men preferred them because they were more patient." By the end of the war, women had flown just about every plane in the American inventory, from fighters to bombers, and many foreign types, like the Hurricane and Spitfire.

The war years witnessed substantial progress in many technical areas, with notable trends in electronics and in rocketry. Although the development of radar progressed apace in the United States and Germany from the mid-1930s onward, British research advanced rapidly as part of a defensive network to thwart German bombers. Early ground radars were bulky and heavy, although they admirably served the RAF in the Battle of Britain in 1939. The British achieved a breakthrough in 1940 leading to smaller, effective radar sets that were installed in English and American aircraft. Airborne radars were used in night fighters in all theaters of the war and offered improved navigational and targeting techniques for both day and night bombing. Another application of electronics involved radio-directed bombs, first used by the Germans against Allied ships in the Mediterranean. The AAF achieved considerable sophistication with its own AZON (azimuth only) bombs. Using a standard 1,000-pound bomb, technicians attached a gyrostabilizing unit and a radio receiver that activated small control surfaces. After launch, the gyrostabilized bomb resisted spinning; in advanced versions, both the rate of descent and the lateral movement could be controlled through radio signals. In both the European and Pacific theaters of the war, the AZON and similar devices proved very successful in precision attacks against targets such as vital bridges. By war's end the U.S. Navy was using a winged bomb, guided by radar, known as the "Bat." During one engagement, a naval plane sank a Japanese destroyer after launching the Bat some 20 miles distant from the target. From weapons like these, engineers gained experience for the development of sophisticated missiles in the postwar era.

After a dormant period of over a century, rocketry reemerged as an effective military weapon. The Allies employed a variety of solid-propellant rocket devices, ranging from projectiles like the bazooka, fired as a single round by an individual soldier, to withering barrages fired from ground batteries or launched from naval vessels. A family of American air-to-ground rockets included the 10-foot-long "Tiny Tim," with a warhead equivalent to a 12-inch naval shell. Carried by fighter bombers, a load of Tiny Tims gave the planes the opportunity to strike hidden, isolated targets with the accuracy and destructive power of a cruiser salvo. American planes also made occasional use of booster rockets to help take off, using JATO (jet-assisted takeoff) units. JATO units boosted heavily laden transports and bombers from soggy airfields and gave an edge of assurance to sluggish naval flying boats taking off from heavy seas. JATO work also led to the development of a significant liquid-fueled rocket, a two-stage army ballistic missile with a solid booster known as the "Wac Corporal." The first-stage booster, adapted from the Tiny Tim, developed 50,000 pounds of thrust. The second stage, filled with nitric acid–aniline liquid propellants, developed 1,500 pounds of thrust, a combination that fired payloads up to an altitude of 43 miles. But the Corporal program did not hit full stride until 1945, and none of these projects compared to the German missile known as the V-2.

The V-2 (from *Vergeltungswaffen-2*, or "weapon of retaliation") had no counterpart in the Allied inventory. The V-2 was 46 feet long, with a diameter of 5 feet, and capable of speeds up to 3,600 MPH to an altitude of 60 miles. By the end of the war, Germany had launched nearly 3,000 of the remarkble V-2s against targets in England and elsewhere in western Europe at ranges of up to 200 miles. With the support of government, private, and university sources for research and development, the von Braun team at Peenemuende solved numerous hardware fabrication problems and technical difficulties (such as the production, storage, and handling of liquid oxygen in large quantity) while developing unique management skills in rocket technology.

By early 1945, it was apparent that the war was nearing its end. Von Braun called a secret meeting of his top staff and reviewed the options: stay on at Peenemuende in the face of the advancing Russian units or try to head south and link up with the Americans. There was no dissent—go south. After regrouping, the von Braun team began plans for contacting the Americans, unaware that the United States was already formulating a program to round up leading German scientific and technical personnel. Better known as Operation Paperclip, the American search gave priority to the von Braun team. With contact established early in May 1944, it still required several months of effort to clear all of the bureaucratic hurdles and prepare the way for over 100 selected German personnel to come to the United States. Finally, von Braun and six top associates arrived at Fort Strong in Boston on September 29, 1945. They brought an invaluable legacy in astronautical research and operational skills. During the first postwar decade, rocketry often took a back seat to exciting developments in jet propulsion and high-speed flight research; thereafter, rocketry came into its own, especially as an element of international leverage in cold war antagonisms.

5

Air-Age Realities, 1945–1955

Restoring balance in the postwar world presupposed plans for the orderly progress of international air transportation. The United States, however, had ambitious dreams of new intercontinental networks, a goal that clashed with that of the United Kingdom, in particular. As other nations recovered from the ravages of World War II, additional airlines entered the market. Fortunately, new postwar organizations of international stature helped to strike a reasonable balance among the contending international airlines. Although advanced, four-engine transports heralded a new plateau in airline service, the major aeronautical changes occurred in military operations, where the rapid acquisition of turbine-engine aircraft initiated the "jet revolution." Experimental high-speed flight research became a major enterprise. The military services also learned to cope with jets in combat, as cold war pressures boiled over into warfare in Korea. The general aviation sector hoped for dramatic growth in the postwar decade, but the light-plane market proved limited. Nevertheless, measured progress occurred, and prospective buyers benefited from a wide choice of both single- and twin-engine designs.

NEW AIR–AGE CONCEPTS

Based on the experience of tactical strikes, strategic bombing, and global transport during World War II, it became obvious that military aviation would continue its significance in the postwar era. A spate of articles, pamphlets, and books by a variety of authors spelled out the devastating role of military aviation in combat and, while expressing hope for a peaceful postwar world, emphasized that military aviation would probably continue to occupy an important niche in the framework of international relations for some years to come. These authors also stressed the growing impact of transport aviation, domestically and internationally, as an aspect of country's total stance in power politics. The Second World War had unquestionably demonstrated the value of air transport in rapid and ad hoc deployment of personnel and matériel to virtually any spot in the world. Many writers argued that an active, technologically advanced civil air network was an extremely important and integral component of a nation's total air power. At issue were the policies required for a healthy international airline, especially in cases where interntional routes were subject to strident, nationalistic competition from competing international carriers.

These were all complex issues, and the search for guidelines and answers attracted the efforts of several prestigious organizations and individuals. In 1944, for example, the Brookings Institution published a two-volume study under the general title *America Faces the Air Age*, written by J. Parker Van Zandt. Volume 1 deals with *The Geography of World Air Transport* and volume 2 discusses *Civil Aviation and Peace*. These books and other postwar assessments repeated similar themes. In the 33d Wilbur Wright Memorial Lecture, delivered in London on May 31, 1945. Theodore P. Wright, U.S. administrator of Civil Aeronautics, entitled his remarks "Aviation's Place in Civilization." Observing that humankind was a society based on communication among its members, Wright stressed the impact of aviation in terms of rapid passenger travel on a global scale. In warfare, he continued, air power had achieved equality with the traditional services, and its destructive role could be a possible deterrent to future conflicts. Hopefully, Wright conjectured, the role of rapid air transport in facilitating international conferences would have a positive influence on world peace. Wright also forecast some of the trends in technical development, including supersonic flight research, advanced radar for navigation, helicopters, and continuous refinements contributing to economy and efficiency.

One of the more interesting responses to the postwar age of aviation and the issues it raised came from a new journal, *Air Affairs*, which published its first issue in September 1946. The roster of editors and contributors drew heavily from academe and from the upper echelons of Washington's civil service and resident power establishment, with a leavening of international aviation spokesmen. The first issues included papers from faculty of the University of Chicago: anthropologist William F. Ogburn, international law and military expert Quincy Wright, and nuclear physicist Harold C. Urey. Well-placed bureaucrats included William A. M. Burden, assistant secretary of commerce, and Edward Warner, former member of the Civil Aeronautics Board (CAB) who was currently president of the Interim Council of the Provisional International Civil Aviation Organization. The board of trustees, chaired by Burden, included Laurence Rockefeller. Themes in three of the articles in the first issue of *Air Affairs* illustrated its general editorial preferences: Ogburn on "Aviation and Society," Urey on "Atomic Energy, Aviation, and Society," and Wright on "Aviation and World Politics." Either in the abstract or in specific detail, succeding volumes of *Air Affairs* addressed issues of the societal implications of expanding aviation activities, the somber implications of warfare in an age of atomic power, and the political implications of living in a world shrunk in time by airlines and totally accessible to hostile air forces. In 1946, Ogburn offered a complete interpretation of the probable impact of the air-age world in *The Social Effects of Aviation*. Ogburn may have misjudged in some respects (limited air warfare in Korea, for example) but forecast many trends in air transport and private flying with accuracy.

SCHEDULED AIR TRANSPORT

American air carriers faced the postwar era of international air routes with a number of issues hotly debated. Pan Am lobbied intensively to enhance its pre-eminent position as America's international airline, a position just as intensively contested by TWA and other carriers with the potential for overseas routes. American aviation leaders clashed with the British in determining which international routes would be served by their respective countries, while dozens of nations on the sidelines sniped at both of them. There was talk of an all-encompassing international aviation organization, but the scope of its activity and regulatory powers remained unsettled.

With the Democratic administrations of Roosevelt and Truman opposed to one airline's dominance of U.S. international routes, Pan Am ultimately had to share its empire with several other carriers, most notably TWA (European routes), Braniff (South American) routes, and Northwest (Pacific and Oriental routes). Agreements on route structures and rates were finally stabilized by the United States and Great Britain through diplomatic channels and maintained by mutual self-interest through the International Air Transport Association (IATA), organized at a conference in Havana in the spring of 1945. Although IATA operated on a voluntary basis, it evolved into a workable tool, at least until the mid-seventies.

Complementing IATA was the International Civil Aviation Organization (ICAO), the result of a convention of delegates from 52 countries held in Chicago late in 1944. Previously, there had been considerable debate in the United States as to the relative degree of control of international air routes. Adolph A. Berle, Jr., an assistant secretary of state, adhered to what was called the "open-sky concept," an approach subject to numerous interpretations but one that generally favored a minimum of nationalistic regulations by the United States, Great Britain, or anyone else. At the other end of the spectrum was a handful of senators who worried that either unrestrained competition or strict regulations would favor weaker airways in the short run and inevitably result in the erosion of American international routes, since they would face stiff competition from heavily subsidized foreign flag lines. Better to have a strong, single, American international airline to compete against the foreign carriers—a position favored by Pan Am. Lloyd Welch Pogue, chairman of the CAB, occupied the middle ground. While opposing monopolistic domination of American overseas routes by one airline, he was unwilling to see American expertise and leadership in international transport flying, developed in World War II, dissipated by blind adherence to open-skies arrangements. Pogue accepted the concept of open skies in principle but felt that specifics should be negotiated on a bilateral basis.

The British, lacking a fleet of advanced, long-range transports, pursued their own goals. During the war, the English aircraft industry had concentrated on fighters and bombers, with the result that the large-scale production of air transports and the operation of Allied air transport networks went to the Americans. When the war was over, the British expected an opportunity to re-establish aeronautical pre-eminence in their traditional spheres of influence. This hopeful assumption ran headlong into American aeronautical expansionism articulated by the congresswoman from Connecticut, Claire Boothe Luce. Her husband, magazine kingpin Henry Luce, championed the "American Century" from his highly influential position as publisher of *Life, Time,* and *Fortune* magazines. The United States, emerging as the world's leading economic and industrial power after World War II, would assume world leadership, Luce said. In aeronautical counterpoint, his wife proclaimed that "American postwar aviation policy is simple: we want to fly everywhere. Period!!!" The British hoped to counter this kind of aerial imperialism through a complicated quota system for the distribution of traffic, regulation of fares, frequency of flights, and amount of subsidies.

Against a background of conflicting proposals and national self-interest, the international civil aviation conference gathered in Chicago on November 1, 1944. For 37 days, until adjournment on December 7, the delegates alternately plunged into rancorous debate or sat around in bored frustration as the Americans, the British, and the Canadians locked horns in clandestine bargaining sessions. Many delegates talked of quitting Chicago and heading home. To keep the conference going, Fiorello La Guardia, the feisty little mayor of New York City and a member of the American delegation, challenged the delegates to roll up their sleeves and keep the conference alive. The savvy Berle had briefed La Guardia for this nicely timed performance, which followed some well-chosen conciliatory remarks by H. J. Symington of Trans-Canada Air Lines.

The problem of fares and flight schedules was left hanging, but other key issues were resolved. The right to navigate the skies of foreign countries, along with certain landing rights at foreign airports, were accepted as the Transport Agreement. The adoption of the International Convention on Civil Aviation, one of the most significant agreements to come out of the Chicago conference, provided for the formation of ICAO, composed of an assembly of all members and an elected council of 21 seats. Until all the member governments ratified the various provisions of the Chicago conference, a provisional group (PICAO) would get things organized for action. Edward Pearson Warner, the highly respected American aviation expert, became the head of both PICAO and ICAO and served with distinction until his retirement in 1957. The operational functions of ICAO were included in a dozen technical annexes, which were eventually worked into the original convention. The annexes spelled out uniform international rules and procedures on communications, traffic control, licensing, aerology, and navigation. After affiliating with the United Nations as a specialized agency in 1947, ICAO became an outstanding factor in safe, regularized international aviation.

Over a period of time, the complicated issues of rates were worked out under the auspices of IATA, organized at a conference in Havana in the Spring of 1945. IATA located its headquarters in Montreal, Canada, where ICAO had also settled, which facilitated liaison between the two groups. The thornier problem of exactly which routes were to be flown, frequency of flights, and interpretations of cabotage (which meant, for example, that a British flight from London across the United States could stop in New York but not pick up passengers bound for San Francisco, although it *could* take on New York passengers headed for Australia or New Zealand) were outlined in the famous Bermuda Agreements of 1946. These bilateral negotiations between Britain and the United States resolved many controversies simmering since the Chicago meeting and, more important, served as workable guidelines for subsequent agreements between the United States and other countries as well as for international negotiations in which the United States was not a party. Following ICAO, IATA, and Bermuda, international civil airline operations were, quite literally, prepared for takeoff.

The principal equipment used on long-distance domestic routes and overseas flights was represented by American aircraft such as the Douglas DC-6, the Lockheed Constellation, and the Boeing Stratocruiser. Although the DC-3 continued to fly for many airlines, both Martin and Convair twin-engine airliners, with higher speed and more passenger seats, began to replace it. Only in a few ways did postwar aircraft differ from prewar planes of the DC-3 type, underscoring the remarkable strides made in the thirties. Briefly, the improvements of the postwar era were a matter of larger size, higher wing loadings and improved flaps, increased engine horsepower and efficiency, availability of higher-octane fuels, tricycle landing gear, and cabin pressurization.

Use of four engines had been pioneered by Igor Sikorsky in pre-revolutionary Russia and appeared in flying boat designs of the mid-thirties. Aside from the Boeing 307, the most significant step in this direction for civil airplanes had been the Douglas DC-4E. Due to shifting airline specifications, Douglas dropped the DC-4E and built a smaller version, the DC-4.

The DC-4 first flew in February 1942, and made its mark during World War II as the C-54 military transport. Early postwar versions, along with military surplus equipment, became important aircraft for many domestic U.S. and foreign airlines. The first major modification, the pressurized DC-6, entered service in 1947. Changes in size, power, and operational characteristics made it the most economical piston-engine transport in airline service, and it was produced in several versions. An even larger and more powerful model of this type, the DC-7, appeared in 1953 and was the first airliner capable of flying nonstop against the prevailing winds on westbound routes in the United States. The DC-7C (the "Seven Seas"), introduced in 1956, was capable of nonstop trans-Atlantic service, eliminating the traditional refueling stop at Newfoundland.

Douglas's chief rival in long-distance, piston-engine airliners was the Lockheed Constellation series. The first Constellation took to the air in January 1943, about a year later than the DC-4, and was immediately pressed into military service as the C-69. The Constellation grew out of an order from TWA, and the plane's specifications quickly prompted purchases by Pan Am. Lockheed evidently benefited from the experience of using Fowler flaps on its Model 14 twin-engine airliner as well as experimental cabin pressurization trials in 1937, since the new Constellations appeared with both features.

For long-distance and transoceanic routes, the graceful, triple-tailed Lockheed Constellation epitomized aerial elegance during the early postwar era. The takeoff of a Connie or similar transport was impressive; the four propeller-driven engines swelled to a deep, melodic rumble like the sustained roll of a battery of tympani. Courtesy of Trans World Airlines.

In the postwar era, the plane's power and range made it a favorite of both TWA and Pan Am on long-range and over-water routes, and Constellations went through a series of stretching and modification. During its production lifetime of 14 years, the plane's passenger capacity, horsepower, and gross weight all increased by as much as 50 percent, and it was one of the earliest commercial airlines capable of crossing the Atlantic nonstop in either direction. In its heyday, the graceful, eye-pleasing Constellation signified the apotheosis of up-to-date air travel.

Perhaps the ultimate in postwar airline travel was available in the Boeing 377 Stratocruiser. The plane evolved as a collection of parts: wing, tail, and landing gear from the B-29; fuselage from the postwar air force C-97 cargo/tanker aircraft; and engines from an improved version of the B-29 known as the B-50. The engines proved to be balky and troublesome, and critics taunted the plane as the best "three-engine" airliner in service. True, the engines were a problem, but the Stratocruiser contributed unmatched luxury to air travel of the late forties and mid-fifties. Seating was unusually flexible, from 55 to over 100, and some planes boasted 28 upper and lower berths for overnight transcontinental and intercontinental runs. Passengers especially remembered the 377 for its wide seats, legroom, and headroom. And the bar/lounge, entered by a spiraling staircase leading to the lower deck, was a sensation for early postwar airline travelers.

The appearance of reliable long-distance transports inevitably made them indispensible to travel by presidents of the United States. Franklin D. Roosevelt's aerial

Pilot, copilot, and flight engineer shared close quarters in four-engined aircraft like the Constellation. En route, the flight engineer (far right) kept a close eye on fuel consumption, oil pressure, engine temperature, and other crucial data. Courtesy of Trans World Airlines.

jaunt to the 1932 Democratic convention was admittedly out of the ordinary. During World War II, the exigencies of international diplomacy called for President Roosevelt's presence overseas, and air transportation not only saved invaluable time in these intercontinental odysseys but also provided more security than did surface transportation in war zones. In January 1943, Roosevelt boarded a Boeing 314 flying boat (Pan Am's *Dixie Clipper*) in Miami and became the first chief executive to make a wartime flight while in office. The voyage also marked the first flight abroad by a chief executive and the first presidential trip to a combat zone since the Civil War. The *Dixie Clipper* flew down the coast of South America and then across the South Atlantic to British Gambia, on Africa's west coast. From there, the presidential party proceeded aboard a C-54 operated by TWA to Casablanca, where Roosevelt met Prime Minister Churchill and other Allied leaders. During the return trip to the United States, Roosevelt celebrated his 61st birthday while flying in the *Dixie Clipper* above the Caribbean.

Following this remarkable journey, aides made plans for a presidential aircraft. The first such plane was a converted B-24 bomber that carried the transport designation C-87. Whimsically christened the *Guess Where II*, the plane never carried the president, although it flew extensively in the course of transporting Mrs. Roosevelt and dozens of other government officials around the world. In November 1943, President Roosevelt sailed abroad a more spacious warship for conferences at Tehran and Cairo, relying once more on the C-54 for rapid transportation within Africa. These trips crystallized the need for a permanent presidential aircraft, one that also would have special appurtenances for the chief executive.

The new plane was a Douglas C-54, selected while under production and finished by a hand-picked crew of assembly workers. Delivered in 1944, it became known as the *Sacred Cow,* a sobriquet bestowed by the irreverent White House press corps. Modifications to the aircraft included a special elevator for Roosevelt, whose bout with polio years before required near-total confinement to a wheelchair. The elevator eliminated the awkward ramps required during prior trips. There were several large compartments within the fuselage, including the presidential stateroom, which measured nearly 8 by 12 feet. The *Sacred Cow* flew a variety of wartime VIP missions in the European and Pacific theaters. Early in 1945, Roosevelt again boarded a U.S. Navy warship for Malta, where he was met by the *Sacred Cow* for the flight to Russia for the Yalta conference. Flying to Cairo after the conference to board ship to return to the United States, Roosevelt made his last plane trip. When Roosevelt died in April 1945, stewardship of the *Sacred Cow* passed to his vice-president, Harry S. Truman.

Roosevelt introduced the concept of presidential air travel, but Truman made it commonplace. During his administration, the frequency of Truman's aerial travels led news reporters to refer unofficially to the presidential plane as the "Flying White House." These journeys not only enhanced the acceptance of commerical airline travel but also facilitated a growing number of international and domestic trips. Domestic flights, if they passed over Ohio, were enlivened by Truman's earthy sense of humor. Ohio was the home state of conservative Sen. Robert A. Taft, a persistent Republican foe. The plane's crew had orders to notify Truman each time a flight crossed the state, whereupon Truman would march off to the lavatory and would then command the pilot to activate the *Sacred Cow*'s liquid waste disposal system, which, in those days could be done in midair.

As improved versions of postwar transports entered service, White House travel experts recommended a replacement for the *Sacred Cow* of World War II vintage. In 1947, a new DC-6, named the *Independence,* assumed the missions of the presidential aircraft. Like its predecessor, it was kept busy as a transport on official business for members of Congress, the military, and foreign dignitaries. In 1953, President Eisenhower commissioned another new plane, the Lockheed Constellation *Columbine II,* named after a similar Constellation that had been assigned to him during his military career in Europe; a year later it was replaced by a larger and more powerful Super-Constellation, *Columbine III.* By this time, the U.S. Air Force had become responsible for a small fleet of aircraft used by the White House and had established a Special Air Missions Unit at Andrews Air Force Base, near Washington. By 1959, Special Air Missions Unit had added jets to the roster of aircraft, opening a new phase of presidential travel.

While the Douglas, Lockheed, and Boeing piston-engine transports made the more glamorous long-distance flights of the first postwar decade, a new generation of

The Independence, *a Douglas DC-6 that served as the offical transport for President Harry S. Truman, inaugurated new dimensions of travel for occupants of the White House. Courtesy of McDonnell Douglas Corporation.*

twin-engine aircraft was developed to replace the venerable DC-3 on shorter runs. Both Martin and Convair produced such planes, but the Convair 240/340/440 series, with pressurized cabins, enjoyed wider acceptance. The Convairs won special favor on intercity routes developed by the so-called trunk lines. Regional carriers, like Allegheny, Ozark, North Central, and others, specialized in service between smaller cities, as well as between them and larger, "hub" airports served by the major carriers. Serving smaller airports with reduced support facilities, the Convairs were designed to be self-contained, with integral boarding stairs and other accessories, and they slowly displaced the DC-3 during the forties and fifties.

The postwar transports produced by Douglas, Lockheed, Boeing, and Convair were pressurized and were powered by a new generation of radial engines burning higher-octane fuel. Also, they all employed the tricycle landing gear. Pressurization not only permitted airliners to fly higher over rough weather that unsettled passengers' stomachs but also improved airplane performance. The reduced density of the air at higher altitudes decreased drag, a condition that increased speed by 17 percent at 20,000 feet, where most of the piston-engines transports operated. But such increases would be possible only if the engines were to develop the same power as they did in the denser air of lower altitudes. For this reason, improvements in superchargers and compression ratios, made possible by higher-octane fuels, were imperative, and the experience of World War II became very significant in the engines used by postwar civil transports. Before the war, airlines generally had resisted the use of 100-octane fuel because it cost nearly twice as much as 91-octane. But large-scale wartime production of 100-octane gasoline for military planes brought down the cost and developed a production capacity that made it feasible for use in postwar airliners. Some experts singled out the fuel factors as the biggest contribution of World War II to the airline business.

The tricycle undercarriage had appeared before World War II, but it was not used commercially until the postwar era. Its use on the new generation of heavier airliners contributed to their efficient operation. Their heavier wing-loadings necessitated longer runways, and the tricycle gear automatically put the wings at an angle of incidence for maximum lift from the start of the takeoff roll. Due to the higher landing speeds of postwar transports, they could also be braked more forcefully without risking a ground loop. In addition, pilot visibility in taxiing these big planes on the ground was improved, and luggage-laden passengers who had scrambled up the aisles of "tail draggers" like the DC-3 enjoyed arriving aboard the newer planes with aisles and seats horizontal.

The decade of the new piston-engine airliners represented a quantum jump from the trimotor transports of the twenties. "In the 20 years between the design of the Ford Trimotor and that of the DC-6," remarked Miller and Sawers, "the cost of carrying a passenger had been reduced by 65 percent, the speed at which he was carried had increased 150 percent, and the distance for which he could travel non-stop had grown eightfold." Operating costs for the larger DC-6, as compared to the DC-3, went down by 30 percent. Much of the difference was attributable to the increase in passenger capacity, from 21 seats to 60 or 70, but there were many other factors: more durable engines and better aircraft maintainability; better management techniques; reduced drag, which raised the 170 MPH speed of the DC-3 to 280 MPH of the DC-6. In short, carrying passengers unquestionably became a paying proposition, and the speed, safety, and comfort of the postwar airliners began to attract passengers in droves.

The makeup of airline clientele began to change. Before the war, an airline trip had still represented something of an elitist mode of travel, and passengers tended to be business executives, movie stars, and others with more-than-comfortable incomes. In the prewar era, one observer wrote, "Stewards gave the kind of service one expected and got on first-class liners . . . porters carried baggage, everything was as it should be and flying was a pleasure." Although the upper strata of clientele and their stylish mode of travel were probably most conspicuous on international routes, they constituted a considerable segment of domestic flights as well. Within a decade after the end of World War II, the airlines were boarding millions of passengers per year, in contrast to the mere thousands of the thirties. Moreover, the clientele represented much more of a cross section of American society, not just the upper crust of the prewar era.

The impressive increase in patronage brought new and different problems for airline personnel. The congestion of travelers at check-in counters imposed new burdens on the procedures for passenger reservations and luggage handling. Added to this was the growing complexity of routes taken by passengers who flew a combination of major airlines and/or trunk lines during a trip. At the end of the trip, impatient passengers fidgeted during longer waiting times for baggage to be unloaded and delivered to luggage claim areas. Passenger complaints swelled from a trickle to a flood. The harassed airlines perceived the computer as a solution to this growing tangle and began making the first installations.

Between the arrival and departure of a flood tide of travelers, the airlines had to cope with their patrons' appetites. Early in the postwar period, many airlines hoped their

Romantic Rio can be yours
This Month

In Rio de Janeiro now, it's summer! And by Clipper, Rio's Copacabana beach is just a week end away from the United States.

Rio's beautiful boulevards run to the ocean's edge. At sea level there's swimming under the bright tropical sun . . . In the flowered hills it's cool enough for golf, tennis, or horseback riding. Brazil has fine hotels, superb food, delightful Latin American color and gaiety.

World-famous Pan American service includes complimentary hot meals served aloft. For rates and information on flights to any country in South or Central America, to the West Indies, Bermuda, Europe, Alaska, or Hawaii, get in touch with your Travel Agent or call the nearest PAA office.

New, four-engined Clippers like this are now being readied for operation out of Miami. Passenger service representatives who speak both Portuguese and English meet you at the airport in Brazil—help you through customs, arrange transportation and check with hotels.

PAN AMERICAN WORLD AIRWAYS
The System of the Flying Clippers

FIRST air service across the Pacific   FIRST plane service across the North Atlantic

A Pan Am brochure of 1946 was designed to entice travelers to the beaches of Rio's Copacabana, "just a week end away." The speed and intercontinental reach of postwar air travel introduced revolutionary changes in American vacation habits. Courtesy of Pan American World Airways.

clientele would grab a bite to eat before departure. On longer flights and oceanic routes, meals prepared beforehand were served en route, either kept hot or reheated in the cramped galley space aboard the aircraft. These warmed-over preparations were generally skimpy in portion, soggy, and tasteless. The repetitious menu of cold cuts, cheese, soup, and other "complimentary snacks" more often than not made travelers long for the more felicitous dining standards aboard passenger trains and ocean liners. When deep-frozen foods became available about 1950, airline caterers responded to them with alacrity. Schedules for food preparation in the caterers' kitchens became less frenzied as complete meals were now prepared and put in cold storage well ahead of time. Reheated in flight aboard the aircraft, the frozen meals provided a reasonable repast, despite snide remarks from die-hard epicures. Forestalling expensive gastronomic wars, with one airline attempting to lure another's passengers with better food and drink, IATA prescribed guidelines for in-flight menus. At the same time, it established standards for airline seats as to headroom and legroom. These strictures pertained to coach or economy class only, however, and first-class travelers continued to be courted with blandishments such as more spacious seating and, on some flights, menus that would have done justice to a first-class restaurant.

Airline service had a discernible impact on the traditional modes of travel, domestic as well as international. The steady decline of rail passenger traffic in the United States undoubtedly owed much to the postwar boom in auto sales and the end of gas rationing. But the longer lines at airline ticket counters signaled increasing inroads into the clientele of passenger trains, especially on long runs and coast-to-coast routes. By 1951, airline passenger-miles were totaling more than Pullman passenger-miles (10.6 million compared to 10.2 million), and the gap continued to widen. The appeal of flying was even more evident in the case of intercontinental travel patterns, where the ocean liner no longer reigned supreme.

Behind these impressive developments in airline evolution a number of significant and interrelated factors were at work. The introduction of modern, multiengine transports was the most obvious, but the formulation and functioning of organizations like ICAO and IATA were fundamental. The prosaic, but essential, procedures for improved airline reservations and food service necessary to handle the growing volume of passengers were indispensible. Within the United States, the web of domestic air routes expanded as a function of federal legislation. The Federal Airport Act of 1950 committed government funds to the development of domestic airways, and airport construction proliferated across the country. In the same year, the Civil Aeronautics Authority opened its first omnirange airways (VOR—very high frequency omnidirectional radio range) station. Although dozens of VOR stations had already been commissioned, specified chains were now designated as controlled airway routes that connected major air terminals. This paved the way for elimination of the four-course radio ranges of the thirties and forties. Cockpit instrumentation now featured calibrated visual information relative to aircraft bearings—a vast improvement over the aural signals that were frequently obscured by static. In one respect, however, American airlines were left trailing in aviation technology. In the thirties, the airlines had taken the initiative in the operational use of advanced designs like the Boeing 247 and the Douglas DC-3. In the postwar era, the major advance came with introduction of jet equipment. In this field, the initiative was taken by the military services and by the British.

MILITARY AERONAUTICS

As speeds of conventional, propeller-driven aircraft edged into the 400–450-MPH range, the appeal of a gas turbine engine became compelling. With a piston engine, the efficiency of the propeller began to fall off at high speeds, and the propeller itself represented a significant drag factor. Although research in gas turbine technology dated from the thirties, when speeds of 300–350-MPH represented norm, a few postwar aerodynamicists already recognized the impending limitations of propeller-driven aircraft at high speeds. The problem was to gain sufficient research and development (R&D) funds for what seemed to be unusually exotic gas turbine power plants.

In England, RAF officer Frank Whittle doggedly pursued research on gas turbines through the thirties, eventually acquiring some funding through a private investment banking firm after the British Air Ministry turned him down. Strong government support finally materialized on the eve of World War II, and the single-engine Gloster experimental jet fighter flew in the spring of 1941. English designers leaned more toward the centrifugal-flow jet engine, a comparatively uncomplicated gas-turbine design, and a pair of these power plants equipped the Gloster Meteor of 1944. Although Meteors entered RAF squadrons before the end of the war and shot down many German V-1 flying bombs, the only jet fighter to fly in air-to-air combat came from Germany— the Me-262. Hans von Ohain, a researcher in applied physics and aerodynamics at the University of Göttingen, had unknowingly followed a course of investigation that paralleled Whittle's work and took out a German patent on a centrifugal engine in 1934. Research on gas turbine engines evolved from several other sources shortly thereafter, and the German Air Ministry, using funds from Hitler's rearmament program, earmarked more money for this research. Although a centrifugal type powered the world's first gas turbine aircraft flight by the He-178 in 1939, the axial-flow jet, more efficient and capable of greater thrust, was used in the Me-262 fighters that entered service in the autumn of 1944.

Independently of the early jet research in England or Germany, some work had begun in America in 1940, but the successful British investigation led the Army Air Force to import a Whittle centrifugal jet engine in the spring of 1941. Copies of this engine, produced by General Electric, were used in the twin-engine Bell P-59A. Flights were cloaked in strict secrecy, so much so that the air force installed a removable propeller on the nose of the plane in an attempt to disguise it on the ground. All of this diversion proved to be a bit overdone, since the P-59A came out second best in mock combat trials against advanced models of the P-47 Thunderbolt and the P-38 Lightning. Still, it was a beginning. The air force already had a more successful jet fighter under development, the Lockheed P-80 Shooting Star, which made its first flight in January 1944. Lockheed delivered over 100 of these fighters before the end of the war, and some were flying with air force squadrons in England and Italy for familiarization purposes.

With their greater speed, rate of climb, and operational ceilings, jet fighters quickly replaced piston-engine types in the postwar air force. The Shooting Star, or F-80 (the postwar air force changed its letter designations from ''P'' for pursuit to ''F'' for fighter), was soon joined by the Republic F-84 Thunderjet and the rakish, sweptwing North American F-86 Sabre.

The striking configuration and performance of the Sabre made it one of the most memorable air force fighters of the postwar decade. In retrospect, it is somewhat ironic that the F-86 sprang from a naval requirement in 1944 for a fighter with straight wings and was eventually adapted for a new air force fighter with the same conventional wing planform. Both versions were designed by North American. As the Allied armies rolled across Europe, American designers eagerly studied captured German aerodynamic data for high-speed jet aircraft with swept wings. The information revealed in these German wind-tunnel experiments coincided with the arguments already voiced by some imaginative aerodynamicists in the United States. With greater confidence in this radical departure from accepted practice, aerodynamicists redesigned the air force version of the F-86 with sweptback wing and tail surfaces. It became one of the most respected jet fighters of its day.

The navy still required a jet fighter capable of operating from aircraft carriers. Putting down a conventional high-performance fighter on the deck of carriers demanded considerable flying skill, and jet aircraft, with higher stalling speeds, presented a new range of challenges. For this reason, North American retained the conventional wing design for its navy fighter, designated the FJ-1 Fury. The navy's first operational jet fighter, however, was the McDonnell FH-1 Phantom, whose design studies predated the North American jet by two years. As the first naval jet designed at the outset for shipboard operations, Phantoms began entering service in limited numbers during 1945. In the summer of 1946, Phantoms completed the first jet fighter flights from a carrier deck and provided important operational experience. The Phantoms and Furies saw brief service before being replaced by the McDonnell F2H Banshee series and the Grumman F9F Panther series. These last two provided the navy with its first mass-produced, high-performance jet fighters, joining active squadrons in 1949.

Despite all the attention and money lavished on jets, the military air services of the early postwar years still found a niche for piston-engine combat planes. In the navy, one of the most successful was the Douglas AD-1 Skyraider. A big airplane designed for torpedo attacks or close air support, the husky Skyraider could operate from rough, front-line airstrips as easily as from a carrier deck. First flown in 1945, the versatile plane remained in fleet service for 23 years and did yeoman service during the Korean War. The air force also kept flying its B-29 bombers and acquired an improved, more powerful version known as the B-50. But larger, more devastating bombers were being developed for intercontinental missions, far beyond the radius of the B-29 and the B-50, and the air force also planned for a high-speed bomber powered by jets.

The first big new bomber was the B-36, built by the Consolidated-Vultee Aircraft Corporation, known as Convair. Plans for the huge plane had begun in 1941, several months before Pearl Harbor, and represented the apotheosis of the air force's prewar preoccupation with long-range strategic bombing. Wartime requirements for production of available combat planes slowed development of the B-36 until 1946, and two more years passed before the bomber went operational. Compared to those of its predecessors, the B-36's statistics stretched the imagination. It had a wing span of 230 feet and a maximum bombload of 72,000 pounds. Defensive armament included 16 20-mm cannon in 8 remote-controlled gun turrets. The 16-member crew (including a relief crew of 5 for long, grueling missions of 12,000 miles) operated the plane from

The North American F-86 Sabre was America's first sweptwing fighter. Sent to Korea to counter the Russian MiG-15, the F-86 not only established air superiority, but also flew missions as a fighter-bomber. Courtesy of Rockwell International.

pressurized compartments—fore and aft—that were connected by a long, pressurized crew-transfer tunnel. The air force finally had an intercontinental bomber capable of operating from bases in the United States. The B-36's complexity and costs made it a center of controversy, and critics charged that the propeller-driven plane was vulnerable to high-performance jet interceptors. Convair later added four gas-turbine powerplants to the B-36's six piston engines to improve its speed and operational ceiling. In a series of simulated combat missions, the B-36 seemed an easy, elephantine target, but when attacking fighter pilots closed in on pursuit curves at high interception speeds, their planes began to flounder out of control at the extreme altitudes where the big bomber flew. A total of 385 B-36s were built and many remained in the air force inventory until 1958.

Still, the air force perceived the need for a jet bomber capable of using the jet plane's inherent speed at high altitudes for its defense. Like the F-86, designers first laid out the wing planform of the Boeing B-47 as a straight wing, until data from German wind-tunnel tests encouraged Boeing designers to change to the sweptwing configuration. The thin, laminar-flow wing section startled test pilots with its extreme vertical deflections in flight, but the design ensured low drag, and the "podded" jet engines, slung below the wing, were precisely situated to damp extreme deflection cycles. The B-47 entered service in 1951. Powered by six jet engines (a pair of engines in each of two inboard nacelles, and one engine in two outboard nacelles), the B-47

The Convair B-36, capable of round-trip nonstop intercontinental bombing missions, represents America's policy of massive retaliation using nuclear weapons. Two pairs of jet engines were added to increase the huge plane's speed and altitude. Courtesy of General Dynamics.

carried a crew of only three and a bomb payload of over 10,000 pounds. It could fly over 600 MPH. Defensive armament consisted of a pair of radar-directed 20-mm guns in the tail. With over 2,000 bomber and photoreconnaissance variants produced, it was the first sweptwing bomber to be built in quantity. The design influenced not only large military jet aircraft but also a highly successful stable of commercial airline jets.

The military's dramatic transition from piston-engine equipment of conventional types to high-speed jets of comparatively unorthodox configurations was notably rapid and successful. Information from captured German jet fighters and bombers and data from German aeronautical research institutes were contributing factors. The Cold War also stimulated development. As evidence of heavy military expenditures and advanced new aircraft designs filtered out of the Soviet Union, Congress responded with unprecedented peacetime military appropriations. New facilities for theoretical and applied aeronautical investigations had already begun to evolve as a result of World War II, and indigenous American high-speed flight research made important strides in the immediate postwar era.

NACA had done important research for the military in World War II and continued this function in the postwar era. A good deal of the work was prosaic but highly practical, such as NACA's research on ice-prevention systems for aircraft, which greatly benefited the military services and civil aviation, and which won the Collier Trophy for NACA investigator Lewis Rodert in 1946. But NACA's failure to stay abreast of England and Germany in the development of jet engines still rankled some air force top brass. General Arnold pushed hard for a separate R&D facility for the air service, and eventually succeeded. The Aeronautical Engineering Development Center at Tullahoma, Tennessee, evolved into a highly sophisticated research complex sprawled over several square miles. Wright-Patterson Field at Dayton, Ohio, had developed into a center for applied aeronautical research during the war, and continued in this vein in the postwar era. There were other military and governmental research centers as well. In addition, NACA and the military air services had funding for a remarkable variety of specialized and esoteric aeronautical research by private firms as well as universities across the country. Supported by this broad spectrum of governmental, private, and university R&D, American military aviation made striking progress.

For the American public, the most eye-catching aeronautical news headlines of the postwar era reported the assault on the sound barrier. The Bell X-1 did it first. Development of this remarkable little plane began in 1943, an interesting cooperative enterprise combining NACA, industry, and the armed services. The impetus came from NACA/Langley aerodynamicist John Stack and Bell Aircraft Corporation engineer Robert J. Woods, who realized that propeller-driven planes had nearly reached the limits of high-speed performance. Supported by the army and the navy, NACA and Bell went to work on a research plane to probe transonic and supersonic flight regimes. Jet engines under development fell far short of the required thrust, so designers built the plane around a rocket engine supplied by an outfit aptly named Reaction Motors, Incorporated. The engine burned a mixture of alcohol and distilled water along with liquid oxygen to produce a thrust of 1,500 pounds from each of four thrust chambers.

The engine was one of the few really exotic aspects of the conservatively designed X-1. Not much was known about the flight speeds for which the plane was intended. On the other hand, there was some interesting aerodynamic information available on the .50-caliber bullet, so the X-1's fuselage shape was keyed to ballistics data from this unusual source. The thin, conventionally straight wings were stressed for 18 times the plane's weight to compensate for extreme loads anticipated at sonic speeds. Since the engine's thrust lasted only about two and a half minutes, the X-1 was designed for a midair launch from a B-29 parent ship. During 1946, Bell test pilots flew unpowered drops to check out release systems and dead-stick landings, followed by preliminary powered flights. Then, air force officers took over for eight special flights, increasing speeds and altitudes in preparation for the milestone mission. On the ninth flight, October 14, 1947, Capt. Charles E. Yeager dropped from the B-29 parent ship at 35,000 feet, opened up the rockets, and zoomed upward toward thinner air just over 70,000 feet, making him the first pilot in history to fly faster than the speed of sound. In recognition of this achievement, the following year Yeager shared the Collier Trophy with Laurence Bell and NACA's John Stack.

The miscellaneous aircraft used in high-speed flight research operated out of Edwards Air Force Base, in southern California. The air base itself was the hub of the Air Force Flight Test Center, an area of 300 square miles of desolate desert terrain in the Mojave Desert. Originating as an air force bombing and gunnery range, Muroc Dry Lake, through its remoteness and expansive, concrete-hard surface, inevitably lent itself to the testing of experimental aircraft and the emergency landings they occasionally had to make. This austere, almost surrealistic desert setting made an appropriate environment for the array of exotic planes and headlined flight records that seemed to make "Muroc" a household word in the forties and fifties.

A succession of X-aircraft, designed primarily for flight experiments, populated the skies above Muroc in a continuous cycle of R&D. Two more X-1 aircraft were ordered by the air force, followed by the X-1A and the X-1B, which investigated thermal problems at high speeds. The navy used the Muroc flight test area for the subsonic, jet-powered Douglas Skystreak, accumulating air-load measurements unobtainable in early postwar wind tunnels. The Skystreak was followed by the Douglas Skyrocket, a sweptwing research jet (later equipped with a rocket engine that would surpass twice the speed of sound for the first time in 1953). The Douglas X-3, which fell short of expectation for further flight research in the Mach 2 range, nevertheless yielded important design insights on the phenomenon of inertial coupling (solving a control problem for the North American F-100 Super Sabre), the structural use of titanium (incorporated in and other subsequent supersonic fighter designs), and data applied in the design of the Lockheed F-104 Starfighter. The NACA kept involved throughout these programs. In a number of ways, the X aircraft contributed substantially to the solution of a variety of high-speed flight conundrums and enhanced the design of future jet airliners—a record of consistent progress aside from the speed records that so fascinated the public.

COLD WAR AND HOT WAR

As the postwar generation of high performance jet fighters, advanced bombers, and other new military designs began entering service, naval and air force operations necessarily underwent modifications. At the same time, the organization of the American military establishment was also revamped, leading to the long-sought independent Air Force. After Pearl Harbor, the idea of an independent air force resurfaced with strong support, but General Arnold advised against it on the wise assumption that such a changeover in wartime would be too disruptive. During 1942–43, however, the Army Air Force officially achieved equal status with the Army Ground Forces and Army Service Forces within the War Department. For nearly two years after the end of the war, a frequently raucous debate ensued over the issues of a unified department of defense, with separate, subordinate departments of the army, navy, and air force. The navy finally accepted the new arrangement after assurances for the continuance and development of naval and Marine Corps aviation. When the National Security Act of 1947 went into effect on July 26, James V. Forrestal became the first secretary of defense and Stuart Symington became the first secretary of the air force.

The United States Air Force (USAF) proceeded to put a good deal of faith in the Strategic Air Command (SAC) and the B-36 intercontinental bomber. Loaded with atomic weapons, the B-36 with its global combat radius was regarded as a potent deterrent to potential aggressors anywhere in the world. Under the tough leadership of

This group portrait displays typical high-speed, research aircraft that made headlines at Muroc Flight Test Center in the 1950s. The Bell X-1A (lower left) had much the same configuration as the earlier X-1, in which Capt. Charles E. (Chuck) Yeager breached the "sound barrier" in 1947. Joining the X-1A were (clockwise) the Douglas D-558-I Skystreak; Convair XF-92A; Bell X-5 with variable sweepback wings; Douglas D-558-II Skyrocket; Northrop X-4; and (center) the Douglas X-3. Courtesy of NASA.

Gen. Curtis E. LeMay, SAC achieved a reputation for high competence and *élan*, but its new bombers and their considerable maintenance problems overseas in peacetime were expensive. These costs were part of the complicated arguments over the B-36 and SAC's role, pitted against the navy's demands for a new generation of aircraft carriers and carrier aircraft with atomic capability. The growing sophistication and complexities of postwar military hardware and the unfamiliar nuances of nuclear deterrent doctrines did not make it any easier for the American public to grasp the issues.

A number of panels, committees, and boards issued statements on defense policies, but the one issued by the Finletter Committee in 1948 seemed to capture the spirit of most of these swirling controversies. Its tendentious title, *Survival in the Air Age*, made it clear that the issues were fundamental. The Air Policy Commission, chaired by Thomas K. Finletter, had been appointed by President Truman in 1947, and issued its report after five months of investigation. The Finletter Report underscored the likelihood that other countries would develop nuclear weapons and the means to deliver them. To deter potential aggressors, the report endorsed more aeronautical R&D

programs to strengthen civil aviation and a 70-group air force. Following some negotiation, Truman and the Congress agreed on 66 groups, still a considerable size and indicative of the postwar era's commitment to the idea of national security as related to air power. The aircraft industry had just begun to gear up to manufacture the 66-group air force when the Korean War broke out.

During the withdrawal of Japanese troops from Korea at the end of World War II, the country provisionally divided into two zones of occupation at the 38th parallel. On June 25, 1950, North Korean military forces armed with Russian equipment plunged across the 38th parallel. Within two days, the United Nations had committed itself to South Korea's assistance, although most of the personnel, arms, and supplies came from the United States. When retreating UN forces regrouped and fought back northward, crossing the 38th parallel and penetrating deep into North Korean territory, Communist Chinese forces moved across the Yalu River into North Korea. The conflict now assumed hazardous new dimensions. If American bombers were to attack traditional strategic targets like factories and supply bases in Manchuria, such raids could touch off another world war. By the same token, atomic strikes at targets deeper in China (or in Korea) could trigger nuclear retaliation from Russia. Under the aegis of the United Nations, the USAF and air units from other nations therefore fought under the rules of limited warfare.

The USAF went to war in the skies over Korea as a totally independent air force for the first time in its history. There was a mild irony in this. The USAF not only had won its independence but also had devoted considerable resources to the build-up of a grand-strategic bomber force. The globe-girdling atomic bombers of SAC were the sum of decades of arguments and dreams. Now faced with its first combat situation, SAC learned that its heavy bombers and nuclear armory were not applicable under the hazy ground rules of limited warfare. Although postwar USAF doctrine had always included roles for fighter planes to win air superiority over the combat zone and for tactical aircraft to fly close-in strikes in support of ground troops, the Korean theater suddenly gave more impetus to fulfilling these requirements as opposed to developing SAC missions. Strategic bombing played a role in Korea but was conducted against limited targets using conventional ordnance dropped by World War II–vintage aircraft, like the B-29 and B-50, operating from bases in Japan. The USAF also discovered that other piston-engine fighting planes of the World War II era were still valuable in the age of jets.

The North Korean air force was soon eliminated and attention centered on tactical air support from USAF planes. In Japan, the USAF had F-80 jets, but their staying time over Korean targets was limited after 100–500-mile flights from Japanese air bases. The jets had other problems. They needed smooth, long runways, and in the early months of the war found it difficult to operate from the rough fields of the rugged, gritty, Korean battlefronts. Moreover, ground-support strikes at low altitudes raised the fuel consumption of jets. Some squadrons surrendered their glossy F-80s for time-worn F-51 Mustangs that could fly from short, rudimentary Korean flying strips and carry a formidable load of rockets or napalm. The mobility of aircraft carriers made them invaluable in this environment, and they ranged up and down the Korean

The first operational jet fighter to join the USAF was the Lockheed F-80 Shooting Star. The group shown here was part of the air force training school for fighter pilots. Courtesy of Lockheed Corporation.

coastline. Navy and marine Panthers and Banshees joined the venerable Corsair and the newer Skyraider, probably the best close-support airplane of the Korean War. Improved air bases within Korea eventually enhanced the usefulness of USAF jets, and the F-80 was joined by the Republic F-84 Thunderjet and jet aircraft from other UN forces. The jets began to play an increasingly important ground-support role and proved to be especially stable "weapons platforms," free of the engine torque and vibrations of piston-engine aircraft.

By the spring of 1951, after massive Chinese counteroffensives, the battlefront had stabilized along the 38th parallel. USAF B-29 raids consistently thwarted Communist attempts to establish major combat air bases within North Korea, and heavy strikes against bridges and generating stations along the Yalu River not only hampered the movements of enemy soldiers and equipment but, severely curtailed electrical power supplied to Chinese factories in Manchuria. From safe havens on the Manchurian side of the Yalu, Russian-built MiG-15 jet fighters of the Chinese air force rose to oppose these devastating attacks. The first jet-versus-jet combat had occurred in 1950, when an F-80 made a successful diving attack on a MiG-15. For permanent control of the air, however, the F-80 was no match for the faster, sweptwing MiG, and the USAF brought in the F-86.

Operating from carriers off the Korean coast, U.S Navy jets like the Grumman F9F Panther provided close air support for United Nations troops. At left are propeller-driven Corsairs, also used in the low-level air support role. Courtesy of Grumman Aerospace Corporation.

The MiG-15, armed with two to four large-bore cannon, appeared to have an edge in fire power over the F-86, and its lighter weight certainly gave it the advantage in rate of climb and agility in combat. The F-86, on the other hand, carried six .50-caliber machine guns with a 60 percent higher rate of fire, increasing the pilot's chance of hitting a fast-moving target. Moreover, a large number of American pilots were combat veterans of World War II and possessed a decisive edge in experience and fighting skills over their opponents. Exact numbers are uncertain, but it is estimated that F-86 pilots shot down MiG-15s at a ratio of at least 10 or 12 to 1. During June 1953, F-86 squadrons claimed 75 MiGs shot down and reported no American losses.

Aerial operations over Korea exacted a price: the Far East Air Force listed more than 1,000 casualties and aircraft lost; the navy and marines lost over 1,100 aircraft, most of them during attacks on enemy ground targets. Balancing these losses was the crippling effect on Chinese and North Korean personnel, transportation, and supplies. Aerial attacks disabled hundreds of tanks and obliterated scores of enemy bunkers and other strong points. Planes demolished thousands of railway and road vehicles and ruined hundreds of bridges and dozens of tunnels. Bombers blew up railroad tracks faster than they could be repaired. Navy Skyraiders torpedoed the floodgates of one dam, raising the water in the Pukhan River below the dam to levels that slowed an enemy advance. There is little doubt that such aggressively accurate air strikes stalled the enemy's offensives and encouraged peace negotiations. It is thought that two carefully planned fighter-bomber attacks in the spring of 1953 proved especially effective in this regard. Within four days of each other, F-84 Thunderjets blasted a huge irrigation dam and a reservoir, breaching the structures and causing widespread flooding. Aside from the flood damage, much of the North Korean rice crop suffered from the loss of controlled irrigation. Following the Thunderjet missions, armistice negotiations moved more swiftly, and the Korean War ended on July 27, 1953.

The Korean conflict demonstrated not only the combat effectiveness of jet aircraft but also the value of the helicopter and transport aircraft. The ubiquitous choppers did artillery spotting and airlifted Marine crews with a rocket launcher (slung underneath) to strategic firing sites. The marines would set up the launcher, fire off a series of withering salvos, and be ferried out again before enemy spotters could lay on a counterbarrage. Helicopters hovered near aircraft carriers to rescue fliers who crash-landed in the sea. They flew emergency supply missions to beleaguered ground troops, snatched up pilots downed behind enemy lines, and saved uncountable lives in medical evacuations from the battle zone. In these and myriad other ways, helicopters proved nearly indispensable, becoming firmly entrenched in the inventories of the army, the navy, the marines, and the air force. The helicopter also became entrenched in American folklore as part of the fictional 4077 Mobile Army Surgical Hospital depicted in the 1970 film *M*A*S*H* and the highly acclaimed television series of the same name.

Other airborne operations in Korea included an unusually successful parachute drop north of Pyongyang in the fall of 1950. Flying C-47s and Fairchild C-119 transports, the air force dropped 2,500 soldiers and 300 tons of equipment in an action that cut off main supply routes of retreating units in the North Korean army. Later, UN forces under General Walker depended primarily on aerial supply during the advance north of the 38th parallel. When Chinese armies halted this offensive late in 1950, the retreating UN forces relied nearly as much on air transports for large-scale aerial evacuation of combat troops from several critical battlefront areas. For the duration of the war, air transport operations continued as a central factor in military planning. Planes like the C-47 and C-119, in addition to performing normal transport roles, conducted aerial resupply missions over the front lines. Other transports, like the four-engine C-54, flew regular cargo and medical evacuation missions back and forth between Korea and Japan. The long-range Douglas C-124 Globemaster performed similar duties on flights between the Far East and the United States.

In the decade after 1945, American military air services experienced unprecedented growth, directly attributable to tensions arising from the cold war and demands of the Korean conflict. With bases girdling the world, the Department of Defense unquestionably commanded a global air force. The numbers of officers and personnel and the remarkable variety of sophisticated equipment at its command required equally sophisticated programs in personnel management and bureaucratic techniques. A continuing R&D program kept the air force involved in a broad spectrum of the physical and

During the Korean War the helicopter came into its own. The Bell Model 47 was highly valued for its work in medical evacuation from front-line areas where the lack of airstrips prevented the use of conventional, winged aircraft. Courtesy of Bell Helicopter Textron.

biological sciences. These developments alone gave the air services considerable influence in national and local politics. Air bases and other installations played a large role in the health of state and local economies. Escalating military sales to foreign allies provided potent leverage for influence, and the presence of sprawling American air bases also influenced regional and local economies overseas. A military presence—domestic and foreign—continued to expand. Even staunch advocates of the military system perceived some dangers in this pervasive phenomenon, as Eisenhower suggested in his farewell address concerning the military-industrial complex in 1961. But aviation, even military aviation, could have benign social influences as well; in an age of complex international relations, some facets of air power offered alternatives other than deadly combat. Three examples illustrate the positive contribution of aviation: climatological research, disaster relief, and crisis diplomacy.

HURRICANES, BLIZZARDS, AND BLOCKADES

One of the more unusual, and apparently foolhardy, uses of aircraft involved hurricane hunting. After tentatively chasing hurricanes to gather storm reports on the periphery, pilots decided they could penetrate the eye of the hurricane itself, obtaining theretofore unavailable data on pressures, storm ceilings, and other meteorological phenomena. The first pilot to penetrate the eye of a hurricane appears to have been a flight instructor named Col. Joseph Duckworth. A specialist in instrument flying, he was sure that a trained pilot could successfully fly in severe storm conditions. In July 1943, weather reports from Texas warned of a hurricane sweeping in from the Gulf of Mexico. Duckworth persuaded a navigator to go along, and the pair took off in a single-engine AT-6 trainer. Severely buffeted by the storm, they flew into the hurricane's eye and home again. Since the base weather officer was so disappointed at having missed this unique experiment, Duckworth repeated the experience for him. During the turbulent flight, the weather officer managed to record some sketchy notes about the hurricane. Clearly, a larger aircraft, able to carry extensive instrumentation, was called for. But it also seemed likely that a larger plane in the violent grip of a hurricane would experience even more structural stress. In what seems a high-risk mission, a trio of air force volunteers, Col. Frank Wood, Maj. Harry Wexler, and Lt. Frank Reckford, successfully flew a Douglas A-20 light bomber into a hurricane—and out again—on September 14, 1944. Organizing a joint program shortly after the end of the war, government agencies and the armed services employed carefully modified aircraft equipped with heavy loads of scientific gear for extensive hurricane research. The navy supplied the PB4Y-2 Privateer, a version of the wartime B-24 Liberator redesigned with a single vertical rudder, and the air force operated the RB-29, a reconnaissance version of the B-29 Superfortress. These planes equipped specially designated weather squadrons based in Florida, Bermuda, and other selected regions. On receipt of weather reports that sent other planes scuttling for the nearest shelter, the hurricane-hunter squadrons prepared for action, deliberately flying into weather that other pilots encountered only in their most fearsome nightmares. Highly useful information evolved from these tumultuous forays, leading to better understanding of hurricane formation, dynamics, and forecasting. Predictions of hurricane strengths and their probable paths saved innumerable lives and millions of dollars in property, particularly in the storm-prone Caribbean and Gulf regions.

Aviation also countered the hazards of other climatological emergencies, airlifting relief supplies to areas stricken by floods, cyclones, or snowstorms. The "Blizzard of '49" represented a major effort. On the first day of the new year, snow flurries appeared in adjoining regions of South Dakota, Wyoming, and Nebraska. By evening of the second day, heavy snowfalls began while temperatures plummeted, and full-scale blizzard conditions prevailed on the morning of January 3. Driven by fierce winds, the blizzard raged for two days, with gusts cutting visibility to only a few feet. When the storm finally spent itself, an area of over 150,000 square miles lay virtually immobilized under an unprecedented blanket of snow. Drifts of 20 to 30 feet blocked railway cuts, and roads were impassable. Thousands of rural families were, quite literally, stormbound inside homes whose windows and doors were blocked by heavily packed snow. Once they crawled out through upper-story windows—or chopped a hole in the roof—farmers and ranchers had to dig out the barns and sheds housing their livestock. But hay for cows and horses lay buried under many feet of snow, and animals in outlying pens and pastures soon began to suffer for lack of forage.

Regional newspapers characterized the blizzard as the worst storm in over a century. Cities and towns laboriously began to dig themselves out, and main railroad and highway routes took days to open. But a state of emergency persisted, as repeated snowfall and recurrent winds created new drifts across roads. In many counties across the three-state region, near-paralysis from the storm lasted until early February. The slow process of recovery heightened the precarious situation on marooned farms and ranches, where food and medical supplies dwindled daily and livestock began approaching starvation. Cut off by towering drifts and snow-choked roads, rural families and herds of animals could be reached only by air.

Units of the army, air force, and National Guard responded to emergency calls from state governors. Military aircraft, like the Sikorsky R-5 helicopter and the Stinson L-5 (a single-engine light plane), equipped with skis, made emergency medical flights and began a vital survey of the stricken countryside. Soon, larger transports were called in. Flying from airports outside the storm region at first, then using cleared airfields closer in, C-47s and other planes of the air force and the Air National Guard dropped tons of animal feed and baled hay to imperiled livestock. The planes also airlifted emergency food and medical supplies to isolated communities. As soon as possible, private pilots dug airplanes out of hangars and attached skis to their landing gear. Dozens of light aircraft crisscrossed the snowy midwestern plains, scouting for emergency symbols fashioned by farmers and ranchers who followed instructions from radio broadcasts.

Carrying requests for food, clothes, and medicine, pilots landed at convenient towns, arranged for the necessary supplies, and flew back to the farm or ranch, landing nearby or dropping packages out of the door on a low pass. Snowbound shopping lists stressed basic necessities, although a grocer in western Nebraska filled one rancher's list that began with two books of cigarette papers, four cartons of Bull Durham for smoking, and one carton of Spark Plug for chewing; these were followed by secondary items like flour, sugar, and so on. Although many of the transactions involved established credit customers at local stores, other deliveries depended entirely on good faith among unacquainted parties. Many proprietors were not paid until roads opened in early spring, when rural strangers appeared with an introduction, a check, and a word of heartfelt thanks.

Authorities coping with the blizzard of 1949 named it "Operation Snowbound." Elsewhere in the world in the same year, aviation played a far more crucial role during missions dubbed "Operation Little Vittles" by flight crews; official military communiqués referred to the flights as the Berlin Airlift. When World War II ended in Europe, Russia and the western Allies divided Germany into separate zones of occupation. Plans for reunification disintegrated as the occupation zones established separate currency systems and administrative organizations in 1947/48. These circumstances intensified friction between Russia and the West, particularly in the case of Berlin, deep within the Russian zone. As the former German capital and one of the great cities of Europe, Berlin possessed unusual psychological and political significance. Since the city was divided into four zones of occupation (French, British, American, and Russian) like Germany itself, Allied ground access to Berlin was restricted to a few rail, highway, and canal routes through Russian-controlled territory.

Allied air routes into the city were also restricted to three narrow corridors. For food, fuel, and raw materials for its industries, Berlin was totally dependent on the ground life lines; air transport was used primarily for passenger travel and high-priority, lightweight cargo. If Russia wanted to isolate the vulnerable city, forcing Allied withdrawal or some political compromise, it only had to cut the major ground routes. On June 24, 1948, it did.

With a civilian population of about 2.5 million people, Berlin had sufficient food supplies for 36 days, it was thought; coal might stretch for 45 days. Gen. Lucius Clay, the American military governor of Germany, ordered airlift operations on an interim basis to stretch Berlin's dwindling reserves of food and fuel. In the beginning, the airlift was seen as a short-term expedient—strictly an emergency measure, like flying serum to Nome, Alaska—something pulled together quickly for an intense effort of very short duration. Meanwhile, Clay and others wanted to send an armed truck convoy to punch through Soviet road blocks and reopen the city. President Truman balked at this proposal, and it soon became obvious that the short-term airlift would become a long-duration operation. It was an interesting proposition, using air power in an active way for a high-stakes international gamble: to maintain the Allied presence in Berlin without resorting to armed confrontation. The question was whether airplanes alone could sustain a city of 2.5 million.

At the beginning of the airlift, planners talked of flying in 500 to 700 tons per day, at the most, for about three weeks. There was no anticipation of continuing the missions into the winter or of regarding the airlift as a potent long-term diplomatic counterstroke. The United States Air Force in Europe (USAFE) rounded up 110 planes, all of them war-weary C-47s whose peeling paint frequently exposed markings from the Normandy invasion or from flying the Hump. Each plane could carry only about three tons; more airlift muscle was plainly needed. Within six days, troop-carrier squadrons flying C-54 transports had arrived from as far away as Texas, Panama, and Alaska. Their payload of around 10 tons made them indispensable, and the gallant C-54 Skymasters became the real workhorses of the Berlin airlift.

Having momentarily solved the airplane problem, USAFE turned to cope with others, each of which threatened to end the airlift in one way or another. The American landing strip at Tempelhof Airport in Berlin consisted of pierced steel segments laid over a shallow base of crushed rubble. Within two weeks it started to give out, and engineers warned that the airfield would be out of commission within 60 days. The situation implied a major construction job, but there was no heavy construction machinery available in Berlin and no apparent way of getting it there in air force transports. The Pentagon found a civilian contractor possessed of unusual skills with an oxyacetylene torch. He was dispatched to the American zone in West Germany, followed by a stream of dismantled bulldozers, graders, and other construction equipment. The man had carefully cut everything into manageable size with his oxyacetylene torch, sent them to Tempelhof by plane, then arrived later himself to weld back together, just as carefully, all of the assorted pieces. Construction workers hurriedly built two new runways at Tempelhof and finished a third at Tegel, in the French sector.

Arriving in Berlin during the 1948–49 airlift, a C-54 has its cargo unloaded while the crew grabs a quick snack from a mobile snack bar. Courtesy of McDonnell Douglas Corporation.

The continual flying, day and night, placed increasing strains on airframes and engines. The USAF ransacked airfields around the world for mechanics and engineers as well as propeller technicians, hydraulics specialists, and several other aviation specialties. Completely new maintenance standards evolved, postponing scheduled down-time in inspection hangars from every 100 hours to 200 hours, and reducing the time for a 200-hour inspection to about 2 or 3 days instead of 7 to 10. When the winter season arrived with murky weather, new advances in electronic Ground-Controlled Approach (GCA) systems allowed aircraft to keep flying. Week in and week out, planes landed at Tempelhof every 3 minutes. Portable snack bars and a meteorology team would hustle out to meet each plane; in 30 minutes the aircraft would be emptied, refueled, and off again. Crew fatigue became a factor as weeks passed under the intense regime of careful instrument flying and strict sequencing of aircraft operations. At Great Falls, Montana, the air force set up a special school to indoctrinate replacement crews in the regimen and flying conditions they would meet in Berlin's winter weather. The planes kept flying, and the tonnage carried into Berlin climbed.

By the end of the first month, incoming cargo totaled 1,500 tons per day, and already plans were in motion to exceed 4,500 tons daily. With the cooperation of the RAF, this figure became reality by the autumn of 1948. The daily average sagged to 4,200 tons in discouraging weather of the winter months, but perked up again with the opening of the new Tegel airstrip. During 1949, renewed efforts by naval, USAF, and RAF crews raised the daily average to 8,000 tons, about four times the city's subsistence level. Some items were not counted in these impressive totals, for example, the candy and gum dropped in "Operation Little Vittles." As pilots roared low over the city on final approach, crew members shoved out hundreds of candy bars and gum packages, each swaying lightly under its own handmade parachute, much to the delight of growing crowds of Berlin's children.

By the spring of 1949, the airlift had supplied Berlin with 80 percent of the food, fuel, and raw materials that had arrived by surface routes before the blockade. On May 12, 1949, the Soviet Union capitulated and ended the blockade, although the airlift continued until the end of September. Allied planes had carried a total of 2.3 million tons of supplies in 277,264 flights into the city. The Berlin Airlift became a remarkable proving ground for air transport. The operation yielded invaluable experience in GCA techniques, air traffic control, and aircraft operation and maintenance under grueling schedules. The airlift also became an excellent example of interservice teamwork and international cooperation with British and French forces. Perhaps most important, the Berlin Airlift demonstrated the effectiveness of air power in postwar diplomacy, and provided an alternative to war.

THE GENERAL AVIATION SECTOR

As early as 1943, when all aircraft companies were backlogged with military contracts and D-Day was still in the planning stages, light-plane manufacturers began looking ahead to a postwar bonanza in private flying. The Department of Commerce estimated the demand for light planes at 200,000 aircraft per year, and even cautious forecasts emphasized the inevitable flood of postwar private pilots from the ranks of 12 million veterans expected to take advantage of benefits from the GI Bill. In addition to ex-air force pilots who would undoubtedly want to keep flying after the war, to say nothing of the thousands of returning GIs who had experienced the fascination of flight on wartime transports, many of the aircrews would probably want flying licenses. After 1945, light-plane advertisers expanded their appeals outside the aviation trade journals and attempted to reach a more variegated audience, from readers of *Business Week* to *Better Homes and Gardens*. These and other periodicals also broke out in a rash of inventive articles for both men and women on the joys of family excursions to any number of alluring places beyond the weekend range of the ordinary automobile. "Texas," promised Cessna Aircraft Company, "won't be much larger than Rhode Island." A notably upbeat article in the February 1946 issue of *Fortune* magazine ("New Planes for Personal Flying") gave an overview of proposed models and planes in production, enthusiastically proclaiming that "this year the buyer of personal airplanes has the widest choice in aviation history." The prices in *Fortune*'s catalog ranged from $1,885 for a Piper Cub to $60,000 for a Twin Beech. The article admitted that though prices were still an obstacle, costs were on the way down.

But costs went up. Early postwar market surveys emphasized a $2,000 price tag for two-place planes and $4,000 for a four-place aircraft. Above those figures, warned the market experts, there would be strong buyer resistance. And *Fortune*'s survey turned up very few planes in those price ranges. Stinson's Voyager, the lowest-priced four-place plane, was tagged at $5,000. The two-place Ercoupe, highly touted for its spinproof characteristics and simplified, co-ordinated controls (eliminating the rudder pedals), was advertised by Macy's for $2,995. Still, everybody set production records in 1946 when the light-plane industry sold 33,254 aircraft, 455 percent over the last prewar annual sales figure. Many industry observers were encouraged. With increased demand, higher volume, and mass production, prices would come down, and even more planes would be sold to all those veterans in flying schools.

Then the market went sour. Sales fell off in 1947, to 15,617 units, skidded to 3,545 in 1949, and plummeted to their nadir in 1951, when only 2,477 planes were sold. The causes were varied. The less expensive two-place planes, with a 95–100-MPH cruising speed and docile flying characteristics, were less than exciting for many ex-military fliers. There were numerous other aircraft options among 31,000 war-surplus aircraft coming on the market, including many Twin Beeches, C-47s, and twin-engine bombers that were modified as executive transports. Novice civilian student pilots comparing the style and comfort of Detroit's new postwar car models found many postwar light airplanes noisy, drafty, and cramped. Manufacturers misjudged postwar production costs; Republic Aviation's four-place Seabee amphibian, supposed to sell for $3,995, retailed instead for over $6,000, and the Beechcraft Bonanza went from an estimated $5,000 to $8,495. These planes were easily grounded in bad weather, and operating costs were discouraging for the majority of families planning to fly on weekend junkets. Moreover, few resort and vacation centers were close to convenient airstrips. In rapid succession, Cold War tensions, the Berlin Crisis, and the outbreak of the Korean conflict meant further damping of civilian light-plane sales.

As the dreams for a vast market and massive production of low-cost light planes evaporated, many companies folded, turned to other product lines, or merged with the few survivors. Among the survivors were the three industry leaders—Beech, Cessna, and Piper—who outlasted the debacle and held on until the aviation market began to revive by the end of the decade. With more realistic notions about the nature of the market, refinement and improvement of aircraft, and development of flying facilities, the light-plane industry managed a slow but stable recovery. Obviously, there was a need for a reasonable number of small, two-place aircraft as training planes, and perhaps for use in utility and pleasure flying. But the vast market envisioned in earlier projections simply was not there. As one acute observer wrote, "Mass-producing Everyman's airplane for a rock-bottom price was not the answer to growth." Instead, the light-plane industry began to focus more attention on designs for the business pilot, producing aircraft with comfort, reliability, speed, and greater range to make them an attractive alternative to other means of transportation. Aircraft that met speed and range requirements, in particular, became available in several cost and performance categories, including a new generation of twin-engine planes. Some of them were under development even as the downswing in the industry was gathering momentum.

The postwar experiences of Piper were typical of many manufacturers. The company completed its 1945 fiscal year in September with $7.7 million in sales. In the space of a few months, the backlog of orders climbed to over $11 million, dealers had to be rationed, and the three-place Piper Cruiser was being sold on the black market at several hundred dollars above the factory list price. Production was increased. When the light-plane market began its violent contractions in 1947, Piper discovered that the industry had overproduced. Many buyers had ordered from two or three companies, then took delivery on the first plane available, cancelling the others. But Piper was also guilty of continuing to market designs that had changed little in one-and-a-half decades. The comparatively small, fabric-covered planes were suitable for sunny-day flights in visual flight rules (VFR) conditions but had limited equipment for night flying or in reduced visibility.

Piper nearly went under. The company survived through Draconian economic measures and a drastic reshuffling of its management. In time, the company found its marketing niche with improved versions of the dependable Cub trainers and a Super Cub variant, whose huskier engines (90 to 135 HP) and agile, short-field capabilities made it popular for a phenomenal range of duties in agriculture, forestry patrol, surveying, mountain rescue, and other roles. The four-place Pacer, with a 150-HP engine, introduced in 1949, became the Tri-Pacer in 1951. Its tricycle landing gear simplified landing and takeoff for the business pilot and its 130-MPH performance was adequate. The series also won praise for well-equipped interiors.

Cessna competed with Piper in the training and sport-plane categories with the 120/140 series, an 85-HP aircraft in deluxe or standard versions. Side-by-side placement of its two seats gave the 120 a tidy arrangement for instructors and student pilots. The fancier 140 also provided side-by-side seating, an arrangement thought by many to be less raffish than comparable seating in tandem. More important was the all-metal fuselage of the 120/140 series, a harbinger of the future for airplanes of this class.

When the postwar sales slump hit Cessna, company managers commissioned a series of market surveys that definitely pointed to a strong market in four-place business aircraft. The company's five-passenger, all-metal 190/195 series, with 245–300-HP radial engines, had momentarily been shelved for development of the 120/140 model. The 190/195 program was quickly reactivated, although Cessna also recognized the need for a somewhat smaller plane. The result was the Model 170, precursor of one of the most successful series of light planes ever built. Developed around the 120/140 design, the four-place 170 entered production with fabric wings and metal fuselage in 1948. Within a year, the 170A, with metal wings, redesigned tail, and a single strut was introduced. This 130-MPH plane was the basis for subsequent modifications, including wing flaps and more powerful engines, until the end of the postwar decade, when the 172 with tricycle gear was introduced. Not only did the all-metal 170A/172 series offer a notable improvement in maintenance over fabric-covered planes, its 500- to 700-mile range and comfortable cabin made it a favorite of business pilots. It became one of the best-selling light planes of the era. Cessna never underwent the wrenching managerial reorganization and financial problem of Piper, although the company occasionally scrambled for liquidity. Cessna found part of the answer in diversification. In 1947, the company took on an air force contract to build wood and aluminum furniture, a venture that took up part of the slack in sales for three years. Cessna also branched out into industrial hydraulics, a profitable operation that was retained.

Beechcraft, like everyone else, entered the postwar era with gusto. Beech management assessed the American love affair with cars and hoped that American families would embrace flying with equal passion. The company had two well-proven products to start with, the Staggerwing and the Twin Beech. Designers realized that the biplane configuration of the Staggerwing, with its wood and fabric construction, symbolized a

The Cessna 170 series featured an all-metal fuselage, making it a trend-setter for planes of its class in the postwar era. Later designs incorporated all-metal wing and tail assemblies as well as a tricycle landing gear. Courtesy of Cessna Aircraft.

prewar hallmark that seemed out of place in the new era of metal monoplanes, and the Staggerwing production line finally closed down in 1948. In the interim, Beechcraft, too, felt the sales drought but weathered it better than its competitors. Demand for improved versions of the Twin Beech held fairly steady, and the company maintained its traditionally conservative fiscal habits. A comfortable surplus carried over from extensive wartime manufacturing contracts. Beechcraft's products cost a bit more, but the company catered to a narrower market than did its competitors, and Beechcraft owners were aggressively loyal. Most important was Beech's early postwar development of a successor to the single-engine Staggerwing. The manufacturer had a reputation for producing high-performance aircraft and shrewdly planned to continue this tradition with a new ultramodern design, while keeping the price to a reasonable $5,000. The new plane was called the Model 35 Bonanza, an apt sobriquet for the sales it eventually generated.

The Bonanza culminated a trend of light-plane advances dating from the depression era. Extensive use of metal skin had marked the high-winged Luscombe series sold in the mid-thirties. Beech's own Model 17 Staggerwing had featured retractable gear. The Spartan Executive of the late thirties, a remarkable design for the time, had incorporated extensive metal work, a low wing, and retractable gear, but a radial engine seemed to mar its otherwise sleek silhouette. The prewar Ercoupe had sported a fixed

In the postwar era, the general aviation manufacturers produced many new designs with advanced aerodynamic features. The "V" tail of the Beechcraft Bonanza of 1946 became a hallmark of this long-lived design. Courtesy of Beech Aircraft.

tricycle gear. The Bonanza's magnetic appeal emanated from Beechcraft's deft melding of these several attributes, including subtle improvements and unique features.

For planes of its class, the Bonanza was the first with electrically retractable tricycle gear. Beechcraft's wartime production experience facilitated the manufacture of an all-metal plane that featured flush riveting. The company also relied on wartime experience to establish more efficient fabrication and assembly procedures. The four-place cabin boasted unusually generous dimensions (extrapolated from a Mercury automobile) that easily accommodated six-foot passengers. The interior featured markedly lower noise levels, tasteful appointments, and an instrument panel laid out according to the latest concepts of industrial psychology. Radio gear and flight instruments for 24-hour and instrument flying were standard equipment. A new, horizontally opposed, six-cylinder engine that produced 165 HP was developed by Continental Motors. And there was that distinctive and unique V-shaped tail. Aerodynamicists had tinkered with V configurations before. During the 1930s, Polish and U.S. glider designers had tried it; occasional British and American wartime experiments had also investigated the idea; but Beech was the first to adopt it. A V-tail configuration decreased the total tail area of conventional elevator and rudder arrangements, reducing both weight and drag. It gained wide acceptance as a Beechcraft hallmark, which is probably why no mass-produced aircraft ever imitated it. The attention to weight and drag extended to other parts of the plane, explaining its notable performance. Early Bonanzas attained a top speed of 185 MPH, cruised at 175 MPH, but landed at under 60 MPH with full flaps. The tricycle gear aided docile touchdown, takeoff, and taxi maneuvers on the airfield. The plane's 750-mile range offered highly efficient business travel. Later models retained the low landing speed, and cruising speed, range, and passenger seating (up to six) improved over the years. The Bonanza introduced a new era in single-engine light planes and in business flying.

The early postwar era also witnessed the introduction of a new generation of twin-engine aircraft. The sleek Aero Commander first flew in 1948 and entered production three years later. The design owed much to a group of Douglas Aircraft engineers who proposed an executive twin-engine plane (or ''twin'') based on the shoulder-wing layout of the Douglas A-20 and A-26 of World War II. This configuration facilitated passenger entry and exit, since the plane sat lower to the ground. Eventually the design group formed their own production company. A pair of 260-HP engines gave the Aero Commander excellent performance at around 200 MPH, with accomodations for five to seven passengers. In 1953, Piper started producing the Apache, a 160-HP twin offering four to five passengers a more sedate cruise of 171 MPH. Additional twins appeared, including a Twin Bonanza with a conventional tail. The Model 18 Twin Beech soldiered on, with better engines and other improvements producing cruising speeds of around 215 MPH for six to ten passengers.

At the end of the decade, twins could generally be grouped into three categories— light, medium, and heavy (i.e., high performance)—based on their passenger load, range, horsepower, and speed. Prospective buyers might look, respectively, at the Apache, Aero Commander, or Twin Beech. The same categories characterized single-engine designs as well, reflected in types like the Piper Tri-Pacer, the Cessna 170, and the Bonanza, respectively. Over the years, each category encompassed numerous additional types, offering buyers a choice of planes for narrowly defined operational requirements and operating environments.

With this new range of postwar aircraft, the general aviation industry began to make its mark. Many skeptics still regarded the industry's products as marginally safe, little planes with questionable reliability over short, inconsequential hops. William P. Odom helped dispel such doubts in 1949 when he piloted a Bonanza from Honolulu, Hawaii to Teterboro, New Jersey, a nonstop flight of 5,273 miles. Three years later, Max Conrad flew the boxy little Piper Pacer on a round trip from the United States to Switzerland to visit his family. In 1954, Conrad ferried a twin-engine Piper Apache to Europe, and light-plane manufacturers began to make regular trans-Atlantic delivery flights to overseas customers rather than shipping the planes in crates for reassembly later. Such flights required extra gas tanks but convincingly demonstrated the range and reliability of light-plane designs and engines. Skeptical prospective customers began to take note. Convenient cross-country flying on instruments became a reality during the decade after VOR units were installed in the early fifties. Although the VOR system had been planned for scheduled airline use, electronics companies like Narco soon developed low-cost, lightweight equipment for the light-plane industry. Private pilots began using the VOR network, giving general aviation aircraft extraordinary new versatility for night flying and operations in messy weather that used to keep them on the ground. Improved, static-free ultrahigh frequency (UHF) radios also came on the market, and

When President Eisenhower began using a twin-engine Aero Commander for short hops, he gave the general aviation industry an important vote of confidence. The American industry also dominated in worldwide sales of general aviation planes; the Commander shown here bears Australian registry. Courtesy of Gulfstream American.

Bill Lear introduced a light-plane autopilot system that eased the strain on pilots whose flights were becoming longer and more frequent. When President Eisenhower's personal pilot began using an Aero Commander to fly the nation's chief executive between Washington and his farm near Gettysburg, Pennsylvania, it seemed to epitomize the maturity and legitimacy of the general aviation sector.

The term *general aviation* in the postwar decade replaced *private flying* in a move to get away from the aeronautical country-club image that many felt the latter term implied. By common understanding in the growing aviation industry, general aviation included all aircraft and operations not related to military flying or the scheduled commercial airlines—a bewildering variety of aircraft and flying activities. The airplanes included not only those manufactured by the light-plane industry, but gliders, home-built planes, and considerable numbers of ex-military planes converted for civil use. These ran the gamut from biplane Stearman trainers to twin-engine Douglas B-26 Invaders and Boeing B-17s. With their big 220-HP radial engines and one of the open cockpits converted into storage space, the Stearmans and other war biplanes were modified for crop dusting and seeding. Beginning in the early 1950s, several dozen Invaders were outfitted with passenger seats and customized interiors to be flown as high-speed executive transports. Others were converted as "fire-bombers" to dump fire retardants on forest fires, to carry parachuting "smoke-jumpers" to forest fires in remote areas, or to operate as spray planes.

The aircraft turned out by the light-plane industry were similarly applied to an assortment of tasks. Airborne farmers and ranchers checked on crops, livestock,

Goodyear airships, or "blimps," began carrying running-copy illuminated signs after World War II. Recent models, like the one shown here, have become familiar sights at major sporting events. Courtesy of Goodyear Aerospace Corporation.

windmills, and fences. Planes were used in cattle roundups and in hunting coyotes and other predators. State and federal agencies flew light planes on forest surveys, fishery patrols, and wildlife inventories; commercial fisheries made increasing use of airborne spotters to increase their catches. Construction and engineering firms made increasing use of light aircraft to transport survey crews, to haul supplies, and to deliver emergency repair parts and drawings to remote work sites. Helicopters, operationally proven in Korea, flew similar missions into areas with no suitable landing strips. The first commercial helicopter certificate in the world went to Bell's Model 47 in March 1946. It became the most-produced helicopter of the time, with wide application in police, Coast Guard, and miscellaneous utility work. Other designs carried a variety of loads, like telephone and electric poles to work crews installing lines in rugged terrain with no roads, and lifted bulky machinery to the tops of office buildings when existing freight elevators proved inadequate. Helicopters and planes alike saved many lives in emergency rescue work and evacuation.

Still, for all these diversified operations, the most extensive use of general aviation involved business flying. The postwar decade finally offered potential customers a variety of aircraft to suit a variety of needs and cost constraints, from the business pilot to the corporate executive who hired a pilot. Products like the Piper Pacer adequately served the insurance agent who conducted business in a 150–300-mile radius of the office; a Cessna 172 offered additional range and speed for a distributor whose business reached into nearby states. Twin-engine planes like the Aero Commander or Twin Beech allowed an east coast industrialist to accompany several other executives on flights to a regional plant over 1,000 miles away, with a company pilot handling the controls while a business planning session proceeded over built-in tables in the passenger cabin. A round of suitable beverages and hors d'oeuvres often provided the finishing touch.

In magazines and newspapers that chronicled the progress of flight between 1945 and 1955, aviation and airplanes in military and civilian roles seemed to monopolize coverage. Astronautics and rocketry, however, quietly continued a measured progress. Events of the late fifties thrust space research into the forefront, and the sixties seemed dominated by issues of the so-called space race between the United States and the USSR. The postwar decade of rocketry can be seen most clearly against the historic launches of the world's first artificial satellites in 1957–58.

6

Higher Horizons, 1955–1965

The 1955–65 era of flight was especially memorable for two major trends: the achievements in astronautics (including Earth satellites, space probes, and manned space missions), and the introduction of jet aircraft into the civil airline fleets. These developments had a pervasive influence on American life and world civilization, as well as on the nature of the aviation manufacturing—or, as it came to be known, the aerospace—industry. The airlines also faced challenges arising from sharply increased passenger demand and in establishing necessary levels of airline safety. The military services integrated a series of aerodynamically advanced fighters and bombers that featured new techniques of fabrication and production. The general aviation industry introduced its own versions of advanced designs, including pressurized aircraft and eye-catching jets.

MISSILES AND ROCKETS

If the vanguard of the von Braun team found the circumstances of their entry into the United States somewhat confusing and disorganized after the war, they found American rocket development in much the same state of affairs. Despite the 1947 National Security Act establishment of a unified military organization under the secretary of Defense, with separate and equal departments for the navy, army, and air force, essential guidelines for responsibilities of R&D and deployment were decidedly fuzzy in the nascent field of military rocketry. As a result, American missile development in the postwar era suffered from interservice rivalry and lack of strong overall coordination, a situation that persisted into the mid-1950s. The air force, with its record of successful long-range bombardment operations during the war, made a strong case for leadership in missile development. But the navy also laid its claim, by producing studies showing the capabilities of missile operations from ships and submarines. The army, in turn, viewed missiles as logical adjuncts to heavy artillery. The air force, however, had initiated long-range missile development even before the end of the war, and this momentum gave them early pre-eminence in that field.

Since American missile technology could not yet produce large rocket-propelled vehicles, the air force at first concentrated on rather large, winged missiles powered by air-breathing turbojet power plants. The air force stable of ground-launched cruise missiles possessed ranges from 600 to 7,000 miles and could carry the heavy, awkward nuclear warheads produced in the early postwar era. Eventually, the Atomic Energy Commission made lighter and less unwieldly warheads available, but the air force pressed on with its cruise missile operations, at the expense of development programs for rocket-powered intercontinental ballistic missiles (ICBMs) like the Atlas. The Navaho project represented the peak of the cruise missile fixation. Weighing in at 300,000 pounds and capable of Mach 3 flights (three times the speed of sound), the Navaho nonetheless never reached operational status. R&D costs came to $690 million before it was cancelled in 1957, when ICBM technology overtook it. The Navaho did make three successful flights, however, and certain aspects of Navaho R&D turned out to be very significant in other areas. The experience in high-speed aerodynamics was applied to other aeronautical research programs, and the missile's inertial guidance system found application in other ICBMs and submarine navigational systems. While the propulsion units for the Navaho used a jet engine for sustained flight to the target, the heavy vehicle was boosted into the air by three liquid propellant rocket engines developing 135,000 pounds of thrust each. Developed by Rocketdyne (a division of North American, later Rockwell International), variants of these power plants were tailored for air force Thor and Atlas missiles, and for army Redstone and Jupiter rockets. The rocket engines for the Jupiter played a highly significant role in the evolution of the Saturn vehicles.

In the early postwar era, while the air force developed cruise missiles, the army acquired expertise in liquid-propulsion rocketry through special projects at the White Sands Proving Ground in New Mexico. At White Sands, von Braun and the rocketry experts from Peenemuende not only made lasting contributions to American ballistic missile capabilities, they also ventured into the arena of space exploration. Besides test-firing a series of captured V-2 rockets for the army's operational experience, the German experts helped coordinate a series of upper-atmospheric research probes. One such project, known as the Bumper series, employed a V-2 as the first stage with a Wac Corporal upper stage, one of which reached an altitude of 244 miles. The last two Bumper launches took place in Florida during 1950 at the Long Range Proving Ground located at Cape Canaveral—a prelude to U.S. space launches of the future. Another major activity included the Hermes program, in which the General Electric Company worked with the von Braun team under Army Ordnance auspices. During Hermes operations, the basic V-2 rocket underwent successive modifications, increasing its performance envelope and payload capabilities and at the same time American contractors' experience in rocket technology. In the course of the program, a number of more or less indigenous American vehicles were flown. Although none of them became operational, they provided a useful exposure to rocket development for government

During the first launch of a rocket from Cape Canaveral, Florida, in July 1950, technicians photograph the lift-off of a captured V-2 topped by a WAC-Corporal second stage. Tests like these helped prepare the way for the space age. Courtesy of U.S. Air Force.

and contractor agencies alike, and one of the concepts, Hermes C-1, contributed directly to the development of the first significant American ballistic missile, the army's Redstone.

As the 1940s drew to a close, the army decided to establish a new center for its rocket activity. Although White Sands remained active as a test range, a facility devoted to basic research and prototype hardware development was needed. A site selection team finally settled on Redstone Arsenal in Huntsville, Alabama. Established in 1941 for the production of various chemical agents and pyrotechnic devices (including small, solid-fuel rockets), Redstone had all the necessary attributes: shops, laboratories, assembly areas, and ample surrounding land to assure both security and space for static-firing tests. Moreover, when compared to White Sands operations, it was more accessible to the Long Range Proving Ground at Cape Canaveral. Von Braun's group from Fort Bliss transferred to Redstone and started operations at the Ordnance Guided Missile Center in Huntsville by the close of 1950. During the Korean War, the new research center was assigned the development of a surface-to-surface ballistic missile with a range of 100 miles. Dubbed the Redstone, its first launch occurred at Cape Canaveral in 1953. The missile became operational in 1958.

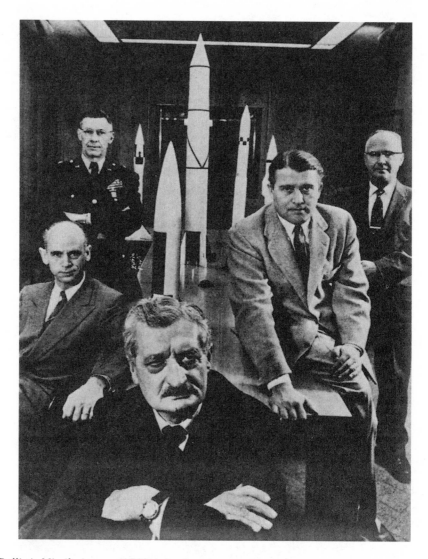

The Army Ballistic Missile Agency (ABMA) developed some of America's most successful early boosters, represented by the models shown behind Hermann Oberth (foreground), *the German rocket pioneer. When the picture was taken in 1956, Oberth was in the United States as a consultant to ABMA administrators. They are:* (clockwise) *Ernst Stuhlinger, one of the von Braun team from Peenemuende days; Gen. H. N. Toftoy, commander of ABMA; Robert Lusser, an ABMA official; and* (seated on the table) *Wernher von Braun, chief of ABMA's Development Operations Division. Courtesy of NASA.*

In the meantime, the accumulated design experience of the Redstone program contributed to a joint army-navy development program involving the Jupiter vehicle—a direct descendant of the Redstone. But the specter of catastrophe if a large liquid-fuel rocket were accidentally to explode on a ship at sea, spewing a hugh volume of volatile propellants everywhere, led the navy to proceed cautiously in the area of liquid-propellant rockets. Late in 1956, naval experts decided to concentrate on solid-fuel rockets instead. This direction of development eliminated the complex logistics and operational difficulties inherent in the deployment of liquid-propellant rockets in seaborne operations, particularly in the case of missiles launched underwater from submarines. The navy gave official authorization to its own strategic missile—the Polaris—and named Lockheed as the contractor early in 1957. Powered by a solid-fuel motor, the Polaris nevertheless borrowed the Jupiter's guidance system, evolved from prior collaboration of army and naval research teams.

The Jupiter experience gave the navy an important boost in the field of vehicle guidance and control while increasing the army's expertise as well. More important was the army's continuation of Jupiter development into a successful intermediate range ballistic missile (IRBM). The army eventually had to surrender its operational deployment to the air force when a Department of Defense directive late in 1956 restricted the army to missiles with a range of 200 miles or less. Even so, the Army Ballistic Missile Agency (ABMA) maintained a role in Jupiter R&D, including high-altitude launches that added to ABMA's understanding of rocket vehicle operations in the near-Earth space environment. This bank of knowledge was to pay handsome dividends later.

During the early 1950s, the Atomic Energy Commission perfected smaller nuclear and hydrogen warheads. These developments eventually doomed the air force's cruise missiles and placed a new emphasis on producing the Thor IRBM and longer-range missiles, like the Atlas. In the process, a new generation of managers from the air force and the Department of Defense struggled to work out financial and management procedures for the arcane world of nuclear-tipped weapons. Successful launches of the single-stage Thor and Atlas occurred in 1957 and 1958, and the air force also began work on an advanced ICBM, the Titan, a two-stage vehicle launched for the first time in 1959. The prime contractors for these vehicles were, respectively, Douglas Aircraft (later, McDonnell Douglas), the Convair Division of General Dynamics, and Martin Marietta's Denver Division. Even as the air force perfected liquid-propellant missiles like these, it kept an eye on the navy's solid-propellant Polaris. In view of the solid-propellant missile's simplicity, low cost, and capability for rapid launch, planners could disregard vulnerable liquid-propellant storage facilities, transfer pumps, and other paraphernalia, and instead plan to emplace new solid-propellant missiles in widely scattered, "hardened" underground silos. By March 1957, the air force was ready to go ahead with the three-stage Minuteman, and Boeing won the prime contract. Boeing and other major corporations, plus a host of subcontractors and suppliers, provided a broadening base for military rocketry (and ambitious plans for space exploration to come). Most of the same contractors also supplied aeronautical hardware to both the military and civil/commercial sectors. For obvious reasons the term *aerospace industry* gained currency in referring to these manufacturers. One indicator of the growing significance of astronautics in national security occurred early in 1959, when the Department of Defense revealed that fiscal year 1960 would initiate extensive integration of long-range missiles into the air force, displacing manned bombers on a major scale.

SPACE RESEARCH

The increasing payload capability of certain military missiles opened the possibility of replacing their warheads and using them as boosters to launch heavy, scientific payloads into space. The United States had already applied its growing expertise in rocket technology to the development of a family of sounding rockets to carry instrumentation for upper-atmospheric research. The navy's Aerobee and the Viking reached altitudes of between 100 and 200 miles. During the period of the International Geophysical Year (IGY) of 1957/58, many nations around the world conducted a coordinated program of sounding rocket launches, including 210 sent up by the United States and 125 launched by the Soviet Union. But the United States had an even more ambitious goal than launching sounding rockets during the IGY. America planned to orbit its first small satellite.

The satellite project began in 1955. In spite of the international spirit of cooperation inherent in 164 programs, there was a strong sentiment in the United States that it should waste no time and should win maximum international prestige with a launch during the IGY. For the booster, a blue-ribbon selection panel from the military and from industry analyzed a list of candidates that included the Atlas, the Redstone, and the Viking. ABMA argued that the Atlas was still untested. The Viking vehicle, its opponents noted, still required a program to uprate its first-stage engines and develop new second and third stages before it could become operational. However, the army's Jupiter C vehicle—a direct derivative of the proven Redstone—appeared to have all the capabilities necessary to launch a satellite. For complex reasons, the committee selected the Viking, arguing that it had been intended from the start as a vehicle for space research and that its development would not impinge on America's young ballistic missile program. But the choice of the Viking, in the context of cold war concerns over international prestige and technological leadership, was a controversial decision. The new program, authorized in September 1955 under the Department of the Navy, was to be known as Project Vanguard.

Although the first stage was successfully launched on October 23, 1957, the first Vanguard with three "live" stages blew apart on the pad, and its successor veered off course and disintegrated before it had ascended four miles. As if these two fiascos were not enough, the Vanguard had already been overtaken by events. The Russians had orbited *Sputnik I* on October 4, 1957. Within the space of four weeks, the Soviet Union then demonstrated that Sputnik was no fluke by launching a second orbital payload, *Sputnik II*, carrying the dog "Laika" into orbit in November. The potent Russian boosters threw a long shadow over the Vanguard. Plans to use an existing military booster gained support once again.

The honor of launching America's first satellite fell to the close-knit group of pioneers who had dreamed of space exploration for so many years, the von Braun team. When the army's Redstone/Jupiter candidate for the IGY satellite was rejected, ABMA assumed a low profile but kept up work that paid off not only for satellite boosters but also for the future of manned space flight. As one ABMA insider explained, von Braun found a "diplomatic solution" to sustain development of the Jupiter C by testing nose cones for the reentry of warheads. Following launch, solid-propellant motors in the second and third stages accelerated an inert fourth stage carrying an experimental nose cone. The nose cones tested ablative protection as they reentered Earth's atmosphere. After successful tests during the summer of 1957, von Braun declared that a live fourth stage and a differently programmed trajectory would have given the United States its orbiter. In any case, ABMA was prepared to put an American payload into earth orbit. Slightly more than four weeks after the launch of *Sputnik I*, the secretary of Defense finally acceded to persuasive pleas from ABMA to put up an artificial satellite, using its own vehicle. Authorization from the secretary for two satellite launches came on November 8, 1957, and the initial launch was set for January 30, 1958, ABMA missed the target date by only one day: a Jupiter C orbited *Explorer I* on January 31, 1958. The unqualified success of *Explorer I* and its successors was due in large part to the existing operational capability of the Jupiter C launch vehicle, the high flexibility of ABMA's in-house capability, and the technical expertise of the Jet Propulsion Laboratory (JPL). The JPL functioned administratively as a unit of the California Institute of Technology but got a large share of its funds through army contracts. The JPL developed the solid-fuel propulsion units for the upper stages of the Jupiter C as well as the payloads for the Explorer satellite. Within the next few months, the Jupiter C vehicles, designated as Juno boosters for space launches, carried payloads into orbit around the moon and the sun.

During the public consternation and political turmoil in the wake of the Soviet space spectaculars, the American government began a thorough reappraisal of its space program. One result was the establishment of the National Aeronautics and Space Administration (NASA) to replace NACA. Created when President Eisenhower signed the National Aeronautics and Space Act into law on July 29, 1958, NASA was organized to ensure strong civil involvement in space research so that space exploration would be undertaken for peaceful purposes as well as for defense. The period 1958–59 was one of feverish activity in space exploration. In the years that followed, dozens of payloads were launched.

Explorer I, the satellite so quickly and brilliantly launched by the von Braun team in 1958, made the first records of micrometeorite activity and led to what was considered the most important discovery of the IGY, the Van Allen radiation belts around the earth. The Explorer designation, like that of most satellite programs, applied to a series of similar satellite configurations; several dozen Explorer satellites had been launched by the mid-1960s, continuing to probe radiation belts, solar winds, interplanetary magnetic fields, and other space phenomena. Beginning in 1959, a series of Pioneer space probes investigated earth-moon trajectories, solar orbits, and extremely long range communications. Lunar exploration progressed much further with the Ranger probes (1961–65). Despite some early failures, the Rangers provided valuable experience in perfecting attitude control systems, solar power and battery power for satellites, component reliability, and other functions. Equipped with television cameras, Ranger spacecraft in 1965 impacted on the moon after yielding the first spectacular close-ups of the lunar surface. Between 1966 and 1968, five Surveyor spacecraft made soft landings

Using a modified version of the Redstone missile, named Jupiter C, the United States launched its first satellite on January 31, 1958. Inset shows the smaller second stage, topped by the pencillike Explorer satellite. Courtesy of NASA.

on the moon and carried special instrumentation to analyze the lunar environment, including the lunar surface. Data from Surveyors, in particular, helped NASA choose the sites for manned lunar landings to come.

Other satellite programs reflected a more terrestrial orientation. The launch in August 1960 of *Echo I* became a widely remembered event because the satellite could be seen by the naked eye. An inflated sphere of polyester plastic about 100 feet in diameter, *Echo I* was a passive communications satellite—a relay point against which messages bounced between different locations on earth. More indicative of things to come, the short-lived (34 days) Project Score, orbited in December 1958, tested a broad spectrum of voice and teletype transmissions and broadcast a Christmas message to the world from President Eisenhower on December 19, 1958, marking the first transmission and reception of a human voice from space. By 1962–63, Bell Telephone's Telstar satellites and RCA's Relay satellites were transmitting television broadcasts for trans-Atlantic and trans-Pacific reception. With their low orbits (a few hundred miles above the earth), they had to be tracked, and signals ceased as they passed out of sight of ground stations. A synchronous orbit, about 22,000 miles above the earth, would keep a satellite in a steady position for continuous handling of radio, television, and telephone signals, eliminating tracking systems and breaks in communication. The Syncom satellites of 1963–64 did precisely that. Built by the Communications Satellite Corporation, better known as Comsat, the Syncom series heralded the commercial status of satellite communications.

Additional satellite launches, like the Transit series of the early 1960s, introduced more accurate all-weather navigation capabilities for aircraft, maritime shipping, and submarines, including the Polaris-equipped subs. NASA's Tiros family (Television and Infrared Observation Satellite) also began operating in the early 1960s. With the launch of *Tiros I*, equipped with special television cameras, in April 1960, a new era in weather forecasting began. *Tiros I* transmissions allowed viewers on earth to see the earth's cloud cover on a global scale for the first time. From the early 1960s to mid-decade, a group of ''observatory'' satellites (the orbiting Solar Observatory, the Orbiting Astronomical Observatory, and the Orbiting Geophysical Observatory) began accumulating new sorts of information. Many satellites in the series were unique. The astronomical satellites, for example, carried special telescopes to observe sectors of deep space perennially obscured by the earth's atmosphere, and also mounted other specialized equipment for experiments in space.

In a remarkably short time, satellites became extraordinarily sophisticated in the kinds of information they could gather, record, and transmit to earth. In the early 1960s, they became an invaluable element of national defense. The role of Transit as an adjunct to the Polaris submarine system is only one example. The Midas (Missile Defense Alarm System) satellite was equipped with delicate infrared sensors to detect the launch of hostile ICBMS. The Samos (Satellite and Missile Observation System), a military reconnaissance satellite, could scan the entire globe. Another group of satellites, the Vela family of the mid-1960s, was designed to detect the explosion of nuclear weapons in space and became operational to support the 1963 Nuclear Test Ban Treaty. Many other types of military satellites were launched during the 1960s, but

The United States has long relied on military ballistic missiles to launch orbital satellites. For the 1964 launch of a geophysical satellite NASA used an Atlas ICBM first stage and a slimmer Centaur second stage. The payload section was enclosed by the aerodynamic shield housing the satellite. Courtesy of NASA.

details concerning most of them remained sketchy. As an indication of the intensity of launches with national security implications, the Department of Defense orbited 70 artificial satellites during 1965 alone, triple the number orbited by NASA in the same year.

Although other nations inaugurated space programs and launched their own boosters and scientific payloads, most public attention fastened on the manned "space race" that developed between the USSR and the United States. Within the first week of NASA's existence in October of 1958, Project Mercury was authorized to put an American astronaut into orbit, and the space agency began negotiations to obtain the necessary boosters and to select candidates for astronaut training. At that time, NASA did not have the resources to develop its own boosters for space exploration. Mission planners reached into the inventory of American ballistic missiles and finalized agreements with the army and ABMA for use of the Redstone, as well as the Atlas ICBM, to be acquired from the air force. In order to check out requirements and systems for manned orbital operations, NASA planned to employ the Redstone for suborbital launches, whereas the more powerful Atlas would be used for the orbital missions. Selections of the first seven Mercury astronauts was announced in the spring of 1959, and work proceeded on the development and testing of the Mercury space capsule. During 1960, the von Braun team (now including dozens of key scientists and engineers and several thousand skilled personnel) transferred from ABMA to NASA, bringing acknowledged skills in the development of rockets. The design and development of launch vehicles became the focus of NASA's Marshall Space Flight Center in Huntsville. With additional major facilities like the Manned Spacecraft Center in Houston, Texas (for astronaut training and manned spacecraft development), the Goddard Space Flight Center, near Beltsville, Maryland (for tracking and data acquisition and for unmanned spacecraft development), and the Kennedy Space Center at Cape Canaveral, Florida (for rocket vehicle preparation and launch), along with existing installations around the nation, NASA prepared to undertake a significant leadership in space exploration.*

Aeronautical research, an important NACA legacy, continued as a significant aspect of NASA's overall program. On the prosaic side, Langley Research Center helped to resolve the phenomenon of hydroplaning, a situation in which the wheels of aircraft landing on wet or slushy runways tended to float on the water's surface, resulting in loss of control. Narrow grooves in the runway to drain the water improved traction and proved successful for highways as well. On the exotic side, the Flight Research Center at Edwards Air Force Base continued an imaginative program of flight testing. Its most dramatic work involved the X-15 program. The rocket-powered X-15, built by North American, explored speeds and altitudes far beyond the capability of any existing aircraft. Briefly explained, the goal was to fly six times the speed of sound in level

*Following the death of President Lyndon B. Johnson in 1973, the Houston facility became Johnson Space Center. The vehicle launching complex at Cape Canaveral had several designations until 1963, when the name Kennedy Space Center was chosen to honor the assassinated president. The Cape itself was also designated Cape Kennedy, reverting to its original name in 1973. MSFC was named after the World War II leader and former secretary of state, George C. Marshall.

President Eisenhower came to Huntsville, Alabama, in 1960 to dedicate NASA's George C. Marshall Space Flight Center. With a Saturn I flight stage looming in the background, Wernher von Braun used the model at left to explain its finer points to the president. Note the display illustrating the many contractors involved in the Saturn program. Courtesy of NASA.

flight and reach heights to be exceeded only by manned spacecraft. Three planes were eventually built, and full-scale tests began in 1960. By the time the program ended in the late 60s, an X-15 had been flown to more than 67 miles above the earth, and high-speed dashes reached Mach 6.7. During high-temperature phenomena tests

The X-15 drops away from the wing pylon of the B-52 that carried it aloft. A Lockheed F-104 Starfighter cruises nearby in its role as the chase plane. Courtesy of NASA.

flights, some sections of the plane became hot enough to turn cherry red. For this reason, the X-15 structure made extensive use of exotic alloys like titanium and Inconel-X, providing invaluable information in machining and manufacturing subsequently used in the design and production of aircraft and spacecraft. In hypersonic flight at extremely high altitudes above the stratosphere, the X-15 provided invaluable data on vehicle control unobtainable by any other means. Such flights required reaction controls (small rocket thrusters) that led to successful control devices for the Mercury spacecraft and later space flight systems. Flight suits for X-15 pilots also supplied the basic design for those used by Mercury and Gemini astronauts in earth-orbital missions.

Early in 1961, a Mercury-Redstone launch from Cape Canaveral carried the chimpanzee "Ham" over 400 miles downrange in an arching trajectory that reached a peak of 157 miles above the earth. The chimp's successful flight and recovery confirmed the soundness of the Mercury-Redstone systems and set the stage for a suborbital flight by an American astronaut. But the United States was again upstaged by the Soviet Union. On April 12, 1961, Maj. Yuri Gagarin soared into flight aboard *Vostok I* and completed one full orbit of the earth to become the first human being to travel in space. Just as the Russians appeared to have overtaken the Americans in the area of unmanned space projects, they now seemed to have forged ahead in manned exploration as well. Although Alan B. Shepard made a successful suborbital flight atop ABMA's Redstone boosters on May 5, this milestone was eclipsed when Soviet cosmonaut Gherman Titov roared into space aboard *Vostok II* on August 6 and stayed aloft for 17½ orbits. Not until the following year did astronaut John H. Glenn win the distinction as the first American to orbit the earth. Boosted by an Atlas ICBM, Friendship 7 lifted off from Cape Canaveral on February 20, 1962, and orbited the earth three times before Glenn rode the capsule to splashdown and recovery in the Atlantic. These and other manned flights proved that humans could safely travel and perform various tasks in the hostile environment of space. Over the next few years, both Russian and American manned programs improved and refined their booster and spacecraft systems, including those for multicrew missions.

Such missions responded to a monumental new challenge. For the United States, it seemed to be a matter of retaining international technological prestige during cold war struggles with the Soviet Union. As the *New York Times* remarked after Gagarin's 1961 triumph, "The Soviet Union won another round last week in the psychological and propaganda war for men's minds." It did not help that, within five days of the Russian manned flight, the United States was implicated in the Bay of Pigs operation, the abortive attempt to topple the communist regime of Cuba's Fidel Castro. Even before the Bay of Pigs affair, President Kennedy had been harrying his advisers, seeking ways to upstage the Soviet Union in space exploration. If nothing else, the Cuban fiasco appears to have given additional impetus to the moon race. After a series of memos between the White House and NASA, it seemed possible that Americans could achieve the first manned lunar landing. As Wernher von Braun, full of gusto, wrote in a memo to Vice-president Lyndon Baines Johnson in late April, "We have a sporting chance." On May 25, 1961, President Kennedy committed the nation to "achieving the goal, before this decade is out, of landing a man on the moon and returning him safely to Earth." In typical rhetoric of the cold war era, Kennedy clearly intended to beat the Russians.

MILITARY AVIATION

Rapid change in astronautics and aeronautics had both positive and negative implications. In time and distance, the world was becoming a more closely knit community. At the same time, aerospace weapons annihilated the protective buffer of distance and gave little time to react to a hostile act. For the United States, the potential threat of a nuclear attack from the Soviet Union prompted not only the build-up of aircraft and missiles but also the development of sophisticated and costly early warning systems. The heavy expense was due to the extent of complex electronic technology required as well as to construction and maintenance costs, since these new defenses were situated in some of the most remote and inhospitable regions of the earth. In the air-age environment of the postwar world, the shortest routes from the Soviet Union to the United States (and vice versa) ran over the Arctic polar region. To detect hostile incoming bombers, USAF established a Distant Early Warning (DEW) Line of radars, extending from the northwest coast of Alaska eastward to Baffin Island. The DEW Line became operational in July 1957; the Soviet Union's Sputnik went into orbit in October. Although the DEW Line continued to serve a vital function in a combined surveillance network, including missiles and aircraft assigned to the North American Air Defense Command (NORAD), Sputnik clearly demonstrated the potential of the growing threat of intercontinental nuclear missiles. Thus, more electronics and more dollars were poured into the Ballistic Missile Early Warning System (BMEWS), which began operations in the fall of 1960 at a base in Thule, Greenland. Additional radar networks, on the ground, backed by radar planes aloft fed information into NORAD's computers and color-coded display screens. In an effort to safeguard this sensitive nerve center from a nuclear blast, USAF spent approximately $100 millon to blast out a rocky bastion deep inside Cheyenne Mountain, near Colorado Springs, where NORAD moved during 1965–66. USAF's growing share of national defense responsibilities was further exemplified by the establishment of its own school, the Air Force Academy. During the summer of 1955, the first class of about 300 cadets entered the academy, which used temporary quarters at an air force base in Denver while a futuristic-looking campus was being built in nearby Colorado Springs. The swift pace of changes in aerospace weaponry gave all military personnel much to study.

In the aftermath of Korea, a new generation of fighters evolved, most of them featuring the sweptwing configuration typical of the MiGs and Sabres that had clashed over the Yalu. During the Korean War, the U.S. Navy painfully recognized that it had no fighter capable of meeting planes like the MiG-15. The Gruman Cougar, a sweptwing variant of the Panther, began to reach carrier squadrons in 1952, but the navy still needed a truly advanced design with supersonic performance. The most important step in this direction was the Chance Vought F-8 Crusader, first flown in 1955. The Crusader's designers set out to build a plane that would fly well above the speed of sound yet possess slow flight characteristics compatible with carrier-deck takeoff and landing. In search of speed the designers used NACA's "area rule" concepts (pinching the fuselage waist) in designing the fuselage to reduce transonic drag. For low-speed control, the designers equipped the Crusader with a shoulder-mounted, variable-incidence wing, permitting the pilot to raise the entire leading edge. This meant that the pilot could make a low-speed landing approach with the fuselage and cockpit almost level for excellent visibility of the carrier deck, while the elevated leading edge developed a high lift. The Crusader was the first carrier plane to fly faster than 1,000 MPH and, flown by a then-unknown pilot named John Glenn, became the first plane to streak from the Pacific to the Atlantic at supersonic speed. Further developed for reconnaissance as well as fighter duties, the Crusader equipped nearly half of marine and naval fighter squadrons at the peak of its operational life.

An even more advanced fighter, and one of the most versatile and successful military jets of the post-Korean era, was the McDonnell F-4 Phantom. The Phantom, originally a long-range attack fighter for the navy, carried a two-man crew and was powered by a pair of high-thrust jet engines that gave it a Mach 2.25 maximum speed. The outer panels of the swept wing were given a pronounced dihedral for enhanced aero-dynamics, and the horizontal tail was angled downward for improved flight control.

The result was an angular, lethal-looking aircraft that became one of the best all-around attack fighters of its day. Following the first flights in 1958, the Phantom was equipped with "blown" wing flaps (using blasts of engine air to augment their lift) for landing. It boasted an elaborate array of electronic paraphernalia for fire control of rockets and guns for assorted weaponry and for miscellaneous combat requirements. As an interceptor, it established over a dozen world records for speed and rate of climb; as an attack plane, its bombload exceeded that of the B-29 of World War II.

The Phantom's impressive performance and versatility caught the eye of USAF officers, becoming the first production naval fighter to win air force contracts. Later versions were ordered by the RAF and by Israel, and a considerable number were sold to other air forces in Europe, Asia, and elsewhere. Although these sales were not the first to NATO, SEATO, and foreign nations, exports of American military aircraft represented a growing phenomenon. The size and complexity of supersonic aircraft and the growing sophistication of their electronic subsystems made them unusually costly in terms of R&D and production. Russia and America, the two major protagonists of the cold war era, dominated in the field of advanced fighting machines, although England, Sweden, and France developed a respectable number of tactical and fighter aircraft, and England produced its own heavy jet bombers in the late 1950s and early 1960s. Still, export of the Phantom and other military types unquestionably represented a substantial share of the American aircraft industry's strength and constituted an influential element of foreign policy in American commitments around the world.

Development of the navy's hot new fighters paralleled a new generation of aircraft carriers, beginning with the *Forrestal* class, the first of the postwar types designed and built for jet aircraft operations. The *Forrestal* took to the seas in 1956, followed by the new carriers *Saratoga, Ranger,* and *Independence* at approximately one-year intervals. Four additional supercarriers were designated by the *Kitty Hawk* class, and the navy's first nuclear-powered carrier, the *Enterprise,* was commissioned in 1961. The *Forrestal* and *Kitty Hawk* class displaced 60,000–65,000 tons and carried 60–90 aircraft, and the 86,000-ton *Enterprise* could handle 100 planes. The big new carriers incorporated a number of innovations, many pioneered by the British, such as angled flight decks, mirror landing systems, and steam catapults, all of which permitted a launch rate of four aircraft per minute. Progress in the design of lighter and more compact nuclear warheads meant that carrier-based aircraft like the Douglas A-3 Skywarrior (a twin-engine shipboard bomber) and the Douglas A-4 Skyhawk (a small, agile, single-engine attack plane) could deliver atomic bombs as well as conventional ordnance. U.S. Navy flattops also carried a variety of specialized planes, such as the piston-engine Grumman S-2E Tracker, for antisubmarine warfare, and the jet-powered Grumman A-6 Intruder, an all-weather and night-flying strike and reconnaissance aircraft. Still other types, with a panoply of sophisticated electronic equipment, were designed for airborne early warning duties and electronic countermeasures.

Like the navy, the USAF pressed ahead with R&D on a new family of military aircraft in the post-Korean era, intended to maintain air superiority against opposing fighters and to blunt the awesome threat of Russia's rapidly growing long-range strategic bomber fleet. The USAF additionally moved to improve its aerial weaponry

Originally designed as a navy fighter, the McDonnell F-4 Phantom enjoyed a long service life, operating aboard carriers from 1961 into the 1980s. Here, an F-4 is launched by steam catapult. Courtesy of McDonnell Douglas Corporation.

for low-level strike aircraft as well as strategic bombers. In terms of fighter and tactical aircraft, the post-Korean designs evolved as the so-called Century series, based on the air force designation numbers starting with the F-100.

The F-100 Super Sabre, first of the Century series, was designed by North American to incorporate lessons learned during combat in the skies over Korea. Since the F-100 was the first U.S. production fighter intended to fly faster than sound for sustained periods, its several innovative characteristics included use of titanium in the airframe because of that metal's lightness, strength, and heat resistance at high Mach speeds. First flights occurred in 1953, and operational units had the plane a year later. The operational regimes of the new supersonic fighters often crossed into aerodynamic conditions that still held nasty surprises. One noticeable feature of the F-100 and many other supersonic aircraft was the low-set horizontal stabilizer, situated to control a condition at transonic speeds known as "pitch-up", a savagely destructive phenomenon that could quickly break up the aircraft. Nonetheless, the world's first supersonic fighter experienced a series of unexplained crashes in 1954, forcing the USAF to ground all Super Sabres. Intensive flight tests at Edwards Air Force Base finally pinpointed unanticipated control problems during high-speed maneuvers, leading to redesigned rudder and tail surfaces and an increased wing span.

The F-100 was not the only plane of the Century series to experience teething troubles. The Lockheed F-104 Starfighter had severe development traumas and finally attained its best reputation (as the F-104G) in the multimission role of tactical support and reconnaissance, rather than the interceptor role for which it was originally designed. Other Century aircraft of note included the Republic F-105 Thunderchief, which could fly into combat with an awesome array of 4,000 different combinations of weapons stores, and the Convair F-106 Delta Dart, derived from the F-102 Delta Dagger, which could initiate a "zoom climb," arching up to 70,000 feet in the thin upper atmosphere of Earth to attack hostile bombers. Aside from their area-ruled fuselages and delta-shaped wing planforms, the F-102 and F-106 were important as the first interceptors produced with no fixed guns, relying entirely on air-to-air missiles. In fact, the evolution of these distinctive aircraft had originated with an air force design contest in 1950 calling for an advanced, automatic weapons system based on homing missiles. Hughes Aircraft developed an electronic fire-control system first, and then the USAF solicited proposals for the aircraft to employ it. The F-106 pilot became the monitor of a complex electronic weapons system. Once airborne, the F-106's digital computer processed data from ground equipment that guided the interceptor toward an intruder until the plane's own radar picked up the target. Then, the aircraft's internal electronic gear tracked the quarry, kept the plane locked on the target, and automatically fired its missiles (such as infrared-seeking homing Falcon rockets) at the optimum moment. Air warfare in the aerospace age became still farther removed from the pilot-to-pilot combat of earlier eras.

The size of such aircraft, the complexities of their construction and manufacture, and their sophisticated and extensive electronic gear made them increasingly expensive. Military planners began to put more emphasis on multipurpose aircraft as a means to control spiraling budgets. These factors influenced the USAF decision to buy the navy's F-4 Phantom (originally designated the F-110 until all the U.S. services adopted a unified numbering system in 1962). Various Phantoms subsequently flew as interceptors, air superiority fighters, strike and close-support aircraft, and reconnaissance planes. Although combat Phantoms generally mounted fixed guns, offensive armament emphasized combinations of rockets and other ordnance. For this reason, the Phantom and similar aircraft carried a "weapons officer" in addition to the pilot, with the former monitoring the weaponry as well as tracking targets with on-board radar and handling electronic countermeasure equipment.

The size, complexity, and sophistication of fighter types carried over to bomber aircraft as well. One of the most remarkable types of the era was the Convair B-58 Hustler, an apt nickname for the world's first supersonic bomber. Features of the Hustler included an area-ruled fuselage, a streamlined delta shape, and four powerful engines that enabled its three-man crew to blaze over targets at Mach 2. The B-58, which entered SAC squadrons in 1960, was conceived as a weapons system, which meant that engines, airframe, bombs, and auxiliary equipment were all developed concurrently as a total concept. A unique feature of the Hustler was a detachable,

The Convair F-106 was one of the first delta-winged fighters to enter service. It featured a "pinched" fuselage. Courtesy of General Dynamics.

streamlined pod, slung under the fuselage, which carried additional fuel in addition to electronic countermeasure equipment, weapons, and reconnaissance gear, depending on mission requirements. Although one of the most advanced aircraft of its type, the B-58 was retired in 1970. Because it was designed for high altitudes and a high-speed dash over the target, the B-58 became vulnerable when high-altitude, nuclear-tipped antiaircraft missiles appeared on the scene. With a comparatively restricted bomb-bay capacity, the B-58 also seemed inefficient for selected missions at low levels to deliver conventional ordnance. The plane proved costly to maintain and operate and possessed a tactical radius of only 1,200 miles. Its place was taken by a slower, larger bomber—the Boeing B-52 Stratofortress—an older plane that still seemed to offer more flexibility in a combat environment dominated by extensive automation, sophisticated electronic offensive and defensive weapons, varied bombloads, and nuclear armaments.

Boeing designed the B-52 as an eight-engine, strategic bomber with a 6,000-mile range that could easily be doubled with air-to-air refueling. The plane's cavernous bomb bays could carry 20,000 pounds, twice the capacity of its immediate predecessor, the B-47. Of the many variants of the B-52s produced, the B-52G and B-52H represented the apotheosis of Stratofortress modification and development. The G type could carry a trio of Quail decoy missiles internally and still maintain its offensive punch with two thermonuclear Hound Dog missiles, one carried under each wing. The Quail and the Hound Dog themselves were like miniature aircraft powered by turbojet engines. The B52H was originally intended to carry an even larger air-launched

The B-52, introduced in 1955, remained operational into the 1980s and was produced in several versions. The B-52H shown here flies over Edwards Air Force Base, the test center originally known as Muroc. Courtesy of U.S. Air Force.

weapon, a rocket-powered vehicle called the Skybolt. Great Britain was intensely interested in the Skybolt as a means of extending the operational lives of its own aging jet bomber squadrons, but development problems and budgetary concerns led the Kennedy administration to cancel it. Britain had to rely on nuclear-armed submarines instead, although the USAF found it feasible to use the B-52 in low-level missions and to equip the plane with other types of advanced weaponry. At high altitudes, jet bombers were easier to detect and subject to attack with defensive interceptors and/or ground-to-air missiles. Hence the appeal of the long-range Skybolt, launched outside the enemy's defensive perimeter. As an alternative, bombers could fly "on the deck," making a low-level penetration to avoid defensive radar as long as possible. USAF planners foresaw this option in the mid-fifties, so that the B-52G and H models were extensively modified to cope with aerodynamic forces expected from low-level turbulence and equipped with terrain-avoidance radar and specialized electronic countermeasure (ECM) equipment. The B-52's capacious bomb bays were fitted with rotary launchers to carry a mix of short-range attack missiles (SRAMs) and subsonic-cruise armed decoys (SCADs). The rocket-powered, supersonic SRAM scooted along at treetop level, whereas the winged SCAD was programmed to fly and appear just like a B-52 on the enemy's radar sets.

To gather information about potential targets and to keep abreast of a potential aggressor's intentions, the USAF depended on a new type of plane for aerial recon-naissance. A curious press quickly tagged the aircraft "spy planes." Details about the planes were scarce until one of them, the U-2, figured in two major diplomatic events, the first in 1960 and the second in 1962. The U-2 had first flown in 1955 but remained cloaked in secrecy until May 1, 1960, when it was shot down over the U.S.S.R. The single-jet U-2 was slow, but its unusually long, tapered wing enabled it to cruise for hundreds of miles in the thin atmosphere of 80,000 feet, a region thought to be safe from Soviet antiaircraft missiles. Cameras and a sophisticated array of electronic gear crammed the plane's fuselage on its clandestine forays over Russia. The U-2's loss came as a rude surprise and was one of the factors that caused cancellation of a summit meeting between Eisenhower and Krushchev. Although the furor surrounding the incident caused Eisenhower to forswear further aerial reconnaissance flights across the USSR, the United States felt compelled to attempt surveillance through satellites and airborne electronic eavesdropping around the periphery of the Soviet airspace. Al-though some U-2 aircraft made high-altitude scientific flights to explore cosmic radiation and clear-air turbulence, the plane's principal role continued to be reconnaissance.

Exact figures remain vague, but out of some 25 U-2 aircraft built by 1978, 9 planes had been lost, presumably to hostile action. The United States lost one U-2 during the Cuban Missile Crisis of 1962. That October, a U-2 flying a surveillance mission over Cuba had returned with photos showing alarming evidence of Soviet surface-to-surface missiles capable of reaching many strategic and industrial targets in the United States. The missile emplacements radically unbalanced the prevailing "status quo," repre-sented a menacing escalation of offensive Soviet power in the Caribbean, and violated understandings that President Kennedy assumed the Russians had accepted. During subsequent flights over Cuba, a surface-to-air missile brought down a U-2, killing the pilot. The confrontation over the Cuban-based missiles brought America and the Soviet Union to the brink of war. The United States put its military forces on alert, SAC bombers maintained a continuously airborne status, and USAF combat units around the world assumed a state of combat readiness. A naval blockade of Cuba stopped further missile deliveries, and Russia agreed to withdraw other missiles from the island. The crisis subsided, but the affair dramatized the continuing role of aerial surveillance (by both aircraft and satellites) and the fragility of global security in a world equipped for destruction of its population through intercontinental jet bombers carrying nuclear weapons, plus hundreds of missiles tipped with nuclear warheads.

The development of military capabilities in the post-Korean era obviously depended much on advances in electronics and miniaturization of components, and this aspect represented a growing chunk of the costs of military aircraft. There were important incremental advances in fan-jet engines (borrowed from the civil sector) that equipped the B-52H. The B-52G and H models also incorporated the ''wet wing,'' resulting in considerable weight savings, increased fuel capacity, and longer range. Integral fuel tanks were not unknown in the 1950s, but the elephantine proportions of the B-52's 185-foot wing spread presented an imposing number of engineering challenges. Excess material in the aluminum wing panels was sculpted away by numerically controlled milling machines, followed by chemical bath to etch metal surfaces to smooth, close tolerances. When the wing was fully assembled, the interior was covered with a special sealant that resisted leaks even when the wings were flexed through an arc of 30 feet in flight. In a constant battle to reduce weight while maintaining safe margins of structural rigidity, aerospace engineers developed ''sandwich panel'' construction, in which sheets of thin metal skin were bonded to lightweight honeycomb cores. This required extensive R&D into materials suited to metal bonding and specialized techniques for ''curing'' the bonded skins to assure their structural integrity. These lighter fuselage and wing panels eliminated the extra weight and riveting associated with conventional skin-and-stringer construction. The Convair B-58 was one of the first high-performance service planes to make extensive use of bonded honeycomb panels, a technique that found increased application in both military and civil aircraft of the era.

The long, grueling missions of military aircraft at extreme altitudes put stringent burdens on aircrew confined to cramped and spartan cockpits. A good deal of time and money was expended in ergonomics, or human-factors engineering. Crew selection included batteries of psychological tests to ensure mental and emotional stability of aircrew subjected to prolonged stress during simulated combat runs and actual alerts—with thermonuclear warheads aboard. Additional research did as much as possible to design cockpit seats, controls, and instrumentation to provide a modicum of comfort and keep the work load under control. Still, the confinement of the cockpits, the heightened tempo of decisions required in supersonic aircraft, the increasing incidence of cues and stimuli from instrumentation—all contributed to crew fatigue as a consistent problem in high-speed, jet-age operations.

Improvements in pressurized cockpits, oxygen equipment, and G-suits necessarily followed the enhanced performance of faster, higher-flying aircraft. But the implicit dangers in bailing out of these high-performance planes loomed as an especially serious problem. A series of air force balloon flights in the mid-1950s (the ''Man High'' tests) took several volunteer officers up to 99,000-foot levels to test high-altitude oxygen apparatus and procedures, and a series of Excelsior experiments perfected parachutes to stabilize free-fall descents from over 70,000 feet. In 1959, Capt. Joseph H. Kittinger, Jr., made the final test leap from an altitude of 102,800 feet, concluding a daring sequence of feats that won him a Harmon Trophy. Other dangers in bailing out also demanded attention in the mid-1950s. Distressed by deaths and injuries from aircraft crashes, and concerned about the effects of wind blast and wind-drag deceleration during high-speed bailout, Col. John B. Stapp planned and developed a rocket-powered sled, running on a track, to test ideas and equipment. Fatalities and crash-induced injuries were reduced by using a redesigned system of body harnesses and nylon webbing suggested after an exhaustive and dramatic series of experiments, in which Stapp himself frequently insisted on riding the sled. Stapp also insisted that he make the even more dangerous wind blast and wind-drag deceleration experiments. Pilots used rocket-assisted ejection seats to escape crippled combat aircraft, but special protection was required at high speeds to avoid severe injury: torn eyelids, lungs and stomach overdistended by the rush of air into the mouth, fracture and dislocation of arms and legs, and possibly fatal G-loads to the body from wind-drag deceleration. Using the rocket sled again, the intrepid Stapp survived repeated, punishing tests and helped to design the special gear required to eject successfully from a supersonic plane. Like Kittinger, Stapp also received several awards for these courageous contributions to aerospace science.

THE AIRLINES

As America moved ahead in the reach into space, a second revolution unfolded in commercial air transportation—the development of jet-powered airliners. In the early postwar years, the British had the lead in gas-turbine engine development, although American progress in aerodynamics and airframe design moved impressively ahead. In time, improved U.S. jet engines consolidated the American lead in jet aircraft design and technology. Frank Whittle's wartime work on jet engines gave the British a consistent one- to two-year lead over American companies for about a decade after 1945. On the other hand, American airplane manufacturers maintained a similar advantage in design over their British counterparts. This was especially evident in the development of advanced, sweptwing military planes like the F-86 fighter and the B-47 bomber, as well as the larger B-52, an intercontinental bomber that first flew in 1952 and began entering service three years later. The increasing experience with jet engines and large jet-propelled aircraft might have led to earlier introduction of jet airliners, except for lingering doubts about jets in commercial operations.

American airline executives remained cautious about the economics of gas-turbine propulsion systems. Jet engines were essentially simpler than piston types, but high running temperatures, requiring costly metal alloy components, raised questions about the jet's reliability and longevity. Although military jets had been operating for years, they were not flown on the tight timetables and virtually continuous operation that characterized airline schedules. Fuel consumption for jet engines remained much greater than for piston engines, implying high operating expenses, and the initial low thrust at takeoff speeds required longer runways. True, the speed of jet aircraft was attractive, and technological advances in engine design would probably resolve many problems, but most U.S. airlines and the transport manufacturers preferred to watch and wait. In the meantime, there was an interesting alternative: the turboprop.

As the name implies, the turboprop used a gas-turbine engine to turn a propeller through a gearbox linkage, using the turbine exhaust for additional thrust in the process. This arrangement was judged to have better propulsion efficiency than the turbojet, better fuel consumption, and less noise than a piston engine. Moreover, the turboprop seemed to be better suited to the speeds (up to 400–450 MPH) predicted for airliners of the postwar era. The British again led the way, developing an outstanding turboprop engine, the Rolls Royce Dart, first ground-tested in 1946 and then test-flown in the nose of a Lancaster bomber in 1947. The first Dart was rated at 900 shaft HP and uprated to 1,480 HP by 1953, with later versions developing over 2,000 HP.

In the United States, airline officials and manufacturers agreed that turboprops were the next step and planned to reequip existing designs, like the DC-6 and the Constellation, with American-built engines. But American engine manufacturers failed to develop suitable turboprops soon enough, and the first such plane to enter service on a U.S. airline was the British-built Vickers Viscount, equipped with four Dart power plants, ordered by Capital Airlines in 1952. (When Capital went out of business later, its Viscount fleet was acquired by United Air Lines.) Lockheed's turboprop Electra did not enter service until 1959.

Meanwhile, the British once more paced airliner development, leading the way with the de Havilland Comet, the first jet airliner. Scheduled jet airline flights began when the *Comet I* inaugurated international service from England to Africa and the Far East in 1952. But after a series of tragic accidents over the next two years, all Comets were grounded. The problem was traced to metal fatigue, induced through microscopic cracks around rivet holes and aggravated by repeated pressurization cycles in flight. Despite this fault, which was eventually rectified, the Comet had been extremely popular with passengers and had proven to be profitable. These facts were not lost on American operators. Pan Am, by 1952, had already placed orders for a larger version of the airplane, the *Comet III*, to have been available by 1956, but the investigation of the Comet crashes upset the timetable. By the time the Comets were ready for service again, U.S. airlines could buy American-made equipment that promised better performance and profitability.

In 1952, the same year Pan Am placed an order for the Comet, U.S. manufacturers decided the time had come to produce an American jet airliner. Several factors converged in the same year to convince American companies that jet liners were feasible. As axial-flow engines improved, their fuel consumption decreased to reasonable levels for airline operations. Industry leaders became especially interested in Pratt & Whitney's J-57, which took to the air in 1952 with the Boeing B-52 bombers. Also, engineers were coming to grips with problems associated with the design of large subsonic planes, including aspects of low-speed control on sweptwing designs. Boeing in particular had amassed considerable expertise during the B-47 and B-52 programs. Military spending as a result of the Korean War yielded profits that Boeing could allocate to commercial jets. Besides, as the B-52 neared production, design teams became available for an airliner project.

Boeing started by designing a jet that would serve as a fast tanker for air-to-air refueling of the B-47 and B-52, and planned to use the same jigs and tools to produce the contemplated 707 airliner. When Douglas presented its competitive design, the DC-8, with a larger cabin for more passenger comfort, Boeing had to rework the 707 to a configuration larger than its tanker. This meant additional costs for new jigs but resulted in a more successful plane that eventually outsold the Douglas rival. Work on the 707 progressed rapidly. The prototype made its first flight in 1954; Pan Am placed

From its first flight in 1954, the Boeing 707 ushered in new standards for airline travel and helped the United States dominate the world market for jet transports. Courtesy of The Boeing Company.

first orders for the plane the next year; and the 707 went into service on the New York–London route in 1958. Boeing thus got its jet airliner into operation ahead of its competitors and captured the lion's share of the market. The DC-8, which first took to the air in 1958, entered service with United and Delta during 1959. Douglas saved time by not building a prototype, though modifications during the production process proved costly. The DC-8 was a popular and successful design, with over 550 planes produced, although 707 production eventually reached the 800 mark. During the early sixties, Convair produced a similar design, the 880/990 series, but production fell far short of the Douglas and Boeing rates, forcing Convair to cease production after completing 102 planes. Meanwhile, a redesigned *Comet IV* went into service with BOAC in 1958 and, in a race for prestige, made the first jet passenger flights across the North Atlantic three weeks ahead of Pan Am. But the American jets boasted better range and speed and in fact dominated jet sales to airlines around the world.

The impressive speed and additional comfort of passenger jets dictated early adoption of a jet transport for presidential travel. A Boeing 707 was delivered in the spring of 1959 and eventually assumed an official designation that was quickly recognized at major international airfields: *Air Force One*. On its maiden flight from Washington to Europe for talks with leaders of Germany, Great Britain, and France, Eisenhower fell in love with the plane and with jet-age travel. Later in the year, the flexibility enabled by *Air Force One* prompted Eisenhower to embark on a news-making tour through the Middle East to India and back, visiting eleven countries in eighteen days. From a yearly average of 120 hours of flying time covering 30,000 miles the president's flight time in 1960 increased by 62 percent (193 hours) and mileage by 262 percent (78,677 miles). When Eisenhower was planning a trip to the Soviet Union in the spring of 1960, the CIA (apparently without the president's approval) installed special cameras in the plane's belly in order to get close-up views of Russian industrial sites, airfields,and other facilities. The U-2 incident, however, forced cancellation of the trip, and the cameras were removed after Kennedy took office in 1961.

Each incumbent made changes to the interior of the presidential plane. Kennedy, for example, commissioned the internationally known industrial designer Raymond Loewy to redesign the plane's interior markings, and which have not been changed. Johnson added an impressive array of communications equipment, along with a large desk that could be lowered to serve as a coffee table. Additional jets joined the White House entourage for the vice-president and other officials; helicopters also came into use. The flexibility and speed of aerial transportation became integral perquisites of the American presidency.

Not only did jets become the preferred mode for presidential travel, but their popularity grew with airlines in the United States and abroad. The big, four-engine jets quickly replaced the piston-engine aircraft on long-distance domestic and inter-continental routes. Once the step was made to introduce jets on these scheduled routes, and they proved to be profitable in operation, a second design trend evolved toward smaller jets for regional service and short-haul stages. The pacesetter in this case was the French-built Caravelle, a graceful plane that entered service on European routes in 1959 and was introduced in the United States in 1961 by United Air Lines. The Caravelle pioneered the "clean wing" design, mounting two engines, one on each side of the rear section of the fuselage. This engine layout (and comparatively lower cruising speeds of 450–490 MPH) contributed to a much quieter cabin, making the plane especially popular with passengers. The clean wing also permitted the in-stallation of flaps along the full length of the wing's trailing edge, which enhanced takeoff performance and control at low speeds. These features influenced the design of a larger and faster American jet airliner, the Boeing 727. For additional speed and capacity, Boeing used more powerful engines and made the plane a trijet, mounting the third engine in the tip of the tail. Using a 707-type cabin, Boeing also kept the overall price at a very competitive figure. The result was an unusually versatile airliner with performance that got it in and out of many airports unable to handle other large jets. The 727 began service in 1964, and in only six years it became not only the fastest-selling jet transport but the biggest-selling, being the first jet airliner to pass the 1,000 sales figure. By the mid-seventies, the 727 was used by about 60 airlines all over the world.

American leadership in worldwide sales of jet airlines was significant. Production of engines influenced employment on the East Coast just as airframe production became a significant factor along the Pacific seaboard. In nearly every state, companies manu-factured components, subassemblies, accessories, and electronic gear, all of which were funneled to the growing jet transport industry. While much the same pattern had been true for military aircraft production, the civilian market was less perturbed by the peaks and valleys of military procurement and represented an increasing share of the

American aviation industry's dollar value. More important, the worldwide sale of American aviation products—to say nothing of spare and replacement parts—became an important contributor to the American balance of trade. Added to this was the prestige factor. At nearly every major airport of the world (with the exception of Communist bloc countries), loading ramps were crowded with American-built planes, even though they flew in the livery of dozens of foreign flag lines. Aviation symbolized advanced technology, implying strength and leadership in world affairs. At international airports, the American presence was much in evidence.

Some dull spots marred the bright picture. Even before jets appeared on the scene, air traffic around major hubs had begun to clog the airway system, creating serious problems in maintaining a smooth flow of landings and takeoffs. Poor weather aggravated the problems. The CAA often seemed ineffectual in coping with difficulties generated by growing air traffic, even though electronic aids like Ground-Controlled Approach (GCA) could have helped.

GCA saw limited use in the final months of World War II and received wide praise during the Berlin Airlift. The system used a pair of radar sets: a long-range set allowed ground controllers to guide pilots into the landing pattern, and a high-resolution set facilitated controllers in "talking" the pilot down to the runway. The GCA system proved highly successful, and the military services relied on it at bases around the world. Paradoxically, while military fliers gratefully accepted GCA, airline pilots balked, refusing to submit approach and landing procedures to ground-based personnel—"clerks," as the pilots called them. Airline pilots preferred a radio method known as the Instrument Landing System (ILS). Work on ILS equipment dated back to the 1930s; but even when the CAA began to equip selected airports with it in 1947, ILS left much to be desired. The equipment used one radio beam (the "localizer") to aid pilots in lining up with the runway; a second beam (the "glide slope") closer in, transmitted a second signal at a shallow angle for final approach. On a dial in the cockpit, a pair of intersecting needles gave the pilot visual references to fly the plane down the glide slope to landing. But the ILS proved sensitive to weather changes, especially snow and rain. At times when pilots needed accurate instrument guidance most, the ILS fluttered and fluctuated maddeningly.

During the postwar era, the CAA never quite overcame airline pilot's stubborn preference for ILS over GCA, which contributed to growing air traffic congestion, to say nothing of delays and inconveniences for airline patrons. Despite administrative changes, the CAA remained in a weak administrative position—buried too far down in the Department of Commerce, which seemed preoccupied with its traditional concerns of business relations, maritime interests, and ground transportation. The CAA was also plagued by budget cuts during the Eisenhower administration's drive for frugality. Aviation services and future planning suffered reduction when air travel reached unprecedented levels of activity. Some dedicated professionals within the CAA managed to cope through careful and efficient administration. Too often, however, cronyism and placidity prevailed in the face of escalating aeronautical needs, particularly when the increased tempo of jet-powered flight operations complicated the situation.

On a nightmarish day in the autumn of 1954, remembered as "Black Wednesday," dirty weather over New York fouled up inbound traffic and kept 300 airliners milling around in the air until air controllers sorted them out. A worse tangle occurred two years later, when spring storms over the East Coast forced planes into layered stacks over New York and Washington. It took control-tower operations 12 to 14 hours to unravel the snarls; some 30,000 passengers throughout the East and Midwest were forced to cancel reservations; many planes diverted to other airports; and Eastern Airlines struggled for three days to restore order to its scattered fleet. These monumental air traffic jams raised ghastly specters of packed airliners colliding in mid-air. Eventually, they did.

The first accident occurred during daylight in 1956. A United DC-7 on a transcontinental route was flying over Arizona at the same time a TWA Constellation assumed the same track and altitude. The two eastbound planes collided at 21,000 feet over the Grand Canyon and plunged into a remote area of the great chasm, killing all 128 passengers. Four years later, a second major tragedy occurred on a stormy winter morning over New York. A United DC-8 inbound for Idlewild Airport struck a TWA Constellation approaching La Guardia Airport. Despite a radar controller's frantic warning, two radar blips inexorably merged, then vanished. None of the 128 passengers survived, and 11 people on the ground also died.

In the shocked aftermath of the Grand Canyon crash, the CAA installed new long-range radar equipment and established more intensive surveillance of air traffic over the United States. Administrative repercussions also followed. By the summer of 1958, a new Federal Aviation Act set up the Federal Aviation Agency (FAA) totally apart from the Department of Commerce. President Eisenhower named Elwood P. Quesada, a blunt, experienced air force general, as the FAA's new chief. The agency assumed sole responsibility for ordering the nation's airspace and acquired new clout in setting air safety standards. The New York collision of 1960 underscored new air traffic control challenges in the jet age. Faster turbine-powered airliners had to maintain lower operating speeds within approach corridors converging on airfields. Accordingly, the FAA mandated slower speeds in terminal areas, beefed up electronic traffic control paraphernalia, and reworked procedures to improve coordination between radar controllers and jet transport pilots. Airline experts claimed that airline travel of the early 1960s surpassed train travel in safety and proved much safer than driving a personal car. Nobody could guarantee complete travel safety, but the airline industry nevertheless amassed an enviable record.

NEW STYLES IN POSTWAR AIR TRAVEL

Air travel began to reflect a host of changes and social effects that had been gathering momentum since the end of World War II. In 1952, air travel still carried the tang of adventure. *Harper's* sent the well-known essayist Bernard de Voto across the country in a DC-6 to chronicle a phenomenon of growing significance for American travelers. De Voto turned in a generally favorable report, complaining about nondescript service on the ground but praising service aloft, where passengers were still treated like celebrities. When Dwight D. Eisenhower assumed office in 1953, domestic airlines were already rolling up three billion annual passenger miles more than railroad Pullman travel. By the end of the year, airliners had decisively pulled ahead as the prime mover for American travelers making trips of over 200 miles. Aviation became the glamor industry of the mid-fifties, attracting extensive press coverage of new routes, new planes, peripatetic personalities, and esoteric changes in fashions worn by steward- esses. Inexorably, the airlines displaced the premier role of ocean liners as the preferred mode of transoceanic travel. In 1958, over one million passengers bound for Europe chose the airways, surpassing ocean liner patronage for the first time; as fast-flying jets entered service, trans-Atlantic air traffic jumped to six million annually within a decade. In a comparatively short time, airlines had usurped the role of both railroads and ocean liners as the principal ''people movers'' in long-distance American travel.

But passengers, having arrived safely on the ground, faced other obstacles. The early postwar surge in aerial travelers caught the airlines unprepared. Airports across the country developed reputations for littered waiting areas, odoriferous restrooms, and packed coffee shops. In the 1950s, airline patrons grew increasingly frustrated with long delays at baggage areas after flights. Jammed passenger terminals frequently seemed the rule rather than the exception; reports of half a million people overwhelm- ing the terminal at New York's Idlewild cropped up with increasing regularity during summer weekends. The pressures of additional passengers and air traffic sparked a wave of airport improvements and new construction, but frustrated officials often discovered, the day after a ceremonious opening, that their gleaming chrome and sheet-glass buildings remained choked with even larger crowds of air travelers. Airline pilots lived continuously with altered runways and torn-up parking aprons as harried airport engineers labored to add landing strips and boarding areas.

The advent of jet transports in the late 1950s meant new, longer runways with thicker concrete for heavier planes. Recently completed construction went under the jack- hammer again. In 1955, Chicago opened O'Hare Field, the world's biggest airport and soon one of the busiest. O'Hare introduced some important innovations, like parallel runways that enabled simultaneous landings and departures. Passenger ramps, with steep stairs and time-consuming maneuvers to roll them out and snug them up to airliners, were displaced by flexible, accordionlike corridors that extended from the terminal boarding area directly to the airplane. Although passengers groused about interminable concourses stretching between ticket counters and boarding lounges, the crowds continued to proliferate. Since new airfields like O'Hare had to be built in rural areas, many miles from the city center, pessimists predicted that the distant site would deter many would-be airline patrons. But as Carl Solberg observed, ''Far from deterring patronage, these distances simply dictate changes in where people met and moved around Chicago, and thereby demonstrated the pulling power of air transport on the nation's social and economic life.'' As the crowds grew, the round of construction at O'Hare and other major fields seemed constant.

Because thronged terminals, baggage delays, and crowded plane cabins left travel- ers fuming, airlines made genuine efforts to respond to passenger complaints and to make the experience as pleasant as possible. At departure gates, the handicapped and parents with young children were invited to board the plane early and get settled before the crush of other passengers flooded the aisle. Frequent travelers could take out memberships entitling them to special lounges furnished with comfortable chairs, divans, bars, secretarial services, copiers, and other blandishments. Once aboard the aircraft, nonsmokers enjoyed reserved sections. If travelers notified the airline in advance, it was possible to receive kosher, vegetarian, or dietetic meals in place of the standard fare. Mixed drinks aboard domestic airliners were still something of a novelty as late as the 1950s, although Pan Am had been serving wine and liquor on international routes in order to keep up with its foreign competitors. Seats were carefully contoured, could be reclined for more relaxation, and included individual controls for regulating the flow of air and for reading lights. Cabin attendants offered an array of current newspapers and periodicals, or patrons could page through the airline's own slickly produced magazine (and, in fact, were invited to take it home). On long-distance and intercontinental flights, most airlines offered other entertainment to make the time fly: aerial pubs with additional snacks and club seating, movies, and audio programs.

Motion pictures and audio entertainment first appeared in the prewar era. After the first airborne movies flickered to life in 1929, TWA revived the experiment in 1934, showing a sound film appropriately called *Flying Hostess*. During the late 1930s, some air transports came equipped with individual radio sets, requiring the pilot to supply information about proper dial settings as the plane passed within range of various broadcast stations. In the postwar era, when the sheer adventure of flying began to wear thin, the long hours in flight between cities and continents were eventually filled with electronic entertainment. TWA generally receives credit for inaugurating a continuing series of movie programs. In 1961, TWA and a firm called Inflight Services joined forces to perfect a projection system and arrange a string of titles for showing. The first to be screened was a Lana Turner film titled *By Love Possessed*. Since then, thousands of travelers have idled away thousands of hours viewing sundry Hollywood epics thousands of feet above the ground. For passengers who preferred music to movies, lightweight headsets carried a variety of programs on a choice of stereophonic channels. Eventually, Inflight Services provided carefully tailored programs for European customers as well as for airlines in the United States, and other movie/music services entered the field. Taped programs were changed at various times during the year, depending on the individual airline's preference. As an audience of one, the airline patron consulted a printed program and dialed the proper channel on a console at each seat to listen. The choice might run to several different music channels featuring individual artists, symphonic programs, country-and-western tunes, Broadway musicals, soothing arrangements of the "easy listening" category, and current popular music, in addition to comedy routines, sports summaries, news broadcasts, and more.

Comfortable, time-saving air travel had a dramatic impact on major league sports, followed by similar influences on college athletics. Isolated use of airlines for team travel occurred in the prewar era, beginning with the Cincinnati Reds baseball team in 1934. The club chartered two Ford trimotors from American Air Lines and flew to Chicago for a date with the Chicago Cubs, although a coach and three players cravenly

On many highly competitive routes, airlines vied with each other to offer special service to in-flight passengers. Such enticements often included lounges like this one in an early 707. Courtesy of Trans World Airlines.

traveled by train. In 1940, having won the Big Ten football title, the University of Michigan decided to fly to California to meet Berkeley in the Rose Bowl. Although it took a trio of DC-3s from United Air Lines to airlift the burly Wolverines to the West Coast, the effort paid off, helping Michigan to a victory margin of five touchdowns. After 1945, professional baseball teams set the pace for air travel. For decades, the pro-baseball network had been circumscribed by the schedules of overnight Pullman travel, which tied clubs to cities up and down the Atlantic seaboard and east of the Mississippi. Airborne conveyance of teams made the migration of old clubs possible and led to expansion of the leagues as well. The Braves abandoned Boston for Milwaukee in 1953. More cataclysmic was the desertion from New York City of one of baseball's most hallowed names—the Dodgers—who decamped Brooklyn for the lush suburbs of Los Angeles. Astounded fans had barely recovered from this apostasy when the Giants also went West, to San Francisco. The introduction of jets made the proliferation of new franchises inevitable during the 1960s. By 1969, both major leagues were fielding 12 clubs, so that cities in the South, Southwest, and Far West boasted of big league teams. The same diaspora of professional sports followed in

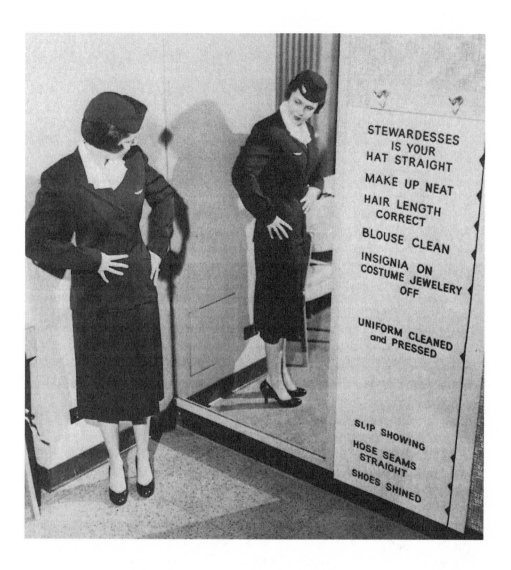

STEWARDESSES
IS YOUR
HAT STRAIGHT

MAKE UP NEAT

HAIR LENGTH
CORRECT

BLOUSE CLEAN

INSIGNIA ON
COSTUME JEWELERY
OFF

UNIFORM CLEANED
and PRESSED

SLIP SHOWING
HOSE SEAMS
STRAIGHT
SHOES SHINED

The job of stewardess was thought by many to be one of the most glamorous assignments open to young women after the war. In addition to youth and physical attractiveness, the job required several specific preflight checks, as shown in this photo from the 1950s. Courtesy of United Airlines.

football, basketball, hockey, and soccer. Similarly, college athletic teams took to the air, making coast-to-coast schedules commonplace.

The postwar generation of airliners expanded overseas business activities and tourism. For Americans who remembered wartime flights in military transports, the speed and cruising altitude of pressurized, piston-engine airliners on long intercontinental routes represented a huge improvement. Immediately after the war, trans-Atlantic flights drew large numbers of American executives seeking new business opportunities in Europe, especially during the development phase of the Common Market when Europeans needed American machinery and products essential to re-industrialization after the war. American corporations also sent executives flying overseas to establish foreign branches or buy into existing firms.

Even though high-flying airliners made transcontinental and international travel less bothersome from the standpoint of airsickness, travelers complained of unusual exhaustion and inability to think clearly for a day or two after reaching their destinations. Faster jets seemed to magnify the complaint. In the late fifties, physicians finally pinpointed the chronic phenomenon as "jet lag." The problem stemmed from the human body's physiological 24-hour cycle, or circadian rhythm. Jet lag accurately described the body's struggle to readjust as air travelers rapidly shifted from one time zone to another. Not surprisingly, the adjustment problem was most acute on east-west routes. Since travel on north-south routes usually stayed within the same time zone, it created less difficulty, even though the flight might be unusually long. John Foster Dulles, a peripatetic air traveler as secretary of state in the Eisenhower administration, reportedly regretted his attempts to conduct summit conferences soon after completing punishing intercontinental flights. According to journalist Marquis Childs, Dulles said that exhaustion from one such long-range diplomatic foray contributed to his controversial decision to cancel American support of Egypt's Aswan high dam, triggering a chain of events that led to the Suez Crisis of 1956 and subsequent strains on U.S. relations with England and France.

Within the United States, the postwar growth of airlines linked the deep South, the Southwest, and the Far West to the rest of the country as never before. Indeed, the statehood of Hawaii and Alaska in the 1950s seemed far more logical because scheduled airlines reduced the barrier of geographical distance to a few hours by plane. Accordingly, air travel was a factor in the decentralization of many corporations and the evolution of conglomerates in the postwar era. Business travel continued to be the principal market for airlines, and the influence of roving executives altered the patterns of urban commerce. In America of the late-nineteenth century, railroads had been the focus for travel and commerce, bringing people to the city center, where manufacturing and other commercial activities had become concentrated. "The city's busy, impressive railway depot," observed historian Daniel Boorstin, became an "inland harbor" and served as a "measure of its commerce and its vitality." The automobile, hastened a process of diffusion in which traditional city centers ceased to be the location of everything commercially significant in city life. The airplane furthered this process. In the post–World War II era, chambers of commerce began to point to the municipal airfield as the measure of commerce and vitality—and the airfield was located on the edge of the city. As a matter of convenience, many firms established office locations close to the airport. Thus, air travel accelerated the diffusion of traditional urban economic patterns by drawing offices, businesses, and even retail activities to entirely new locations. Air transport, according to Boorstin, also made intercity travel a "pseudo-experience." Business people arrived at an airport out in the countryside, conducted their meetings, found entertainment, stayed the night, and departed, having spent the entire time within the confines of the airport or its immediate vicinity without having seen the central city itself.

The phenomenon of postwar air travel also dramatically changed traditional vacation habits. In the 1920s, families of reasonable means along the Atlantic seaboard planned summer excursions to traditional locales like Cape Cod, Atlantic City, and Cape May. Trunks were sent ahead and the families followed. Well-to-do Californians might retire to Catalina Island, and other Americans with time and money visited established spots along the Gulf Coast, Great Lakes, or Rocky Mountains. The very wealthy from West Coast states might journey to Honolulu by Matson Line steamship, while their counterparts along the East Coast boarded stately liners like the *Ile de France,* bound for the Continent. In either case, such time-consuming journeys were mounted only every few years or so. Travelers would spend the entire summer, at the least, to tour abroad since a three-week vacation dedicated to a European trip made little sense if two weeks were consumed in sailing over and back. Many still considered these excursions to foreign climes as risky undertakings. William Allen White noted that any citizen from Emporia, Kansas, who had ventured forth to Europe did so with the earnest concern of all Emporians for a safe return. At home or abroad, vacation excursions cost a lot of time and money, which posed problems for many pre–World War II families of blue- or white-collar status.

Postwar trans-Atlantic flights made such trips perfectly feasible. For those with the money and the inclination, the arrival of jet airliners in the late 1950s and early 1960s permitted occasional weekend forays to London or Paris, or perhaps a long weekend on the ski slopes of Switzerland. The "jet set" had arrived. More important, the spread of domestic as well as foreign routes brought an unprecedented degree of democratization to travel. Secretaries and hourly workers could find reasonably priced package tours to Miami's beaches, Las Vegas's gaming tables, or Vail's ski slopes, and crowded the same aircraft with jet-setters and high-rollers bound for the same sites for the same pursuits. The airlines and travel agencies also mounted aggressive advertising campaigns that lured an astonishing cross-section of the American public on bargain-priced excursions overseas. Traditionalists carped that many of these junkets insulated travelers from the foreign culture they had come to see by housing them in Holiday Inns or American-style hotels built by the airlines themselves. Still, there were new experiences to be savored, and 14-day package tours whisked middle-class Americans off to Brussels and Vienna, even though their incomes and vacation allowances came nowhere near the requisite standards of the prewar Grand Tour. In 1970, when foreign travelers numbered five million per year, only 3 percent were seafarers.

Never before had so many ranged the world in such numbers in search of pleasure, culture, and knowledge. Low-cost charters took thousands of college students abroad for study and for the fun of it. Economical package tours lured their parents to locations hitherto thought too bizarre to contemplate. By the 1970s, so many tourists had booked air passage to the Galapagos Islands that Ecuador threatened quotas in order to protect flora as well as fauna. Midwesterners who had motored to the Black Hills of South Dakota now flew off to the Seychelles, tiny dots in the Indian Ocean—hardly a locale that Emporians or anyone else would have envisioned as a vacation destination in the 1920s. International travel became a major component of world trade, and many foreign nations identified American tourism as a major export. For the United States, tourism abroad had to be listed as a major import as American dollars winged overseas, adding to the American balance-of-payments problem.

Not only did postwar air travel carry people in search of culture to distant countries, it also facilitated the flow of cultural artifacts and events. Individual musical artists from overseas, for example, were able to perform in the United States in greater numbers and with far greater frequency than ever before. World-renowned orchestras not only played Carnegie Hall in New York City but also appeared in Buffalo, St. Louis, Des Moines, Omaha, Denver, and parts west. In cultural counterpoint, Americans concertized around the globe, maintaining schedules that had been impossible in an era of steamships and rail travel. Similarly, international art exhibits, lectures, conventions, and symposia became the norm rather than the exception.

GENERAL AVIATION

Except for a downturn in the U.S. economy that jolted sales in the early 1960s, the trend in the general aviation sector after 1955 moved upward. The product lines of the "Big Three" (Beech, Cessna, and Piper) expanded to include a wide variety of new types, from specially designed agricultural aircraft to advanced twins. In fact, the demand for general aviation aircraft was so strong that new firms entering the market in specific aircraft categories were able to compete effectively with the established companies. With a broad array of aircraft types that often benefited from advanced aerodynamic and fabrication techniques of the era, U.S. manufacturers dominated the world market for general aviation planes.

In terms of production volume, Cessna emerged as the leader in 1955, when the company not only outsold its rivals but delivered more new airplanes (1,746 commercial and 24 military) than any other company in the world—a distinction that Cessna maintained year after year. By far the company's most popular plane was the Cessna 172, which appeared in 1955. The chief features differentiating it from the Model 170 were its squared-off tail and its tricycle landing gear for improved ground handling. In 1960, the plane acquired a sweptback vertical tail, and subsequent models were known as Skyhawks. Cessna had produced more than 11,000 of the 172/Skyhawk series by 1966. The four-place Skyhawk types, with a respectable cruising speed of over 125 MPH, also formed the basis for several higher-powered and faster variants. These high-wing designs were all comfortable, efficient aircraft, and their rugged construction made them popular with ranchers and farmers as well as with business pilots. Cessna continued production of a series of light twins, including the unique Model 336 Skymaster. The Skymaster had its pair of engines mounted in tandem at the front and rear of the fuselage and carried its tail assembly on two booms extending back from its high wing. The placement of both engines to provide center line thrust made the plane much safer to handle in the event that one engine went out, where con-

Using convenient rivers and lakes, airplanes like this float-equipped Cessna 185 frequently represented the only means of travel in remote regions of North America and in other parts of the world. Courtesy of Edo Corporation.

ventionally designed aircraft, with an engine on either wing, developed strong torque characteristics that required skillful piloting.

The Piper Company came up with its own odd-looking plane in the Pawnee, a new concept for agricultural aircraft, or ag-planes. Most ag-planes of the era were converted from existing types, especially castoff World War II biplane trainers like the Stearman. Agricultural flying required tight maneuvers at low level, a demanding operation that led to disturbingly frequent crashes. Pilots flying these heavily loaded, open-cockpit planes had little protection, and fatalities mounted. During the late 1950s and early 1960s, the 5,000 ag-planes registered in America experienced one accident for every 2,600 flying hours, and one death for every 20,000 hours. When the Pawnee started flying, the figures significantly improved to one accident per 5,000 hours, and one fatality per 90,000 hours. Much of the credit for the Pawnee went to its chief designer, Fred Weick, and to Howard (''Pug'') Piper, who pushed its development. The pilot sat high above the fuselage in a reinforced, enclosed cockpit that protected him in crashes and reduced exposure to toxic chemicals. The plane was designed so that the low wing, engine, fuel tanks, and chemical hopper would act as buffers in accidents, further protecting the pilot. After Pawnees went into operation about 1960, the plane accounted for six of every ten ag-planes sold in the United States, and export versions were at work in 91 countries. In Florida, Piper claimed the Pawnee helped wipe out pests and enabled the state to grow corn for the first time in its history. Elsewhere, the Pawnee and others like it (Cessna, Grumman, and Rockwell International designed and produced other ag-planes) fertilized crops and sprayed them to liquidate voracious insects and spike noxious weeds. Ag-planes sprayed livestock and poultry for ticks and mites, flew to control plagues and epidemics, and dropped fish to restock lakes and streams in remote areas. Specialized aircraft like these became an integral part of general aviation's product line.

Like Cessna, Piper refined its line of light, twin-engine aircraft, and continued to manufacture the single-engine, high-wing Super-Cub as a versatile utility plane. On the other hand, Piper moved away from a stable of high-wing designs, developing the low-wing Cherokee trainer and a new series of high-performance, single-engine aircraft, including models with retractable landing gear. In this respect, the company began to compete with the Beechcraft Bonanza, although the most talked-about new plane in this category was the Mooney Mark 21. Besides giving it a distinctive, forward-raked rudder, said to enhance aerodynamic efficiency, the Mooney's designers also paid close attention to streamlining the trim, four-place speedster. The M-21 derived a maximum cruising speed of 180 MPH from its 180-HP engine. Although the Piper Comanche and the Beech Bonanza were faster than the Mooney, their 260- to 285-HP engines raised their purchase price and reduced their fuel economy. With its snappy performance, sporty looks, and low operating costs, the Mooney outsold every other retractable in its class for several years in the 1960s and reflected a growing diversity in the general aviation manufacturing industry.

An especially notable trend within the industry involved the continued development of "heavy twins," incorporating advanced equipment like pressurized cabins and turboprop engines. Even more outstanding was the development of jet aircraft. The postwar development of light twins, while increasing the pilot/passenger accommodations for up to six or eight seats, still put everyone in cabins of comparatively limited size. For true comfort during flights of increasing duration, larger cabins were unquestionably more attractive, and larger planes could more easily incorporate creature comforts such as food, drink, and "sanitation compartments." Moreover, larger aircraft enhanced "club seating," comfortably arranged chairs and folding table surfaces for business conferences en route. Large corporations had been using such executive aircraft for years, beginning with the venerable prewar Model 18 Twin Beech (which remained in production through 1969—a remarkable 37-year span) and continuing with retreaded airliners like the DC-3 and the Lockheed Lodestar. But these aging stalwarts were declining in both efficiency and appeal. The market seemed to be there, just waiting for the right new planes.

In many ways, the Grumman Gulfstream I represented the ultimate of the new breed of heavy twin, designed and marketed with the top-level corporate executive in mind. Deliveries began in the spring of 1959; 60 planes were in corporate or personal service by the end of 1960, and over 180 had been sold by the spring of 1966. The executive version of the Gulfstream I carried 10 to 12 passengers and a crew of two in luxurious comfort at speeds of 350 MPH at over 30,000 feet—the first of the new heavy twins to combine turboprop engines and a pressurized cabin. Its opulent interiors and customized appointments set the standards for executive flying. As in the case of Mooney, the fact that the general aviation market had attracted a successful new competitor was in itself a significant event. Beechcraft, who had dominated this end of the market for so long with the Model 18, responded with another cabin-class twin at a different point on the scale. Deliveries of the Model 65 Queen Air had begun early in 1959, but this plane, compared to the Gulfstream, reflected a more conservative approach. The 230-MPH Queen Air mounted conventional reciprocating engines and seated 7 passengers plus the pilot in an unpressurized fuselage. Although its size and performance seemed less impressive than that of the Grumman, the Queen Air was still important for the concept of the heavy twin corporate aircraft built for a broad spectrum of clients who aspired to executive-class flying comforts. Larger, improved models followed, capped by the pressurized King Air turboprop series, seating 7 to 10 passengers, introduced in 1964. Beechcraft thus equipped its dealers with a comprehensive range of executive twins that sold in numbers running well into the hundreds by the end of 1965, when the company had orders for over 150 King Airs alone.

Mounting sales of aircraft in the heavy twin class demonstrated that corporations were willing to pay from several hundred thousand dollars to over a million dollars for high-speed executive flying. The obvious question was whether there was a similar market for jets. Jets were likely to be much more expensive to buy and to operate, thereby limiting sales. Potential manufacturers were understandably leary because they anticipated high R&D costs and worried that a presumably limited market might result in considerable losses—especially if competitors got into a scramble for sales. Although Cessna had developed the T-37 jet trainer under an air force contract, none of the traditional general aviation manufacturers (Beech, Cessna, and Piper) had any experience in developing a passenger jet for the civil market. The first step toward this new plateau, as with the advanced heavy twin, again came from a company outside the circle of traditional general aviation manufacturers.

The Lockheed Corporation, an aerospace giant, developed the JetStar as a private venture with one eye on an air force requirement for a four-engine light jet transport and crew trainer. The first prototype took off in 1957, followed by a second in 1958. After three years of prototype testing, initially with British-built engines, production aircraft were fitted with Pratt & Whitney power plants of 2,400 pounds of thrust each, and the first delivery went to the Continental Can Company early in 1961. Later versions of the JetStar, with engines of 3,000 pounds of thrust and more, cruised over 550 MPH above 33,000 feet. Lockheed hoped that its JetStar, as the first executive jet available, would generate several hundred commercial sales in addition to substantial military orders. But its price tag of over $1.5 million discouraged most corporate prospects, and military purchases proved disappointing. Although 71 planes (including 53 commercial sales) had been delivered by mid-1966, the company remained far from the break-even estimate of 300 aircraft. The North American Sabreliner, a twin jet similarly developed for military and/or corporate use, also available by 1961, had experienced similarly sluggish sales. Many skeptics felt justified by the low sales

The Beechcraft King Air reflected an important trend toward executive aircraft with pressurized cabins. The King Air design was extrapolated from the Queen Air series. This photo shows the King Air Model 90 on its maiden flight in 1964. The nose boom carried test instrumentation. Courtesy of Beech Aircraft.

figures, arguing that a large market for executive jets simply did not exist. Others felt the market had not really been tapped, and that the answer was a less expensive airplane. One of the most insistent of these jet enthusiasts was a maverick named William P. Lear.

After World War II, Bill Lear had developed an innovative company that pioneered a number of electronic devices and electromechanical equipment. In 1949, he received the prestigious Collier Trophy for having designed the first successful autopilot. During the mid-1950s, Lear made a reputation in the corporate aircraft business by converting Lockheed Lodestars into long-range executive transports called Learstars. By 1959, he was already thinking about a radical new executive jet that would be fast, efficient, and small. Controlling the size would keep the price down, and making it speedy would appeal to corporations who saw it as a time-saving business machine, not a luxury liner. The configuration was laid out while Lear was living in Switzerland by a Swiss engineering team using the existing wing design from a cancelled Swiss fighter-bomber, the P-16. The wing design of the P-16 appealed to Lear, and its thin airfoil section, requiring the use of wing-tip fuel tanks, gave the new jet a fighter-plane look. Starting with the P-16's wing and certain other systems, Lear hoped to control costs by using additional off-the-shelf hardware and by subcontracting for various parts and assemblies on a worldwide basis. But work went slowly, and the tangle of

A savvy sales strategy helped make the sleek Learjet synonymous with air travel for modern sophisticates. Lear's plane appeared in the background for numerous ads featuring stylish goods and services and also provided transportation for the high-style operations of the private investigator in Our Man Flint, *played in the motion picture series by James Coburn. Courtesy of Gates Learjet.*

international suppliers kept schedules in a turmoil. Lear abruptly moved the whole project to Wichita, Kansas, in 1962. The move not only put Lear closer to primary American markets but also offered the production skills and technical base needed for the Learjet, as the plane was called. In Wichita, Cessna's prior work on the T-37 jet trainer helped to provide a nucleus of suppliers, production managers, and factory workers with experience in small jet aircraft. The city arranged attractive tax benefits for Lear's company and had low labor costs. A new plant went up in remarkably short order, and the first Learjet streaked off the runway in the autumn of 1963.

Lear's fleet executive jet carried two pilots and six passengers at over 500 MPH to altitudes of 45,000 feet, statistics considerably above those of its competitors. In addition to its speed, the Learjet's operational altitudes allowed flying in a region free of headwinds, turbulence, storms, and other aircraft. Though the plane's cabin appeared somewhat cramped in comparison to those of heavy twins and other corporate jets, its performance and efficient operational envelope were compelling. Moreover, its price tag amounted to about half those of the Lockheed JetStar and the North American Sabreliner. By the autumn of 1964, only a year after the Learjet's first flight, the first production aircraft went to a corporate owner, with additional deliveries and sales passing the 100 mark in just 12 more months. As the first jet specifically designed for executive use, the Learjet, in the words of one NASA publication, "revolutionized business and corporate aviation." With a snug, though well-appointed, cabin, Bill Lear's sleek speedster offered superior performance at an unbeatable price, and the plane's sales paved the way for additional executive jets from foreign as well as American manufacturers. The era of executive jets had truly arrived.

The advent of pressurized executive aircraft and high-flying corporate jets marked a new plateau of sophistication for the general aviation sector that was paralleled in the development of specialized interest groups and trade associations. The Aircraft Owners and Pilots Association (AOPA) had appeared as early as 1939. In addition to the promotion of recreational and business flying the AOPA also took an increasing interest in state and federal legislation. As aviation became more specialized so did its organizations in the postwar era. For example, the National Business Aircraft Association (1947) looked after the interests of the corporate aviation community, while the General Aviation Manufacturers Association (1970) represented a more explicit lobby group in terms of national legislation and public information. Over the years, a variety of other organizations proliferated: operators of antique aircraft; flying farmers; flight school executives; flying dentists; crop dusters; flying accountants; flying police; flying undertakers, and more. Collectively, all of these groups reflected the increasing role of general aviation in American society, although the organization and activities of AOPA, NBAA, and GAMA in particular reflected the institutionalization and maturity of the general aviation sector.

7

From the Earth to the Moon, 1965–1975

Like so many eras of history, the years 1965 to 1975 seem full of paradoxes. It was a decade in which the conflict in Vietnam dominated headlines. As the war intensified during the late sixties and early seventies, American military aviation reached new levels of technical proficiency, teamwork, and hardware. Air power proved to be a potent weapon, but one that left a battered nation and a ravaged countryside in its wake. Bombing raids and other aerial attacks helped feed a growing antiwar sentiment, not only in the United States but elsewhere in the world, even among traditional American allies.

At the same time that the fighting in Southeast Asia was dividing national alliances, air transportation continued to shrink time/distance relationships around the world, creating a more closely knit international environment. The advent of jumbo jets stimulated domestic as well as foreign travel, and air cargo operations also reached record levels. Regional and commuter operations proliferated. The general aviation sector introduced improved aircraft featuring turboprops and pressurization for added efficiency and convenience in business and corporate flying.

In space exploration, the Apollo program achieved a stunning success with a manned lunar landing mission. It was a program conceived and carried out in the cold war ideology of besting the Soviet Union. At the same time, America and the USSR initiated talks that led to a unique mission in which spacecraft of the two nations made a rendezvous in space.

MILITARY AVIATION

The position of the Strategic Air Command (SAC), with its panoply of missiles and bombers, represented only one aspect of flight as an element of foreign policy. During the cold war era, with the United States as one of the principal "superpower" protagonists, air transports frequently had occasion to airlift troops and supplies to distant quadrants of the globe. Flare-ups dictating an American response generally occurred with little or no warning, and emergency supply requirements needed prompt attention. Since seaborne troops took too long to transport and deploy, the airlift capability of the Military Air Transport Service (MATS) and its successor, the Military Airlift Command (MAC), became an important element of American diplomacy. MATS originated in 1948, when the Department of Defense decided to consolidate air transport duties as a matter of economy and convenience. Long-range transports of the army and navy formed the basis of MATS squadrons, with the air force acting as the operating agency under Department of Defense jurisdiction. Although both the army and the navy eventually equipped themselves with short-range tactical transports, MATS/MAC aircraft remained the principal means of large-scale global airlift.

The Berlin Airlift illustrated the expediency of MATS airlift capability but also underscored the need for specifically designed military transport aircraft. The C-47 and C-54 workhorses of the Berlin Airlift dated from civil airline designs of the prewar era, converted into military transports as a result of emergency wartime needs. Even enlarged cargo doors in the side of the fuselage failed to expedite freight operations and parachute drops of both soldiers and supplies. Medium-range, twin-engine transports like the Fairchild C-82 Flying Boxcar and its improved descendant, the C-119 Packet, were laid out in the early postwar era with military requirements uppermost in mind. The tail booms extending back from the engine nacelles left the aft fuselage section unencumbered so that troops and bulky cargo could easily be loaded and unloaded. Either paratroops or military cargo could conveniently be air-dropped through the capacious rear cargo doors. Additional freight and troop transport designs entered military service in succeeding decades. Moreover, the air force pushed development of larger transport planes capable of intercontinental airlift missions in the cold war environment.

When the Douglas C-124 Globemaster entered MATS units in 1950, the plane's name reflected its intended role. Oversized clamshell doors in the nose enabled rapid handling of artillery, vehicles, freight, and troops. The piston-engine Globemaster also represented the last of the conventionally powered large transports, since its successors followed the example of air force bombers by using gas-turbine power plants. The powerful Douglas C-133 Cargomaster, equipped with four turboprop engines, entered MATS squadrons in 1957. With a range of 4,400 miles carrying a full load, it added a new dimension to military reach. Although originally configured to haul IRBM and ICBM surface-to-surface weapons, the C-133 could carry an appreciable array of military equipment. Another four-engine turboprop, entering service just a few months earlier, proved even more serviceable. Lockheed designed the C-130 Hercules as a medium-range transport carrying up to 92 troops or myriad freight loads. The broad, deep fuselage and rear-loading integral cargo ramp gave the plane unusual cargo flexibility; its rugged construction and ability to fly from short, rough fields gave it unusual versatility. The plane's reliability afforded usage in forward combat areas and made it highly regarded for flights to remote global trouble spots where only a few modern airfields were available. In short, it became the leading transport in "brushfire'' conflicts around the world, prized by all branches of the U.S. military and by numerous foreign air forces.

Since speed persisted as the limiting factor in turboprop transports, jet-propelled designs eventually appeared. Two Lockheed aircraft dominated the air force inventory: the C-141 Starlifter and the C-5 Galaxy. Entering service in 1963, the Starlifter carried 154 troops on flights of 6,000 miles. The Galaxy, billed as the world's largest airplane, could carry 345 combat troops on two decks, or a pair of M-60 main battle tanks, or 125,000 pounds of miscellaneous cargo to destinations up to 8,000 miles away. Production deliveries of 81 Galaxy transports began in 1969. For all the controversy resulting from severe cost overruns, and a spate of trouble involving structural shortcomings (especially the plane's gigantic wings), the C-5 provided the American armed services with a truly remarkable logistical potential to any continent on earth. The broadening diplomatic significance and capability of this global outreach found new expression in a revised nomenclature for MATS in 1966 when it became the Military Airlift Command. Many MAC aircraft played a role in the Vietnamese conflict. They complemented seaborne cargo during the early stages of the conflict, but their capability for rapid aerial delivery of high-priority military cargo and troops constituted a major service throughout the duration of hostilities. The big transports also performed a singular role in medical evacuation, transporting wounded personnel to hospitals in the Pacific area as well as in the United States.

In Vietnam, combat aircraft of the army, navy, and air force played a leading, if controversial, function in a war that itself became highly controversial. The U.S. government claimed to be extending aid to South Vietnam, allegedly under attack from a Communist (North Vietnamese) power in the early 1960s. Many critics of American participation in the affair argued that the conflict sprang from internal problems—a civil war—in which case the use of air power would necessarily be limited, since the identity of friend or foe would always be difficult. In a guerilla-style conflict, fought in a dense jungle environment, easy identification and attacks on conventional ''enemy'' targets like bridges, roads, and villages presented dilemmas. Friend and foe alike suffered casualties and experienced severe economic disruption. These issues seemed especially perplexing in the early years of American air operation south of the demarcation line. As the American commitment escalated, the air force and navy attacked conventional industrial targets within North Vietnam. In the meantime, air power played a prominent role in the jungle war with heavy strikes against enemy troops wherever they were reported. Although fixed-wing aircraft continued to play a central role, helicopter operations evolved as a major feature of American air warfare, particularly in airborne troop transport of army ground forces, and the army increasingly used armed helicopter gunships in tactical assaults.

The first American forces in Vietnam, serving in an ''advisory'' capacity, looked to air support from a rather motley collection of propeller-driven planes, including the venerable Douglas B-26 Invader of World War II vintage. The single-engine Skyraider of Korean War fame equipped naval and Marine Corps units as well as squadrons of the South Vietnamese air force. These planes, though slower than jet-powered counterparts, nevertheless seemed less vulnerable to ground fire than the jets that appeared in

A Republic F-105 fires a devastating salvo of rockets. Armed with bombs as well as rockets, the F-105 was heavily utilized for bombing and for low-level attack during the Vietnam conflict. Courtesy of Fairchild Republic.

later years and could do a better job of spotting and attacking targets on their own. But their numbers were finite; more and more high-speed jets with equivalent bomb capacity were pressed into service, especially after 1963, when the beleaguered South Vietnamese government tottered through the Diem assassination and the ineffectual regime that followed. President Lyndon Baines Johnson strongly reacted to increasing Viet Cong actions, especially after the famous (if disputed) attack on U.S. naval ships in the Tonkin Gulf in 1964. The Republic F-105 Thunderchief bore the brunt of ground attack missions. One of the heaviest single-engine jets in the USAF inventory, the F-105 could carry nearly as much tonnage as a B-17 of World War II. Backed up by the F-100 SuperSabre and the F-4 Phantom, these fighter-bombers delivered a devastating amount of fire power on enemy targets, using guns, rockets, bombs, and napalm.

It is ironic that these finely honed supersonic fighters were so widely employed in the ground attack role, making attack runs at half of their supersonic speed capability. And even at half speed, they flew so fast that ground references often appeared as a blur, making it difficult for their pilots to pinpoint a target. The attacking jets might spot and hit a specific target area, but getting in low to the ground took the planes out of their intended environment of high-speed, high-altitude combat. Over the Vietnamese jungles, a high-performance jet whose entire fuselage was crammed with miscellaneous operational gear became suprisingly vulnerable to small-arms fire. Random rifle hits that brought multimillion dollar fighters to ruin made the war an expensive one for the United States. As in Korea, it became necessary to employ slower aerial spotters to guide attacking aircraft. The job fell to the forward air controller (FAC), flying planes like the Cessna 0-2, derived from the Cessna Skymaster. In a risky air environment, the FAC marked targets with smoke rockets and directed the jet fighter-bombers to their target areas by radio.

Air-to-air combat was the exception, not the rule, for American pilots in Vietnam. When these duels took place, they were generally brief, since aircraft on both sides usually relied on air-to-air rockets, which were rapidly expended. Unlike the Korean War, where jet dogfights had been largely determined by machine guns and gun sights, aerial duels in Southeast Asia pitted radar-directed fighters against each other, and the planes fired electronically guided rockets. The virtual absence of an enemy air force showed up in U.S. figures, which listed only 86 enemy aircraft shot down within a two-and-a-half-year span between 1965 and 1968. Most of these were MiG-17 and MiG-21 aircraft, bested by American F-105s and two-seat F-4s. Such statistics do not mean that the United States escaped with minimal losses. Through 1967, the United States lost some 3,000 fixed-wing types and helicopters. Although about half this loss was attributed to assorted operational factors, the remainder was the result of combat action, including ground fire and surface-to-air missiles (SAM). SAM defenses were formidable, although American pilots learned countertactics, like making final target runs at low altitudes beneath effective SAM trajectories and using sharp turns to throw off SAM tracking gear. As time went on, the USAF made increasing use of electronic countermeasure (ECM) equipment to throw off SAM guidance equipment.

The most vulnerable aircraft in Vietnam were the helicopters operated by the army; 4,200 were lost by 1971. To be sure, this high attrition rate can be traced to the nature of the helicopter itself. The huge rotors produced a large amount of torque, placing great strain on the rotor housing and machinery that controlled the pitch of rotors in ascent and descent, and under combat conditions, strict maintenance schedules were not always possible. But 45 percent of helicopter losses were the result of hostile fire, including antiaircraft rockets. Still, the choppers played a major role in a difficult war; in fact, part of the reason for such high losses stemmed from their high utilization rate: the army reported as many as 18,000 combat sorties for each helicopter that was lost. In addition to their obvious flexibility, the choppers also provided a strong element of surprise. Infantry, artillery, or both could be shifted in the midst of an engagement, startling the enemy by the sudden appearance of fresh troops or subjecting them to an artillery barrage from an unexpected quarter. In the dense jungles, helicopters constituted an instantaneous logistics capability and eliminated long, vulnerable supply routes to remote outposts. By the same token, rapid evacuation of wounded personnel kept down combat deaths and became a strong morale factor. Helicopters in Vietnam also evolved well beyond the role of aerial transport, becoming "gunships" armed with a lethal variety of automatic weapons and rockets. The principal aircraft of this type was an armed version of the Bell UH-1 "Huey," dubbed the "Huey-Cobra" in the gunship version. Like conventional transport choppers, the gunships provided remarkable flexibility and surprise in combat as well as fire power that frequently added

Combat troops jump from a hovering Bell UH-1 (nicknamed "Huey") during operations in the Vietnam jungle. The U.S. Army increasingly relied on helicopters for both mobility and firepower. The legend on the nose reads "Headhunters." Courtesy of Bell Helicopter Textron.

a decisive advantage in sharp encounters with the enemy in close jungle fighting where conventional artillery or other fire support was unsuitable.

This sort of pinpoint concentration of fire more developed in the case of even larger gunships. The venerable C-47 (based on the DC-3 of World War II), rechristened the AC-47, became an airborne concentration of automatic weapons. Aptly nicknamed "Puff the Magic Dragon," the AC-47 typically carried three rifle-caliber GE Miniguns, each firing up to 6,000 rounds per minute. With all three guns aligned through the port windows of the plane, the AC-47 could unleash a withering stream of fire while circling above the target. Troops discovered that the "dragon ships" were especially effective in breaking up enemy night attacks, and additional transport aircraft types, ranging all the way up through the AC-130H Hercules, were adapted for the gunship role. With their capacious cargo holds, planes like the Hercules carried varying installations of high-intensity search lights, flares, infrared sensors, radar, and formidable fire power which could include four Miniguns in addition to four 20-mm Gatling rapid-firing cannon. The AC-130H had a 105-mm howitzer.

This kind of unparalleled fire power in the tactical arena had a counterpart in strategic bombing capability. Although F-105 and similar types of fighter-bombers were employed, often accompanied by larger, specially equipped "pathfinder" RB-66 planes for rader bombing, America's premier long-range strategic bomber, the B-52, found extensive usage in a conflict that resembled a guerilla war more than a conventional engagement. By early 1969, over 100 B-52 aircraft had accounted for 1,800 sorties per month from neighboring Thailand, from Okinawa, and from Guam, some 2,000 miles from the battle zone. Designed to deliver nuclear weapons over high-priority strategic targets, the B-52 instead dropped huge clusters of 750-pound bombs in "spoiling attacks" over suspected enemy staging areas and supply routes. By and large, such raids represented an uneconomical use of the huge B-52 bombers. As air-power historian Robin Higham remarked, it "amounted to air-freighting high explosives over great distances."

Although half the bomb tonnage dropped in Vietnam was carried by fighter-bombers like the F-105, the sight of B-52s disgorging dozens of bombs highlighted the controversy over costs, efficacy, and morality. Since the bombers so frequently struck at will-of-the-wisp guerilla concentrations, it meant they unintentionally hit territory that could be "friendly" as well. Thousands of high-explosive bombs left deep, uncounted craters, which eventually filled with water where malarial mosquitoes bred. Bomb splinters lodged in trees around the impact areas caused a recurrent problem for the timber industry, since many hours each day were lost while saws were resharpened at lumber mills. As Higham wryly observed, "This was hardly the way to win over the populace." An even greater controversy raged over the use of defoliants. The densely packed jungle trees created a leafy canopy that shielded enemy supply routes and made it difficult to detect supply dumps, troop movements, and ambushes. The air force inaugurated an ambitious aerial herbicide program, lacing hundreds of square miles of countryside with defoliants. Denuding the jungle, it was argued, would deprive the enemy of an effective cover. During subsequent spraying missions, the defoliants decimated the trees in rubber plantations generally imbalanced the ecology of vast areas within the country, forcing a halt. This particular program remains one of the most controversial undertakings of the Vietnam conflict. Other means were eventually employed to detect enemy movements and supply routes, using increasingly sophisticated electronic devices instead. But the ecological damage still persists after many years.

Assessing the efficacy of air power in Vietnam is not an easy task because of the nature of the conflict, the political overtones, and the emotional crosscurrents it generated. Higham, writing before the end of the conflict, nevertheless highlighted a number of the major themes. In a combat environment more akin to guerilla war, jets were not as efficient as helicopters or other slow aircraft in attacking small, isolated, slow-moving targets. Moreover, the jets remained vulnerable to small-arms fire. Combat pilots and planners were irritated because many strategically crucial targets, such as those in the Hanoi area, were off limits until later phases of the war, and even then were frequently restricted so as to encourage ongoing peace negotiations. As in Korea, the most appealing targets remained off limits, inside the borders of Communist China. The extensive use of F-105 fighter-bombers and similar combat planes in long-range raids over North Vietnam was interesting since the aircraft required aerial refueling, an advantage made possible by virtually complete American control of the skies. But U.S. air forces may not enjoy such a luxury in future engagements. The same can be said for the unusually wide use of slow-moving helicopters over Vietnam, since future encounters may involve an adversary whose own air arm could severely curtail helicopter operations with concentrated ground fire and sophisticated ground-to-air rockets. While strikes launched from aircraft carriers were useful in the early stages of the conflict, their cost effectiveness and utility began to diminish as more airfields became available within Vietnam and neighboring Thailand. If anything, the navy realized its need for slower, though more long-ranged and heavily armed, attack planes, like the F-8 Crusader.

For the most part, USAF and naval aircraft in Vietnam represented a prewar inventory. One exception was the A-7 Corsair II, outstanding tactical fighter delivered to USAF and naval combat squadrons during the late sixties. It featured a continuous-solution navigation and weapon delivery system, coupled with an all-weather radar bomb delivery capability. However, the basic configuration owed its ancestry to naval versions, like the F-8. As a truly innovational combat plane, the controversial F-111 stands out. The plane evolved in the early 1960s as a multirole aircraft ranging from strategic bomber to tactical fighter—a tall order, and one that created a rash of development problems. The air force viewed the bomber version, FB-111A, as a Mach 2 replacement for the B-58 Hustler and early models of the B-52. The tactical version had to meet air force specifications for Mach 2+ at high altitudes, supersonic dash in low-level attacks, ability to operate out of primitive airstrips in forward combat zones, and outstanding flight characteristics through all the varied flight regimes. A variable-sweep wing design seemed the most feasible route, and, even though experimental work in this regard dated back to the X-5 of the 1950s, a mass-produced combat plane with a swing-wing intended to fly as a strategic bomber or terrain-hugging tactical fighter promised unique design and production challenges. Despite the tangle of compromises facing design engineers, Secretary of Defense Robert McNamara pushed the F-111 program as a cost-effective solution to discouragingly high appropriations for separate aircraft in various combat duties.

General Dynamics Corporation became prime contractor for the F-111, and acrimonious debate swirled around the plane and its manufacturer from beginning to end. For the contractor, continuous R&D snarls led to successive cost overruns, driving cost estimates far above the initial figure. But the first deliveries of the operational F-111A began in October 1967, and the FB-111A entered service two years later. Equipped with different engines for longer range, the bomber's extended wing span of 70 feet was 7 feet longer than the fighter's, and the bomber's gross weight of approximately 100,000 pounds exceeded the fighter version by about 9,000 pounds. Obviously, there were many differences in electronic gear and weapon loads; nevertheless, the two versions were nearly identical in appearance and overall dimensions.

In 1968, the air force dispatched a clutch of six F-111A planes to Vietnam to wring out the complex swing-wing craft in a combat environment. In little more than three weeks, half of the planes were lost, and an additional crash back in the United States prompted air force authorities to ground the aircraft. The Vietnam crashes were not

With wings partially swept back, the variable-sweep F-111 cruises at high altitude. After a disappointing combat debut in Vietnam, the F-111 performed effectively. Courtesy of General Dynamics.

combat losses, it turned out, but were traced to weld failures in the tail control mechanism. In fact, objective analysis of this abortive combat experiment later indicated outstanding performance in terms of high target accuracy after long flights deep into enemy territory without the need for refueling, top cover, and ancillary ECM air support. Following structural modifications and other work, the F-111A returned to Southeast Asia during late 1972 and compiled an enviable combat record. McNamara was at least partially justified.

The peculiarities of the Vietnamese war diminished the impact of heavy American bombing. The Ho Chi Minh trail continually frustrated U.S. military planners in Vietnam, since repeated heavy attacks on this important enemy artery seemed to disrupt the flow of food and munitions only momentarily without ever effectively cutting the supply line. The flow of soldiers and supplies from North Vietnam into the South did not appreciably dwindle; troop movements in 1965–66 rose from 4,500 soldiers per month to 7,000. Electronics played an increasingly important role in attempts to solve the recurrent problem of how to spot enemy troops and equipment beneath the jungle canopy and as a means of reducing the costly loss of highly trained pilots and multimillion dollar aircraft to ground fire. The electronic answer to this problem masqueraded under the code name of "Igloo White," a system designed to inform air forces of convoy activities and their locations. In practice, airplanes first scattered small electronic canisters across the Ho Chi Minh trail and other known truck routes. The sensitive canisters could detect ground vibrations caused by a truck and send a signal strong enough to be picked up by radio aboard picket aircraft. After confirming the geographic location, an air strike was called in on the target area.

Reconnaissance operations employed sophisticated electronic devices for mapping, tactical surveillance, and location of enemy radar sites. But the North Vietnamese began to deploy their own array of sophisticated electronic weaponry. By the late sixties, air attacks against industrial targets in the North had to be pressed through an increasingly effective screen of radar and fire-control centers that backed up a multiplicity of guided missile emplacements and ranks of antiaircraft guns. Many analysts rated this defensive gauntlet as tougher than anything American pilots had faced in World War II or the Korean War. To knock out these formidable defensive ramparts, the air force fell back on weapons such as the somewhat dated Bullpup guided missile, but aircraft still had to loiter around the target area to control the missile, exposing themselves to counter-fire. The air force urgently required an accurate "stand-off" weapon, one that allowed the plane to launch its missile from a longer range, then retreat to a safer position.

The Shrike, an antiradar missile with such a capacity, had been introduced in 1965 but with disappointing results. Intensive R&D improved the Shrike's effectiveness by 1968, and the missile scored consistently by automatically homing in on enemy radars. For attacks on other types of targets, American military aircraft began mounting the Maverick, guided by television signals, and the HOBOS (homing bomb system). The United States hurried development of a new family of "smart" weapons, which homed on the target by laser light or at extreme ranges (40 miles or more) by relying on television guidance. Continuing their efforts to reduce plane and pilot losses, the U.S. military began using remotely piloted vehicles (RPV). Several different models saw action in Vietnam; the standard RPV design evolved from a target drone known as the Firebee I. Dimensions varied, but the basic configuration had a wingspan of 13 to 14 feet and a length of 26 to 29 feet. The vehicle used small jet engines rated from 1,700 to 2,700 pounds of thrust, giving top speeds of over 700 MPH. As utilized in Vietnam, operational versions were air-launched from modified C-130 Hercules aircraft and carried out many photographic and/or electronic reconnaissance missions over North Vietnam. The Hercules parent ship carried all necessary gear for command, tracking, and data relay. For recovery, the RPV either descended by parachute or was picked up in midair by a helicopter. Some of the drones reportedly tested air-launched weapons, such as the Shrike and the Maverick air-to-ground missiles.

During the winter of 1972, the USAF mounted especially heavy raids against Hanoi and the port of Haiphong, hoping to pressure the North into a final cease-fire; but armistice talks, already in progress for several years, dragged on. The raids were costly; about 15 B-52 bombers and crews were lost to improved SAM missiles. Although air power partisans contended that the severe damage done to industrial and port facilities in the North hastened the final cease-fire agreements, President Nixon's historic journeys to China and to Russia during 1972 seem to have significantly improved the climate for peace and negotiations, concluded in 1973. After U.S. troops pulled out of the country, fighting between North and South continued until the South Vietnamese regime collapsed in the spring of 1975.

CIVIL AVIATION

While the military role of aviation demonstrated its destructive fire power in Vietnam, the role of civil aeronautics attained new significance in American society. In the numbers produced, as well as in dollar value, the figures for manufacturers of airliners and general aviation aircraft climbed to record levels. Although a contraction occurred during 1970–72, largely as the result of increased fuel costs related to Arab-Israeli conflicts in the Middle East, the aerospace industry quickly rebounded. The value of deliveries for airline transports rose from $1.2 billion in 1965 to $3.8 billion in 1975, and general aviation revenues in the same period went from $318 million to $1 billion, the first time that general manufacturers had reached such an exalted total. Although the general aviation industry continued to be recognized as a separate sector, the changing nature of the air transportation market brought many models of general aviation into scheduled air transportation operations.

By the late 1960s and early 1970s, two additional types of scheduled airline passenger services had emerged as increasingly important components of the air travel network: regional airlines and commuter airlines. Regional operations (as opposed to major trunk airlines emphasizing long-distance routes) began late in World War II, when the CAB launched a program to encourage short-range local and regional services. Known at first as "feeder lines," they provided air service between smaller communities and funneled passengers into major cities served by the trunk lines. The early feeder lines flew a hodge-podge of equipment like the twin-engine Lockheed Electra and twin-engine Beechcraft Model 18. After the war, surplus DC-3 transports provided a reliable, economical plane. During the 1950s and 1960s, some consolidation occurred, so that multistate, or regional, airlines emerged. The new organizations purchased twin-engine Convairs converted to turboprops, along with new designs like the turboprop Fairchild F-27. Several companies eventually equipped themselves with jets like the DC-9 and Boeing 737. By the mid-1970s, regionals had come to represent a network that carried 27 million passengers to over 450 cities.

Another level of scheduled services also developed momentum during the 1960s: the commuter airlines. Only a dozen or so operated in 1964, but over 200 companies were offering service by the early 1970s. Some of the commuters owned a half-dozen aircraft or fewer; equipment ranged from single-engine light planes and twin-engine equipment from the general aviation sector to conventional piston-engine twins like the venerable DC-3 to new types of small turboprops. The Canadian-built de Havilland DHC-6 Twin Otter became a widely used commuter aircraft. Originally designed to the demanding specifications of bush pilots, the Twin Otter's fixed gear and rugged construction made it a reliable, low-maintenance airplane for cost-conscious commuter airlines. Moreover, the plane's short takeoff and landing (STOL) characteristics proved valuable for many smaller city airports and allowed the plane to operate from short runways within terminal areas of busy hub airports. Although commuters offered direct, regular service between smaller communities, their routes also included flights between these smaller communities and major airline hubs. One such operation in the mid-1970s, Metro

The 1960s marked a period of intensive airline competition. Changing styles in fashions for cabin attendants were introduced with much fanfare, the better to catch the interest of vacationers as well as business travelers. These stewardesses model the trendy "A-line" look of the era. Courtesy of United Airlines.

Commuter airlines became one of the fastest growing segments of passenger transportation during the sixties. Metro Airlines, of Houston, Texas, operated DeHavilland Twin Otters between Houston Intercontinental and smaller population centers. Photo by J. Pamela Culpepper.

Airlines of Houston, Texas, offered up to 50 daily flights between Intercontinental Airport, on the north edge of the city, and Clear Lake City, some 50 miles away on Houston's southern edge. Although the flights included some passengers continuing from other points, the major share of Metro's patrons on this particular route originated from NASA's Johnson Space Center and its various contractors, as well as from the heavy concentration of petrochemical plants and associated industries.

The demand for air transportation from smaller cities grew partly because convenient railroad passenger connections declined in the face of interstate highway construction and the continuing momentum of auto travel. Nevertheless, there were many instances when infrequent flight schedules to and from smaller cities convinced many individuals and companies to buy their own business planes. Moreover, new standards of performance and convenience available in general aviation planes of the era frequently put them on a par with the scheduled airlines.

By the mid-1960s, the appearance of turboprops, pressurization, and jets set the basic pattern for general aviation development over the next one-and-a-half decades. Although American manufacturers continued to dominate the market, the expanding worldwide demand provided openings for foreign designs. The Japanese penetrated the American market with the Mitsubishi MU-2, a stylish, medium-range twin turboprop assembled in the United States from imported components. Similarly, the market for executive jets increased. After Learjet paved the way for volume sales, British Hawker Siddeley H.S.-125 and French Dassault Falcon executive jets enjoyed considerable popularity. In the late 1960s, Cessna began marketing its early Citation series of jets, with performance and costs a step below some of the competition. In short, executive jets, like turboprops, now offered a range of prices and performance.

The elegant Grumman Gulfstream I continued to set the pace for pressurized turboprops. Its size (24 passengers in high-density seating plans), performance (348 MPH at 25,000 feet), and cost (between one-half million to one million dollars) kept it in a premier class, so that other manufacturers successfully produced a variety of less expensive but versatile aircraft. After Beech introduced its pressurized turboprop, the King Air, in 1965, Piper and Cessna also begun to pressurize their twins and add turboprop power plants. By 1974/75, buyers could choose from over two dozen aircraft offering various options of pressurization, turbocharged engines and/or turboprops, ranging from $80,000 to $900,000.

Even if potential customers found the price tags too high or the performance still too limiting, a bewildering array of choices still remained. A two-place Cessna 150 with fixed gear sold for $9,400 and cruised at 117 MPH. Nearly four dozen other single-engine, four-place designs, with fixed or retractable gear, were priced from $14,000 to around $53,000, with speeds up to 200 MPH. An array of piston-engine twins carried six to eight passengers at about 250 MPH and cost from $55,000 to over $200,000. The least expensive jet, Cessna's Citation I, was priced at over $700,000; over a dozen different manufacturers offered designs that could easily cruise at 550 MPH and cost $3.5 million. Exact prices for the higher performance singles and twins were always hard to define, since an airplane's total cost increased as a function of added instrumentation and, in the case of large turboprops and turbojets, the degree of opulence specified for customized interiors.

In addition, there was a variety of sport planes and sail planes in the general aviation community, along with a considerable number of float planes, amphibians, and agricultural aircraft. Helicopters began to proliferate, partly because of their inherent utility and partly because of the Vietnam conflict, which increased production, stabilized prices at acceptable levels, and created a large population of experienced pilots as the war wound down. Due mainly to complexities of rotor design and construction, helicopters remained expensive to own and to operate, with five-place models listing at over $130,000 in 1975 and larger craft at over $630,000.

Buyers were willing to accept the high prices of airplanes and helicopters because they offered the means to save time and money. The patterns of their utilization generally followed those established in prior decades, and general aviation became even more appreciated during the late seventies and early eighties as a result of airline deregulation (see chapter 8). A dynamic aspect of the nation's air transportation phenomenon, general aviation continued to change and grow, just as air cargo and airline travel did.

From hesitant beginnings in the late 1920s and slow progress in the 1930s, commercial air cargo also made great strides in the post–World War II era. Wartime experience in long-range cargo operations helped; more important was the postwar availability of dozens of surplus military multiengine transports. Their scarred, spartan interiors required no refurbishing for hauling freight, and numerous operators formed companies to fly these war-weary castoffs. Not many cargo airlines survived. Running the old military transports frequently called for extensive overhauls and costly maintenance. Locating profitable cargoes and patching together efficient route structures

The Boeing 727 became the best selling airliner of the jet transport era. The Falcon 20 jet (foreground), made in France, competed with American designs in the expanding market for executive jets. Federal Express, pioneer of the overnight, small package freight service in 1973, successfully blended the capabilities of some 74 aircraft to serve a variety of large and small cities across the United States. Courtesy of Federal Express.

was never easy, and fierce competition among the fledgling freight lines eventually took its toll. A handful of survivors began to show a profit, and a few even used jets when they became available. The scheduled passenger lines, sensing lost revenues, began to pay more attention to cargo services on their normal passenger routes and began to operate their own all-cargo services. The volume of air cargo, including freight, mail, and express, comprised a very small percentage of the total ton-miles moved by all forms of transportation. But many commodities were best moved by air freight; the florist industry, the fashion industry, and others became extremely dependent on air transport. Many standard production-line items also went by air, since rapid transport could reduce inventories, cut warehousing costs, and simplify handling procedures. Between 1965 and 1976, the cargo volume moved by certicated airlines escalated from 2.3 billion ton-miles to 5.4 billion ton-miles. The cargo lines as well as regional and commuter lines benefited from the second-generation jet transports of the 1960s and from improvements in navigation and more efficient air traffic control.

The unqualified success of new turbine-engine airliners like the Douglas DC-8 and the Boeing 707 quickly moved former industry skeptics to increase the passenger capacity of these planes in order to capitalize on the traveling public's enthusiasm for jet-propelled travel. New plans envisioned capacious aircraft with a carrying capacity that would have been thought totally impractical only a few years before. If not revolutionary in an aerodynamic sense, the big new jets were certainly revolutionary in size. Along with the aerial behemoths came a new stable of smaller jet transports to service intermediate routes and smaller communities. Taken together, the second-generation jets generated a new revolution in passenger travel, becoming the major intercity mass transit carrier on domestic as well as international routes.

The first step toward bigger jets came when Douglas decided to increase dramatically the number of passenger seats in the DC-8. Anxious to recoup its traditional dominance in the airliner manufacturing industry, Douglas set out to surpass upstart Boeing in this field. Attempting to save time and money, Douglas decided to stretch its basic DC-8 into a new "Super Sixty" series. The first of these, the DC-8-61, took to the air in 1965. The plane's fuselage included an added segment to make it nearly 37 feet longer than its predecessor, and the new version seated up to 259 passengers, compared to a maximum of 189 for early DC-8 models. The passenger capacity seemed unprecedented for the time, and the passengers' first glimpse of the cabin aisle, which seemed to stretch forever, was not quickly forgotten. In stretching the DC-8, Douglas realized an unanticipated advantage. The wing sweep and its location on the fuselage permitted the fuselage of the "dash-sixty" series to use the basic DC-8 wing manufacturing jigs and landing gear hardware. When Boeing engineers sat down to consider similar changes to the 707, they discovered than an elongated passenger cabin, given the Boeing's wing sweep and landing gear location, would have the tail scraping the runway each time the plane rotated into the necessary angle for takeoff. The Boeing people realized that workable modifications to a stretched 707 would require too much redesign and engineering to be cost-effective. Why not build an entirely new plane? They did, and thereby triggered the "Jumbo Jet" generation.

Boeing entered its 747 design studies with momentum carried over from the company's proposals for an oversized military cargo plane, a project that went to Lockheed as the C-5A. Although Boeing was still thinking about stretched versions of the 707 during 1965, the legacy of the military design studies obviously contributed to rapid progress of the 747 design once the company decided to go ahead with the project early in 1966. Pan Am, impressed by the new plane's specifications, placed the first orders (for 25 aircraft) in April of that year. Meanwhile, Boeing had to build a sprawling, cavernous new assembly building for the plane near Renton, Washington. The first ceremonial rollout of a new 747 occurred in the autumn of 1968, followed by production deliveries in December 1969. The logistical aspects of the accommodations for some 490 passengers in a dense-seating arrangement posed new problems for passenger cabin staffs. Serving meals—and cleaning up afterward—became a carefully orchestrated enterprise, especially on long-distance flights, when passengers received two, or sometimes more, meals en route. Obviously, the oversized passenger sections of a 747 changed the intimate and comradely travel arrangements of the prejumbo era. On the other hand, the 747's long double aisles offered ample opportunity for passengers to stroll and stretch their legs, and first-class travelers could avail themselves of an upper-deck lounge that boasted, among other amenities, a small piano. For all its size, the huge plane held its own with contemporary airliners in speed, capable of cruising at 590 MPH.

The Boeing 747 gave credence to the descriptive terms "jumbo-jet" and "wide-body" jet. The escalation in size is evident from the 707 and the 747 shown here. The 1960s also marked a period of rapid growth in air cargo services. Special freight containers were designed to fit the curved fuselage of jet transports. Courtesy of American Airlines.

Boeing buttressed this achievement with continuing improvements to the distinctive Boeing 727 trijet, designed for short- to medium-stage lengths and for service to and from smaller airports. Variants offered airlines a choice of planes seating anywhere from 100 to over 160 passengers in different first-class/tourist-class arrangements. But the 727's strongest appeal lay in its apparently complex high-lift wing system. Boeing engineers had to meet some apparently contradictory requirements calling for a high cruising speed (around 600 MPH) but a low landing speed and short takeoff run for the shorter runways at smaller airfields. Boeing evolved its own high-lift wing system, with an arresting array of leading-edge and trailing-edge slots and flaps for enhanced lift at lower speeds. Fully deployed, the high-lift system made the wing seem to come apart in a myriad of metal slabs, leading to a standard joke about the 727 in which the captain asks the copilot to put the wing back together again after taking off. For all of its seeming complexity, the high-lift system turned out to be very reliable, with acceptable maintenance costs.

Like the 727, the Douglas DC-9 represented a design that opened many new cities to the convenience of jet travel. A twin-engine plane, the DC-9 featured a self-contained boarding stair that eliminated expensive ground facilities at small airports. The DC-9 also represented a new formula to cope with the extraordinarily high costs—and risks—of jet airliner development by a single manufacturer. The Douglas Aircraft Company* enlisted the support of de Havilland of Canada, who assumed a good share of production risks by building the DC-9 wings, betting on future sales of the Douglas jet to offset its own initial development and production costs.

By the late 1960s, air travel growth had prompted the airlines to consider acquiring additional types of larger wide-body transports having a high density seating for about 300 passengers but suitable for short- and medium-haul duties. American Airlines issued such specifications in 1966, prompting both McDonnell Douglas and Lockheed to develop new airliners. Originally, both companies had considered designs for a twin jet, but both eventually opted for a trijet configuration, which gave their airliners a potential range and safety margin for transoceanic routes. The high costs of developing the new transports forced both companies to seek risk-sharing partners. In the context of the 1970s, this increasingly meant a foreign partner, since the American airliner market by itself did not promise the sales necessary to amortize development and production costs. Linking up with an overseas partner presented mutual advantages: the U.S. manufacturer kept its own financial risk under control and secured a more promising entree into foreign markets; the overseas partner could retain national prestige, keep employment stable, stay abreast of advanced aeronautical technology, and reap a good profit in the event export sales developed as hoped.

In the cases of McDonnell and Lockheed, the foreign marketing strategies reached out in different directions and in different ways. Douglas stressed foreign as well as domestic options to underwrite the DC-10's costs and development. Lockheed's venture found support in the United Kingdom, where Rolls-Royce in particular saw potential profits in supplying its advanced RB211 turbofan engines for Lockheed's wide-body design. Development troubles plagued the new engine, however, and progress of the L-1011 TriStar began to falter. Despite some talk of replacing the British engines with American power plants, Lockheed stuck with Rolls-Royce. The engine manufacturers had recently been nationalized by the Labour government, who promised Lockheed unusually favorable prices, delivery arrangements, and interest rates as a means to maintain employment levels in the British aerospace industry. From Lockheed's viewpoint, installation of Rolls-Royce engines gave the American manufacturer a favorable position for selling the TriStar to British Airways, the nationalized airline. The problems finally worked themselves out, and the L-1011 developed into a successful airliner. But the Lockheed story held particular significance for the airline manufacturing industry because of the trend toward dependence on major multinational ventures in the development of costly new aircraft.

By the early 1960s, serious consideration of the first supersonic transport (SST) had begun, based on collective improvements in civil air transport technology (including airframes, aerodynamics, electronics, and power plants) and reinforced by high-speed flight research on large aircraft like the B-70 (an advanced bomber that was test-flown

*A 1967 merger created the McDonnell Douglas Corporation, of which the Douglas Aircraft Company became a subsidiary.

Aerospace exports represented a significant amount of American foreign trade. The Lockheed L-1011 TriStar, designed to use Rolls-Royce engines built in Great Britain, also represented a growing trend toward international risk-sharing in aircraft development. Courtesy of Lockheed Corporation.

and eventually cancelled). The exceptionally high R&D costs prompted the U.S. government to play a large part in underwriting SST development, guided by the Federal Aviation Agency (FAA). Following a lively and competitive design review concluded in 1967, Boeing won the contract for the airframe and General Electric won the engine contract. Engineers had planned first flights for the SST prototype by late 1972 or early 1973. Development proceeded in an atmosphere of international competition, since Russia had its own SST under development and a joint British-French SST project, known as the Concorde, was well advanced. The original American design, unlike its foreign counterparts, featured a variable-sweep wing configuration, but continuing technical problems led to a fixed, wide-span delta wing similar to those of the Russian and Concorde aircraft. Technical difficulties were not the only troubles to hinder the American SST project, however. The program encountered growing opposition from environmental groups and cost-conscious critics both in and out of Congress. Alarm about the "sonic boom" factor, causing damage to buildings and windows—to say nothing of people's nerves—finally led the FAA in 1970 to restrict future supersonic cruises to overwater routes. There was also rising alarm about adverse effects in the upper atmosphere; exhaust wastes from SST engines at extremely high altitudes, critics said, would reduce the ozone layer. Ozone layer degradation, critics continued, would permit more ultraviolet rays to reach the earth's surface, possibly increasing skin cancer and retarding the growth of plants. In the face of

The big new jet transports featured double aisles. In the DC-10 first-class section, seats were installed in pairs, with generous space allotted to each seat. Compare this to the interior of the Boeing 247 on p. 102. Courtesy of McDonnell Douglas Corporation.

growing opposition, the House of Representatives voted down additional funds for the SST in 1971, bringing the American project to its close.

For the United States, a country that prided itself on its technological leadership, particularly in the aviation field, scrubbing the SST was an unprecedented step. Since the Concorde and Russian SST programs continued, some aviation figures in the United States charged that America might lose its aeronautical advantage to foreign countries, and still others gloomily forecast a dangerous loss in export funds as American carriers bought a European SST, rather than the reverse. Following successful Concorde flights in 1969, a protracted R&D schedule followed before the plane entered scheduled foreign service early in 1976 (Paris–Rio de Janeiro and London-Bahrain), soon followed by routes to New York and Washington, D.C. But rising fuel costs and technical reasons discouraged sales to the United States and other foreign flag lines. (Braniff in the United States leased a Concorde for service on its Dallas-Washington, D.C., route in 1979, flying the plane at subsonic levels). The plane was generally popular with passengers, and its noise levels proved reasonable for operation from other U.S. airports. Nonetheless, no additional sales developed. The Concorde production lines shut down in 1979 after only 16 planes had been built. Even though the Concorde failed to develop high sales and corner a new international SST airline market, the multinational venture nevertheless underscored the trend toward such collaboration. Boeing's careful cultivation of foreign manufacturers for its new airliner development in the 1980s illustrated the viability of the multinational approach in a high-cost, high-risk industry.

Just as American leadership in civil aeronautics persisted, despite the growth of foreign competition and international ventures, astronautics in the United States began to reflect the same patterns.

SPACE EXPLORATION: APOLLO-SATURN

Confidence in manned spaceflight operations from the Mercury missions (1961–63) encouraged NASA as the space agency looked toward multicrew flights in the future. The Russians again led the way in such missions with the flight of *Voskhod I* in 1964 (a three-man crew). A Russian cosmonaut performed the first "space walk" during the *Voskhod II* mission in 1965, the same year that NASA began its own series of two-man launches with the Gemini program. Using a modified Titan II ICBM as the booster, the first Gemini mission blasted off from Cape Canaveral on March 23, 1965. The Gemini program, which continued into the winter of 1966, included the first American space walk as well as highly important rendezvous and docking techniques. The maneuvers required to bring two separate orbiting spacecraft to a point of rendezvous, followed by docking, helped to pave the way for more ambitious manned space missions. Plans for multicrew space stations and lunar exploration vehicles depended on these rendezvous and docking techniques as well as the ability for astronauts to perform certain tasks outside the protected environment of the spacecraft itself. The successive flights of the Mercury-Redstone, Mercury-Atlas, and Gemini-Titan missions were progressive elements in a grand design to launch a mission to the moon and back.

Against the background of Mercury and Gemini developments, work was already progressing on the Apollo-Saturn program. The spacecraft for the Apollo adventure evolved out of the Mercury and Gemini capsule hardware, and other R&D was directed toward the new technology required for a lunar lander and associated systems. A similar effort involved the development of an entirely different family of boosters. Heretofore, NASA had relied on existing launch vehicles requisitioned from the armed services—the Redstone missile, along with the Atlas and Titan ICBMs. For manned lunar missions, a rocket of unusual thrust and lifting capacity was called for—literally, a giant of a booster.

To reach the lunar surface, NASA planned a gigantic, three-stage launch vehicle known as the Saturn V. The evolution of such an advanced vehicle required the fabrication and launch of several developmental rockets, known as the Saturn I and Saturn IB types. Both of these were two-stage vehicles, with lower stages propelled by kerosene/liquid oxygen engines and the upper stage by advanced liquid hydrogen/ liquid oxygen engines. A series of ten Saturn I launches during the early 1960s verified the safety and reliability of clustered rocket engines and liquid hydrogen technology. Five Saturn IB flights tested the capability of the Command and Service Module (or CSM, which carried the three astronauts into lunar orbit and back to the earth) and the Lunar Module (or LM, which carried two of the three-man Apollo crew to the lunar surface and back to the CSM). The development of the three-stage Saturn V proceeded on a parallel schedule.

These complex problems required a far-flung managerial enterprise. At Marshall Space Flight Center (MSFC), Manned Spacecraft Center (MSC), Kennedy Space Center (KSC), and other NASA centers, NASA conducted many in-house R&D projects related to the Apollo-Saturn scheme, although final development and production of flight hardware was carried out by primary contractors and an elaborate network of subcontractors and suppliers in virtually every state. For example, Grumman (on the East Coast) produced the LM and North American Rockwell (on the West Coast) built the CSM. A multitude of components came from subcontractors and their suppliers across the country. Aerojet-General supplied the rocket engine for the service module, whereas other engines came from Rocketdyne in partnership with Space Technology Laboratories (LM descent-stage engine) and from Rocketdyne in partnership with Bell Aerospace (LM ascent-stage engine). The roll-call of familiar manufacturers for other major spacecraft components ran from AC Spark Plug to Elgin Watch to RCA to Westinghouse. In the case of the Saturn V, Boeing served as prime contractor for the first-stage booster, North American Rockwell for the second stage, and McDonnell Douglas for the third. Main engines for each of the stages came from Rocketdyne. Similarly, the list of Saturn suppliers represented nearly every state in the union and covered the alphabet from Aeroquip of Jackson, Michigan (couplings and hoses), to Chicago Rawhide Company (seals), to Philco of Philadelphia (radio equipment), to the Wyman-Gordon Company of the North Grafton, Massachusetts (forgings).

The evolution of the Saturn class of launch vehicles mirrored many of the challenges of the Apollo program for lunar exploration. The Saturn vehicles were unique because they included the first rocket boosters designed and built specifically for manned launches, and because their remarkable size meant that nearly everything associated with them had to be built or scaled from the start to unprecedented proportions. Even with a solid technological foundation, it was the process of scaling up that so often created problems and costly delays in the Saturn program.

Compared to operational ballistic missiles of the 1960s, the massive scale of the Saturn V launch vehicles seems awesome. The Thor, a standard USAF intermediate-range ballistic missile (IRBM) of the era, had a diameter of 8 feet and stood 65 feet high. Powered by a single rocket engine of approximately 165,000 pounds of thrust, Thor's lift-off weight came to about 110,000 lbs. By contrast, the Saturn V boasted a diameter of 33 feet and towered 363 feet above its launch pad. Powered by a cluster of five engines, the first stage of three-stage Saturn V generated a total thrust of 7.5 million pounds, with a lift-off weight of more than 6.1 million pounds. Trying to interpret these awesome statistics in comprehensible terms, press releases frequently noted that the Saturn V was as high as a 36-story building and weighed more than a good-sized naval destroyer. The powerful first stage had its own impressive figures. Its girth was ample enough to permit three moving vans to be driven, side by side, into the tanks. The oxidizer tank alone could take on enough liquid oxygen to fill 34 railroad tank cars (or 54, depending on which press handout one read). To force fuel and oxidizer to the engines, which consumed three tons of propellants per second, the pumps of the first stage generated the power of 30 diesel locomotives. Getting three men from the earth to the moon and back again required impressive machinery.

The first steps toward the development of the Saturn rockets had been taken by the Department of Defense, who foresaw potential applications for a very large booster to launch defense-related hardware into space. Possible payloads included reconnaissance satellites, space stations, and similar cargo. In 1958, the department's Advanced Research Projects Agency (ARPA) contracted with the Army Ballistic Missile Agency (ABMA) for an unusually large rocket stage to demonstrate the feasibility of a booster having a thrust of 1.5 million pounds. The ABMA development group, with the von Braun team as its core, proceeded on the assumption that this demonstration stage was to be static-fired only and was not considered a flight vehicle. For this reason, the demonstration booster emerged as a "bargain-basement" design, an unwieldy-looking vehicle that seemed to be the very antithesis of a sleek, Buck Rogers spaceship.

Preliminary paper studies for the booster soon focused on the use of clustered engines to achieve such tremendous thrust, since no single engine of that power was yet available. Limited funds put a premium on inventiveness to keep costs under control and also prompted ABMA designers to use "off-the-shelf hardware" rather than to develop expensive new gadgetry. This design approach characterized the entire Saturn program. Still, the size and complexity of unusually large Saturn components presented a number of unique R&D challenges. Many problems cropped up, and many millions of dollars were eventually required to achieve success.

The original big booster, known as the Saturn I, emerged from the drawing boards as an awkward compromise. For the engine cluster ABMA decided to use an S-3D propulsion system, a proven engine used to power IRBM weapons like the Thor and Jupiter. But first, each engine had to be uprated from a thrust of 165,000 pounds to 188,000 pounds. This proved to be no simple task, since the more powerful engines developed an unhealthy combustion instability that required a redesign of the propellant injectors. The uprated engine, newly designated as the H-1, also experienced shortcomings in several other areas, each of which necessitated costly redesign and testing. For all the immensely sophisticated research facilities and computer techniques, the successful H-1 engine program, as in so many aspects of Saturn hardware development, frequently came down to a question of "cut-and-try."

As engine development progressed, it occurred to ABMA planners that if the engines were to be clustered, why not cluster the Saturn tankage as well? For this reason, the Saturn I's first stage became a bundle of eight separate Redstone tanks surrounding a larger Jupiter tank. Engineers at ABMA were not terribly worried about the complications in valves and plumbing that this arrangement entailed, since the original ARPA order had called only for a static-firing test to demonstrate the efficacy of the cluster concept. But there were those who still insisted that it was not possible to get all eight H-1 engines ignited at the same time and keep them running in concert. When ARPA suddenly decided to go ahead with flight tests, the ABMA rocket team had little choice but to cling to the original bundle of tanks and engines. Still, many observers remained skeptical about the awkward-looking Saturn I. As the time for firing tests drew closer, there were many pointed remarks about "cluster's last stand," von Braun recalled.

Launches of the two-stage Saturn I increased NASA's confidence in tankage, engines, and cryogenic propellants like liquid hydrogen, setting the scene for the first manned missions of the Apollo program. Courtesy of NASA.

On April 29, 1960, all eight engines of the test booster roared into life during a successful trial firing of the big rocket cluster, and the first launch occurred on October 27, 1961. In the meantime, ABMA had been transferred to MSFC, and important decisions were reached concerning upper stages for the clustered first-stage booster—

decisions that determined the nature of future Saturn configurations as well. During Decembr 1959, NASA and Department of Defense representatives, including von Braun, had organized the Saturn Vehicle Team (known as the Silverstein Committee for its chairman, Abe Silverstein). Among other things, the Silverstein Committee decided to rely on liquid hydrogen (LH$_2$) as the fuel in upper stages of the Saturn boosters, to develop a series of multistage rockets, and to follow the "building-block concept," whereby each succeeding vehicle would utilize a new upper stage while incorporating already proven stages from its predecessor. By 1963, the tangle of paper studies, changing vehicle requirements, and nomenclature had been boiled down to three basic, large launch vehicles: Saturn I, Saturn IB, and Saturn V. These vehicles generally followed the building block concept, and all utilized LH$_2$ as fuel in the upper stages.

The decision to use LH$_2$ was somewhat controversial. As a fuel, LH$_2$ was especially appealing because of its high specific impulse, a basic measure of rocket performance. Compared to a conventionally fueled engine (using a grade of fuel known as RP-1, similar to kerosene), a propulsion system burning LH$_2$ could increase the engine's specific impulse by 40 percent. Still, many propulsion system engineers at MSFC regarded LH$_2$ with a jaundiced eye. Its exceedingly low temperature (423° F) caused embrittlement of many materials generally used for tankage, pipes, and valves, and it was viewed as a tricky fuel to handle. Based on research at NASA's Lewis Research Center and elsewhere, Silverstein, the Lewis Center's former director, took the lead in convincing the other committee members to go along with LH$_2$. It was a landmark decision, since utilization of high energy propellants like LH$_2$ in Saturn's upper stages became a key factor in developing efficient and successful booster vehicles.

Fortunately, an LH$_2$ engine, the RL-10, was potentially available for use in Saturn I upper stages. Developed by Pratt & Whitney, the RL-10 had originally been intended for an LH$_2$ fueled upper stage booster called the Centaur, developed under the direction of the air force. Using a cluster of RL-10 engines, NASA planned its first LH$_2$ upper stage booster. In the process, McDonnell Douglas, the S-IV contractor, had to cope for the first time with the problems of producing large, unitary propellant tanks. Unlike the complex, clustered tankage of the Saturn I first stage, upper stages were single units, but their dimensions created a host of new problems in fabricating and production. Special, oversized jigs and welding fixtures had to be designed and built. Since the propellant tanks carried thousands of gallons of cryogenic propellants under pressure, all of the unusually long welded tank seams had to be absolutely flawless. Designers incorporated a common bulkhead to separate liquid oxygen and LH$_2$ within the tank, eliminating the extra length and weight of two separate propellant tanks. But in order to gain this length and weight savings, the size of the common bulkhead required new techniques of fabrication, metal bonding, and insulation. New facilities for assembling oversized stage components had to be built, and static test stands had to be constructed. The challenges of designing and producing the S-IV stage, in short, reflected many of the complexities faced by each of the stage manufacturers of the Saturn program.

Using some of the experience gained from its Thor IRBM for the air force, McDonnell Douglas resolved a number of production problems encountered on the

S-IV stage and used the legacy of experience from the S-IV to build the slightly larger S-IVB. The S-IVB also differed in that it had a single engine, the more powerful J-2 produced by Rocketdyne. Applying the building-block concept, the S-IVB first flew as an upper stage on the Saturn IB two-stage vehicle, and eventually as the final stage of the Saturn V three-stage rocket. The J-2 engine was additionally incorporated on the second stage of the Saturn V vehicle. Thus, by the mid-1960s, hardware for the upper stages of the Saturn series had begun to reach operational status, and development of engines and tankage for the lower stages had begun to accelerate.

The first stage of the Saturn V was powered by a cluster of five F-1 engines, also manufactured by Rocketdyne. The F-1 was a legacy of an air force experimental program dating to 1955. Although the F-1, like the H-1, was a conventional propulsion system that burned RP-1 fuel, its awesome thrust of 1.5 million pounds for a single engine generated many developmental problems of commensurate magnitude. Combustion instability nearly killed the engine at one point, and in the process of getting the F-1 to work in 1965, one engineer characterized engine development as a "black art." After innumerable changes, he explained, one finally had an intuitive feeling that the engine was going to function, even though it was not entirely clear which modification—or combination or modifications—finally eliminated the problems. The F-1 was also the center of an early controversy over the number of engines in the first-stage cluster: four or five. Although the fifth engine initially added weight, its extra thrust eventually provided the margin of power to boost the Apollo spacecraft to successful missions. As von Braun mused years later, he was certainly grateful for the fifth engine, since every time he got a progress report on the design of the spacecraft "that damn LM had gotten heavier again."

Inheritors of a strong tradition stressing the army arsenal custom of in-house R&D and manufacturing, the von Braun team at MSFC was reluctant to let contractors do all of the work. The initial five S-IC first-stage boosters were produced by MSFC before Boeing took over as prime contractor. This effort was not made solely to salve the egos of MSFC's highly trained work force but also to maintain skills to back up contractors when they got in trouble and to provide NASA with a "yardstick" in monitoring performance and cost curves for hardware produced by the contractors. By and large, this approach proved effective and was one factor in keeping the Saturn program on schedule and within reasonable cost. It was easier for MSFC to follow the S-IC stage more closely, since its final production took place at the Michoud Assembly Facility, a federally owned manufacturing complex near New Orleans. Moreover, Michoud was not far from the Mississippi Test Facility, where S-IC and S-II stages were test fired prior to shipment to Kennedy Space Center for launch. Partly because MSFC monitored the S-IC stage so closely, pouring its accumulated knowledge and expertise into the effort, the brawny first-stage booster progressed through design, manufacturing, and testing with remarkably few problems. Such was not the case with the S-II, the second stage of the Saturn V.

The S-II stage contractor was North American Rockwell. In some respects, it seems that the S-II should have progressed with only a few problems, since the engine cluster used the J-2 model (proven on earlier flights of the S-IVB), and booster design could benefit from experience accumulated by S-IV/IVB fabrication and production, including the common bulkhead arrangement. Instead, the S-II became the problem child of the Saturn V stages, threatening to upset the entire booster program. The S-II dimensions were larger, and scaling up inevitably generated new difficulties. Part of the imbroglio seems to have developed from an overload on North American's management, burdened not only with the S-II but also with the Apollo Command Module and with all of the Saturn's main engines. At the same time, the S-II got caught in a weight-reduction program, as the Apollo payload began to add pounds. Since the S-IVB third stage was already in production, little could be done to reduce its weight, and the peculiarities of orbital trajectories made it more compelling to shave weight from the S-II stage rather than the S-IC first stage.

As the S-II shed poundage, the stresses on its structure and systems increased, so that a rash of test failures resulted. Late in 1965, the situation became so critical that some top-level managers at MSFC muttered darkly about the possibility of taking the S-II stage contract away from North American and giving it to somebody else. Before this dire step was taken, North American vigorously overhauled its managerial organization and appointed some new executives. Manufacturing techniques, especially welding technology, were refined with the aid of technical assistance from MSFC and the rocky progress of the S-II program finally evened out by 1968.

The process of managing the seemingly overwhelming number of individual parts, components, and subassemblies for each engine and each stage called for unusually stringent management procedures. The Saturn Program Office at MSFC, which had major responsibility for this task, placed a good deal of emphasis on what it called the Program Control Center, a room crammed with constantly updated progress charts, duplicated by identical control centers at each major contractor plant. Equipped with microphones and projection screens, control centers facilitated executive meetings with simultaneous talks and visual presentations linking different locations across the country.

Among the numerous managerial tangles that inevitably cropped up, one of the most intriguing involved logistics. While launches occurred in Florida, manufacturing and testing of rocket stages took place in Alabama, Mississippi, Louisiana, and California. Since the size of the Saturn booster stages ruled out conventional transport by rail or truck, NASA acquired a flotilla of barges and seagoing vessels specially equipped to transport and protect its elephantine rocket stages. En route to Kennedy Space Center, for example, the S-IC, as well as lower stages of the Saturn I and IB, floated by barge down the Mississippi across the Gulf, and the S-II sailed from California by ship via the Panama Canal. The S-IVB had a more unique mode of travel, arriving by air freight. The carrier, appropriately dubbed Aero Spacelines, flew converted Boeing Stratocruiser airliners. In order to accommodate the S-IV and S-IVB stages, the company added a huge, bulbous structure to the original fuselage, giving rise to a new sobriquet for the aircraft: the "Pregnant Guppy" (which carried the S-IV) and the larger "Super Guppy" (which carried the S-IVB).

By air and sea, each stage safely arrived for launch from Kennedy Space Center following full-duration firing tests by NASA/contractor test teams. Despite the tests,

To carry oversized but vulnerable space hardware and keep the Apollo program on schedule, NASA relied on the Super Guppy, a radically modified Boeing Stratoliner. The unique plane carried various cargoes like the instrument unit segment of Saturn rockets, as well as Apollo command modules and upper stage rockets of the Saturn launch vehicles. Courtesy of NASA.

however, NASA had originally planned to fly each stage of the Saturn V in successive launches: the first stage only "live," with inert upper stages; then first and second stages in separate live launches; then all three stages live. It was a conservative approach, consistent with the step-by-step tradition of the experienced rocket engineers from ABMA and MSFC. Then, in an effort to save time and money, NASA headquarters decided to launch the initial Saturn V "all up,"—each stage live. The idea came from George Mueller, NASA's associate administrator, who took steps to implement it during 1963. MSFC received the new directive with considerable consternation. After a frequently rancorous debate, MSFC finally agreed in 1964 to go along with what was thought to be a calculated risk. The first Saturn V launch, known as AS-501, was a stunning success on November 9, 1967. Still, as von Braun confessed later, he had remained doubtful until the mission was completed.

During 1968, a second launch of the Saturn V developed some unpleasant surprises: severe longitudinal vibrations ("POGO," in astronautical slang), and mysterious failures in some J-2 engines. Left unchecked, the POGO effect might destroy a booster in flight; engineers solved the problem by "detuning" complementary resonant frequencies between the first stage and its F-1 engines. The mysterious engine failures baffled NASA and contractor slueths for several weeks, until special tests in a vacuum chamber uncovered vibration problems in one of the J-2 fuel lines. During standard engine trials in an engine test cell at sea level, a build-up of frost reduced excessive vibrations in the fuel line. In the vacuum chamber, which duplicated the arid atmosphere of space, vibrations went unchecked until the fuel line developed tiny cracks that triggered an automatic engine shut-down. A simple modification to the line halted the failures. Chastened, but still determined, NASA renewed its schedule for a series of manned launches.

GIANT LEAP: MANNED MISSIONS

Development of manned spacecraft, vehicles for landing and takeoff from the lunar surface, self-contained spacesuits for lunar excursions, and a variety of scientific experiments brought a host of difficulties no less frustrating than the production of the Saturn V. Building the manned capsules for the Mercury and Gemini missions provided a mass of useful data, but the three-man Command Module for Apollo missions created new demands, new problems, and new solutions. Missions would last longer and extend further into space, which meant increasingly sophisticated instrumentation and life-support systems. Returning to earth, the three-man Command Module needed to make precise reentry maneuvers in order to survive its fiery plunge back into the earth's atmosphere. Managers and engineers often worked themselves to exhaustion to solve one dilemma, only to discover a frustrating problem elsewhere that caused another slip in the schedule. The most sobering delay occurred in the wake of a tragedy that took the lives of three astronauts in 1967. At Cape Kennedy, astronauts Virgil Grissom, Edward White, and Roger Chaffee entered a Command Module to run through a series of tests in a simulated flight environment. Something, probably an electrical short, touched off a flash fire in the oxygen-rich environment of the closed and sealed Command Module, killing the three men in a matter of minutes. Stunned by the accident, NASA launched a painful but thorough reassessment of its manned space program that resulted in a redesigned atmospheric system, in addition to numerous other changes in the Command Module itself. The lag in the program also provided an opportunity to reconsider and revise other aspects of the launch vehicle and the spacecraft.

Using the Saturn IB launch vehicle, *Apollo 7* was the first manned shot to test the modified three-man Apollo Command Module in earth orbit in 1968. With renewed confidence NASA made a bold decision to fly *Apollo 8* as a manned mission, and to reach out much further into space, using the Saturn V to put the Apollo CSM into a lunar orbit. Launched on December 21, 1968, the crew entered earth orbit, then fired the Saturn third stage again to speed the spacecraft away from the earth on a lunar journey. Astronauts Frank Borman, James A. Lovell, Jr., and William A. Anders were on their way to the moon. As one NASA historian wrote, they became the first humans to view the earth as a receding globe, "a mix of blue oceans, brown continents, and white clouds that was startling against the blackness of space." Two days later, the crew of *Apollo 8* also became the first humans to have left the earth's gravitational influence for that of a different body of the solar system. Following the successful return of *Apollo 8,* two more manned launches of the Apollo–Saturn V verified operations of the LM in space. Everything seemed ready for the momentous mission of *Apollo 11.*

Each of the flights preceding *Apollo 11* had vindicated one or more of the several steps required in a complete Apollo-Saturn mission to the moon. After launch, the CSM and the LM entered an earth parking orbit, still attached to the S-IVB third stage of the Saturn V. The first two stages had already fallen away into the Atlantic Ocean as their propellants became depleted during the ascent trajectory. After a final check to verify all systems, the S-IVB reignited and propelled the CSM and LM out of earth orbit and into the lunar trajectory. The CSM separated from the third stage and performed a docking maneuver with the LM to continue to the moon. The depleted third stage was jettisoned into a separate trajectory to orbit the sun or crash onto a remote area of the lunar surface. After the CSM/LM entered lunar orbit, two astronauts transferred into the LM for the descent maneuver, while the third astronaut continued to orbit in the CSM.

Following exploration of the lunar surface, the ascent portion of the LM blasted off to rendezvous and dock with the orbiting CSM. The astronauts transferred their lunar samples to the CSM and deorbited for the trip back to earth. Recovery of the three astronauts required separation of the Command Module from the Service Module, with the three astronauts protected in the Command Module during a fiery descent through the earth's atmosphere. At the appropriate altitude, parachutes deployed to slow the rate of descent of the Command Module before it splashed into the Pacific Ocean recovery area. Ships and helicopters in the vicinity proceeded to finish the recovery operation.

Thus, most of the critical sequences had been demonstrated in the series of missions leading to *Apollo 11.* But the descent to the lunar surface and the ascent back into lunar orbit were maneuvers to be performed for the first time. Despite all of the previous flights and hours of practice in several different types of simulators, the mission retained a sharp edge of risk.

Among the various devices for training astronauts in moon-landing techniques, the lunar landing training vehicle seemed the most awkward. With its open, tubular framework, a maze of wires, exposed propellant tanks, and convoluted pipelines for fuel and oxidizer, it immediately brought to mind the improbable gadgetry of Rube Goldberg. Astronauts dubbed it the "flying bedstead." Firing an array of vertical and horizontal rocket thrusters, the ungainly trainer could lift off and then make the kind of descent maneuvers necessary for a lunar landing. It was notoriously unstable; Armstrong had to eject from one of the free-flying trainers before it crashed in the spring of 1968, and a NASA engineer bailed out of another floundering vehicle a few months later. But its free-flight characteristics made the flying bedstead useful, and Armstrong made repeated flights during an intensive three-day practice session in June 1969, a month before the *Apollo 11* mission.

With the object of its flight eerily suspended in the twilight, an Apollo/Saturn V space vehicle sits poised on Pad 39 at Kennedy Space Center. Courtesy of NASA.

On July 16, 1969, Neil Armstrong, Edwin Aldrin, and Michael Collins boarded the Apollo-Saturn rocket for the *Apollo 11* mission—the first human voyage to the lunar surface. On July 20, four days after the Apollo-Saturn rocket had thundered away from Kennedy Space Center, Armstrong and Aldrin descended toward the eerie, silent lunar surface aboard the LM, code-named *Eagle*. The 12-minute flight seemed interminable for the millions who followed the radio and television broadcasts. Aldrin scanned the instruments, calling out the altitude and fuel supply as Armstrong controlled the descent engine and watched the stark terrain rise to meet them. He spotted the landing site but realized that massive boulders lay scattered in their path. Armstrong manually fired the *Eagle*'s thrusters, carrying them past the boulder field. About 15 feet above the surface, the *Eagle*'s rocket engines kicked up a sheet of dust, making it difficult to see, and, at the same time, the craft began to drift rearward. Armstrong corrected the flight path and concentrated so intently on maintaining control during the last phase of the landing that he did not immediately perceive the moment of contact when probes beneath the landing pads first touched the surface. Then the *Eagle* settled, and Armstrong switched off the engine. "Houston, Tranquility Base here," he announced. "The *Eagle* has landed." Obviously relieved, the control center radioed back, "You got a bunch of guys about to turn blue. We're breathing again."

After a check-out of the *Eagle,* followed by a short rest period, the two astronauts suited up to exit and explore the moon. Television equipment, attached to the LM, transmitted live coverage to breathless viewers around the world and caught Armstrong's statement: "That's one small step for man, one giant leap for mankind." Aldrin and Armstrong set up several scientific experiments on the lunar surface, gathered 46 pounds of lunar soil and rocks for later study and analysis, and rejoined Michael Collins, who had been orbiting the moon in the Command Module *Columbia.* A safe return and descent to earth, a triumphant reception, and the *Apollo 11* mission became history.

But it was only the first in a series of manned lunar excursions. Subsequent missions carried additional, refined scientific instruments, although *Apollo 13* had to abort its lunar landing following an explosive failure of critical electrical systems. In 1971, the *Apollo 15* mission began the first of three second-generation Apollo missions, utilizing the Lunar Roving Vehicle (LRV), which dramatically extended the range of the astronauts' explorations as well as their staying time on the surface. With the advantage of the LRV, each of the *Apollo 15, 16,* and *17* missions during 1971 and 1972 yielded results that scientists declared were three times as valuable as any of the prior lunar landings. The *Apollo 17* mission of December 1972, the last in the series, did not mean the end of scientific data from the moon. Years later, scientific instruments and experimental paraphernalia left by the Apollo missions still sent billions of data bits from various locations on the lunar surface, adding to our knowledge of the moon and our universe.

The Skylab program provided an unusually practical environment for human-directed scientific research and observation in space. NASA and contractor personnel began in the early 1960s to plan conversion of an S-IVB stage into a combined laboratory and workshop in earth orbit. This eventually became the Skylab, about as

After the first lunar mission, NASA planners feared possible contamination of the Earth environment, so the astronauts were relegated to a "mobile quarantine facility." Armstrong, Collins, and Aldrin are shown with President Nixon aboard the carrier Hornet. *Courtesy of NASA.*

large as a medium-sized, two-bedroom house, providing a true "shirt-sleeve" environment for the astronaut crew and allowing them to live and work without cumbersome spacesuits. The launch of the spacecraft in the spring of 1973 used the first two stages of the Saturn V booster to put the Skylab (the converted third stage) into orbit. Unfortunately, a protective micrometeoroid and heat shield ripped loose during the launch, and one of the two solar-cell wing arrays that supplied power to the crew area also tore away. The mishap prevented the deployment of the remaining solar wing and increased the interior temperatures of the Skylab to dangerous levels. Launched aboard a Saturn IB, the first three-man Skylab crew boarded the spacecraft and erected a large parasol device to deflect solar rays and reduce interior temperatures. A difficult "space walk" by two of the astronauts freed the remaining solar wing, and the Skylab soon began to respond to the additional electrical power generated by the wing's solar energy cells.

The first crew spent nearly a month aboard the Skylab before returning to earth. In spite of early difficulties in readjusting after prolonged weightlessness, the astronauts returned in fine shape, and physicians endorsed plans for two succeeding manned

Astronaut James Irwin salutes during the Apollo 15 *mission, the first to employ the lunar roving vehicle, seen at right. Courtesy of NASA.*

missions of two to three months each. The second Skylab crew spent 59 days in orbit, and the third trio stayed a record 84 consecutive days in space (a mark since surpassed by the Russians) before splashdown on February 8, 1974. One of the major contributions of the Skylab program was convincing proof that manned missions could operate for extended periods of time in the weightless environment of space.

In addition to providing information about the crews' adaptability to prolonged space missions, the Skylab carried a wide array of sophisticated instruments and equipment to gather data on the earth's environment and resources; the sun and the Solar System; and possible industrial processes that would be enhanced by performing them in the environment of space without the perturbing factors of the earth's gravity. The Skylab also carried a select group of miscellaneous experiments submitted by secondary-school students all across the United States—a remarkable opportunity for many young scientists. Some of the more interesting astronomical work done by the Skylab included extended observations of an unusual period of solar flare activity in

The Skylab provided an unusually large area for "shirtsleeve" working conditions in space. The successfully deployed solar panel is seen to the left; the emergency solar shade is visible above the main section of Skylab; the telescope mount and its four-way array of solar cells is above left. Courtesy of NASA.

1973 and, late in the year, unparalleled observations of the newly discovered comet Kohoutek. Extensive photography and analysis of the earth and its environment yielded significant new knowledge of the ocean, weather and climate, pollution, and natural resources.

Production of the new boosters for space flight required specialized manufacturing techniques and assembly fixtures like those at the right. First stages of the Saturn I and IB boosters used a cluster of tanks, as shown in the photo. This seemingly awkward assemblage of tanks nevertheless remained operational until 1975, when it flew during the Apollo-Soyuz Test Project. Courtesy of NASA.

As it continued in orbit, empty and unused after the third crew's departure, the Skylab became increasingly troublesome. NASA had hoped to reach the Skylab with a shuttle mission and boost the space station into a long-term orbit. But delays plagued the shuttle program while the Skylab's orbit began to degrade; its rolling and tumbling threatened to bring it into an uncontrolled reentry. NASA finally reacquired communication with the recalcitrant Skylab during the spring of 1979 and stabilized its attitude. Eventually, orbital decay led to a reasonably controlled reentry over the Indian Ocean, although chunks of the Skylab survived a fiery descent through the earth's atmosphere and plummeted into remote sections of Australia. The incident became a reminder that the increasing frequency of spacecraft in earth orbit created dangers, as humanity's own debris fell from space.

During the summer of 1975, NASA collaborated with Russian officials in launching the Apollo-Soyuz Test Project (ASTP). In view of the strong competition in space between the United States and the USSR, sharpened in recent years by Vietnam and other cold war tensions, ASTP stood out as a dramatic example of international good will and cooperation. ASTP involved separated launches of the two-man Russian Soyuz spacecraft and the three-man American Apollo spacecraft (which used a Saturn IB for launch). After rendezvous and docking, the two crews opened the hatches of their spacecraft, exchanged triumphant handshakes, and visited each other's quarters. This marked the final use of the Apollo-Saturn hardware in manned missions. In the future, NASA planned to employ a reusable, multimission vehicle called the Space Shuttle.

Aerospace Perspectives, 1975 – 1983

The American aerospace industry continued to be a strong member among U.S. manufacturing enterprises, although European consortiums began to build airliners that strongly challenged American firms for a share of the lucrative international market. Within the United States, bargain fares attracted additional thousands of air travelers as deregulation took effect late in the 1970s. When many airlines suspended service to smaller cities and on less lucrative routes, the commuter airline operators, along with general aviation manufacturers, found new, lively marketing opportunities. But rising fuel costs, among other factors, eventually influenced a decline of passenger traffic as well as aircraft sales.

The continuing moves and countermoves of the cold war stressed the utilization of electronic warfare equipment for military aircraft and missiles and for satellites used in navigation and surveillance. At the same time, a growing family of satellites for communications and other peaceful purposes profoundly influenced the daily lives of citizens everywhere. A successful launch of the Space Shuttle brightened hopes for expanding peaceful uses of space, although the national defense roles proposed for future shuttle development symbolized the duality of aerospace technology in a contentious world.

CIVIL AVIATION

In 1978, the 75th anniversary of the Wright's first flight, the airline industry issued any number of self-congratulatory press releases. On the surface they seemed full of statistical trivia, but the numbers did demonstrate the extraordinary evolution of air transport. When airlines had begun comprehensive service in 1926, mail had constituted the principal cargo. During the first full year of operation, only 6,000 passengers had bought airline tickets. By 1955, approximately 41 million passengers were using the airlines, about the same number that used Chicago's O'Hare Airport in 1976. During the late 1940s, only 10 percent of the American adult population had flown in an airliner; 63 percent had done so by 1977, and the figures kept moving upward. During 1978, airline travelers numbered an estimated 275 million. The number of flights available progressed at a kaleidoscopic rate. The airlines had listed 700 daily flights in 1932, a figure equivalent to the number of flights per hour in 1978.

A new surge in statistics came in 1978, with President Jimmy Carter's encouragement of deregulation. Taking its cue from Carter, the Civil Aeronautics Board (CAB) eased its constraints on routes and fares. Airlines trimmed marginally profitable routes, scrambled for lucrative new routes into hub cities, and offered travelers an enticing array of discount fares as the competition for passengers intensified. Air travel and airline revenues reached record levels. Most of the passenger activity involved large trunk lines serving major cities, but regional and commuter services grew as well.

For the twenty-nine major national and international airlines (as defined by the CAB), 1979 was the peak year for passengers and cargo: 317 million passengers, 5.9 billion freight ton-miles. Total operating revenues in 1979 amounted to $27.2 billion and reached $36.5 billion by 1981. But fuel costs and the sluggish economy had begun to take their toll, and the airlines recorded a net operating loss of $421 million in 1981. Revenue passengers fell to 286 million, and cargo was also off, at 5.6 billion freight ton-miles. Another reason for the airlines' slump was a strike called by the Professional Air Traffic Controllers Organization (PATCO) during the summer of 1981. PATCO sought shorter working hours, higher salaries, and earlier retirement, arguing that high stress factors of their jobs justified their demands. The FAA, citing a federal statute that made it illegal for government employees to strike, sent dismissal notices to controllers who stayed off the job. Using supervisory personnel as flight controllers, the FAA maintained about 50 percent of scheduled flights, and restored 75 percent of flight schedules within several days. Newly trained operators brought the situation back to normal by 1982/83, although the reduced schedules and long delays reduced the profits of the airline industry. But despite the sagging statistics, the airlines continued to play a vital role in the national economy.

Within the air transport industry, increased utilization of electronic systems paralleled a similar trend in national defense. In the civil sector, one of the pressing issues involved the mounting numbers of airline flights. The higher speeds of jet airliners, coupled with their higher frequency of airport arrivals and departures, unquestionably increased the volume of air travel and, consequently, the burden on air traffic controllers. By 1970, the airlines were already carrying 150 million passengers annually, a figure three times that of a decade earlier. As for the frequency of flights, forecasters of 1967 had estimated 13 million flights per year within a decade, but that figure became a reality even before the 1960s ended, and the FAA faced the prospect of 28 million flights per year by the mid 1970s. During this hectic growth period, en route air traffic continued to be handled by an increasingly cumbersome radar system that had to be updated manually during the progress of any given flight. A controller's radar screen might show several moving "blips," each signifying a plane in flight. Each blip was identified with a plastic marker, or "shrimp boat," carrying an identification code and abbreviated flight data; additional data could be had from a nearby information board containing ranks of more detailed "flight strips" for each plane being tracked. If an en route flight requested or received instructions for a different cruise speed or altitude, the information had to be entered manually on both the flight strips and the shrimp boats. Repeated changes on a hectic day kept traffic controllers busy monitoring blips and keeping flight information updated. Controllers not only had to keep up with the traffic represented by the shrimp boats in front of them but also were responsible for flight strips representing planes scheduled to enter their sector—information that came through via telephone or teletype. Some days could easily drive a controller to distraction.

Automation seemed the best answer to the FAA, especially for the en route portions, and the agency worked with IBM to perfect a program called En Route Stage A, an interrelated array of radars, computers, and computer programs to ease the controllers' workload. As in the old system, contiguous American airspace remained divided into 20 air route traffic control centers, each responsible for guiding a plane through its territory and handing off aircraft to other centers or to an airport controller. Though centers varied, a typical territory might cover a chunk of airspace measuring 500 by 600 nautical miles. Under the new system, high-speed, high-density traffic came under realtime control that incorporated automatic radar tracking and processing of flight planes. Controllers continued to make basic decisions concerning aircraft movement but had at their disposal a superior visual display and data information system. The display screen dispensed with mere blips, showing instead a computer-generated symbol of greater clarity, accompanied by an "alphanumeric data block" with relevant flight information. This flight information, entered into the computer system before takeoff, was continuously and automatically updated on the display screen, eliminating the persistent clutter of shrimp boats. The new system immensely simplified the crucial action of handing over a flight from one center to the next, as well as to airport controllers.

During the 1970s and 1980s, navigation and landing systems also became increasingly sophisticated. A modern jet airliner might carry several different navigational aids, depending on the size of the aircraft and the requirements of its route. It became commonplace for the larger transports to draw information from their own inertial navigational system (based on a combination of interrelated gyroscopes and accelerometers) and to receive signals from navigational satellites in orbits far above the earth. More and more transports were also equipped with their own airborne weather radars, permitting flight crews to detect storm cells and take action to avoid dangerously turbulent weather.

Great strides in safer landing techniques accompanied this growing electronic sophistication. Radar and other equipment allowed control tower personnel to give pilots accurate information as to their flight path in relation to the proper runway, as well as a plane's height and distance from the airfield. On final approach, pilots learned to rely on a combination of electronic and visual lighting equipment. The instrument landing system (ILS) continued to be a standard landing aid. But any pilot liked to see as much of the runway as possible, and an elaborate coded lighting system not only marked the runway and its boundaries but also provided a "visual approach slope indicatory system." In the final moments of approach, one string of lights offered a visual confirmation of distance to the threshold, while a bank of red and white lights allowed the pilot to see if the plane was above or below the correct angle of approach. The International Civil Aviation Organization (ICAO) standardized worldwide electronic and lighting systems for airports and established categories for cloud ceiling and visibility. In foul weather, aircraft had to have the proper equipment to operate from airports with landing aids for the established categories. By the late 1970s electronic controls aboard the airplane could automatically guide an airliner down a precision approach, bringing the plane within one-half mile of the runway, where the pilot took over. Refinement of this equipment controlled the aircraft all the way to touchdown, permitting totally automatic landings in dense fog.

Continued electronic advances resulted in a new type of precision landing equipment, called MLS, for microwave landing system. Existing ILS equipment required planes to approach in single file over a long distance, and remained prone to "multipath effects," extraneous signals caused by reflection from other planes, buildings, or local terrain. MLS was designed to eliminate multipath effects and to offer increased flexibility through a variety of electronic approaches converging on the runway. Several nations, including the United States, developed MLS equipment over the course of a decade, but international consultations led to agreements for a standardized system to be chosen by a multinational committee within the ICAO. The choice narrowed down to competing British and American concepts. As acrimonious controversy erupted when the British charged that test results had been misleading in favor of U.S. equipment. International prestige, as well as many millions of dollars in sales, sharpened the issue, eventually settled in 1978 when new tests gave the nod to American equipment.

The MLS debate represented only one facet of increasing international competition in the aerospace industry. The highest stakes involved hundreds of jet transports due for retirement from the world's major airlines during the 1980s. By the end of the 1970s, some 1,200 jet transports out of 4,800 in service outside of Soviet and Soviet-bloc countries were approaching retirement age. Since the sales for new airliners in the 1980s were estimated at $80 billion, this high-stakes market generated unusually strong interest among aerospace firms on both sides of the Atlantic. The British, French, and German governments in about 1965 had held the first talks leading to a plane called the A-300 Airbus, a wide-body, high-density transport for short- to medium-stage lengths. Powered by Rolls-Royce or General Electric turbofans, the 300-passenger Airbus competed with the DC-10 and L-1011 over many routes. In the spring of 1978, Eastern Airlines announced its intention to purchase a fleet of 23 A-300 Airbus transports for a total price of $778 million. Although European transports had been purchased by American carriers in the past, this was the largest deal ever negotiated by an American airline for a non-American aircraft, even though Eastern's planes came equipped with General Electric engines. Because American transports traditionally dominated the Western World's airline fleets and aerospace exports had recently outpaced other American manufactured exports in contributing a favorable trade balance, the Eastern

Headed for a landing, a DC-10 sweeps over an instrument landing system (ILS) array. Increased use of electronic systems not only enhanced safety but also increased the airlines' ability to keep schedules despite unfavorable weather. Courtesy of Federal Aviation Administration.

purchase sent a ripple of concern through American aerospace boardrooms. This European coup was shortly followed by an announcement that a Pan Am purchase of Lockheed TriStars included specifications for British-built Rolls-Royce engines. Balancing these European successes later in 1978 was the news that Boeing's proposed 767 jet transport, similar to the Airbus in design and capacity, had won a $1.2 billion order from United Airlines. Orders of this magnitude permitted Boeing to plan deliveries, prices, and financing to compete with other designs.

The wide-body 767, seating from 211 to 289 passengers, depending on the cabin layout, featured spacious seating with a twin-aisle arrangement. Another new Boeing transport, the 757, used a single-aisle arrangement in a narrower fuselage seating about 185 passengers. The 767 entered service in 1982, followed by the 757 a year later. Both planes incorporated nearly identical two-place cockpits, designed to eliminate the flight engineer and save considerable money. In compensation for the displaced engineer, advanced cockpit instrumentation relied on a cluster of cathode ray tube displays and computers to call up information at the pilots' command or to indicate potential problems. The planes also used newer, fuel-efficient engines with computerized fuel management systems and incorporated lightweight composite materials in many parts of the wing, tail surface, and fuselage. Collectively, improvements on the new 767 and 757 were intended to reduce the seat-mile cost by as much as 20 percent in comparison to other transports in service. As one means to create a better climate for international sales, Boeing planned on cooperative international risk-sharing. The Italian manufacturer, Aeritalia, had a 15 percent share of airframe design and a similar proportion of development costs. Aeritalia's airframe work, in the form of advanced composite structures, also gave it an important new capability for other aerospace projects. But Airbus Industrie, the European consortium, countered with advanced airliners of its own, and aggressive European marketing kept the international sales competition at a very keen level.

The general aviation sector attained remarkable levels of active aircraft and continued to function in innumerable utility roles. A comparison of aircraft categories in the United States emphasizes the growth of general aviation. By 1980, there were about 26,000 military aircraft in the inventory of the U.S. armed services. Scheduled airlines in the United States listed about 2,300 transports in domestic service. By contrast, the American general aviation fleet numbered some 200,000 aircraft, and production in 1980 reached approximately 13,000 units per year including both fixed-wing aircraft and helicopters. Millions of people flew to their destinations in general aviation aircraft. Agricultural aircraft treated over 250 million acres of crops each year and planted practically all of the nation's rice crop. Thousands of lives were saved annually by emergency flights of helicopters and planes. Even though fuel costs and the recession also hit general aviation, cutting production to 11,000 units in 1981, general aviation activities continued to be significant. Within a multifaceted society, a myriad of wings were at work in useful and unique ways.

AEROSPACE INDUSTRY TRENDS

By the 1980s, the roster of principal airframe manufacturers were showing the strain of adapting to postwar trends in the aerospace industry. The staggering costs of producing complex new aircraft and spacecraft—both military and civil—had forced stalwart companies to diversify, reorganize, or find new corporate partners. During the early 1960s, the Glenn L. Martin Company had merged with the Marietta Corporation, so that the new entity, Martin-Marietta, continued to do business, balancing its aerospace component with production of chemicals and construction materials. At Douglas Aircraft, the company found itself with a backlog of company orders for DC-8 and DC-9 aircraft, but production delays and difficulties in securing adequate interim financing forced the company to seek a partner. McDonnell Aircraft, in St. Louis, had won extensive military contracts but had been unsuccessful in breaking into the civil airline sector. After courting Douglas for several years, McDonnell finally succeeded in acquiring the West Coast firm in 1967. McDonnell Douglas thus emerged as a balanced aerospace company, producing airliners, military aircraft, and space hardware.

North American also went through a merger. The company had been successful in winning military business and also had acquired major contracts in NASA's space program during the 1960s. But management fretted about overdependence on the government as a customer and actively sought a nonaerospace partner. The solution appeared in the form of a merger with Rockwell Standard, a manufacturer of automotive components who operated plants in 14 countries. North American Rockwell was formed in 1967, later assuming the name of Rockwell International. Other companies, with varying success, also tried mergers, diversification, or a balance of civil and military orders.

Diversification and a balance in products, however, did not always mean stable balance sheets. Lockheed achieved dominance as the supplier of air force transports but also wanted to revive its position in the airliner market. The company had already entered the fields of helicopters, electronic equipment, space hardware, and shipbuilding. The latter venture fared badly and had lost over $15 million by the mid-1960s. Lockheed's return to airliner development in the early 1970s seemed to be more successful, although the company had to overcome considerable difficulties. As noted earlier, plans to use Rolls-Royce engines for Lockheed's TriStar created snarls when the British manufacturer experienced higher development costs, thus delaying delivery schedules; the company's efforts to meet competitive American pricing triggered still higher losses. Rolls-Royce went into bankruptcy, and the imbroglio required Lockheed to negotiate new contracts with airline customers.

Moreover, Lockheed had serious corporate troubles in other quarters, because its C-5 Galaxy transport program had resulted in large financial losses as well. Even though Rolls-Royce was quickly reorganized as a nationalized firm, Lockheed nearly went under. An emergency government loan of $250 million was passed in Congress in the spring of 1971. The special legislation indicated the significance of the aerospace industry; Lockheed's survival affected not only a number of key national defense contracts but also employment rates in several congressional districts in widely scattered areas of the country. Lockheed eventually recovered, and its TriStar won praise for quiet, efficient operation. Unfortunately, the worldwide recession caused airlines to cut back on purchases, and in 1982, Lockheed reluctantly announced its decision to phase out the TriStar production line within two years.

The airlines went through similar phases of reorganization. During the early 1950s, the numerous smaller lines that had sprung up after World War II had either disappeared or consolidated into larger organizations. A second wave of consolidation occurred during the late 1960s and early 1970s, giving rise to larger regionals like Allegheny in the Northeast and Hughes Airwest in the West. Regional lines like Delta became major carriers by absorbing smaller entities like Chicago & Southern in 1953 and Northeast in 1972. The impending move for deregulation in 1978 gave new impetus to merger plans for the early 1980s. Allegheny renamed itself USAir late in 1979 and soon spread into new territories. North Central Airlines joined with Southern Airways during 1979–80, becoming Republic Airlines, which then acquired Hughes Airwest. As a result, Republic and USAir were elevated to the status of major carriers. In 1980, Pan American World Airways took over National Airlines, giving Pan Am a long-sought domestic route structure to feed into its global network.

Freed from the route constraints that had been imposed by the CAB, regional and major airlines alike launched new routes as soon as possible. Consequently, orders for new airplanes shot up and the airlines shouldered new expenses in the process of enlarging terminals, adding reservation desks, and hiring new personnel. At the same time, fuel prices climbed sharply, and high interest added to a rising tide of red ink. Carriers engaged in a no-holds-barred struggle for passenger business, slashing fares on major routes in the United States and offering substantial reductions on many overseas flights. Frequent travelers acquired bonus points that could add up to a free intercontinental trip. During 1982, several airlines offered two seats for the price of one; another company began to give its passengers S&H Green Stamps; still another gave free trips to kids who turned in proof-of-purchase seals from cereal boxes. The permutations of calculating the least expensive trip became monumental, driving travel agencies to the point of despair.

The airline industry lost hundreds of millions of dollars during this hectic period. Fuel cost, recession, and overexpansion took a continuing toll, leading to the total collapse of Braniff International in 1982, and forcing many other carriers to furlough personnel, retire aircraft, and stretch out delivery dates for new planes like the 757. Although the airline picture looked somewhat brighter in 1983, as the recession flattened out, interest rates decreased, and fare wares abated, the improved climate came too late for airlines like Braniff and manufacturers like Lockheed.

The general aviation sector also experienced changes that gave the corporate lineup a different look by the 1980s. The generally prosperous 1960s had been good for the Piper Aircraft Corporation, generating record sales (up by 20 percent, to $106 million for 1968–69 alone) and a healthy balance sheet. Piper was ripe for a takeover by one of the conglomerates that had become active during the decade. Early in 1969, Chris-Craft Industries tried just that. A well-known manufacturer of recreational boats, Chris-Craft had branched out into synthetics, chemicals, and broadcasting. The Piper family fought back, wanting to maintain managerial autonomy, but eventually realized that they would have to find a merger partner that could outbid Chris-Craft and also promise operational independence. The winner in this case was Bangor Punta, a Fortune 500 conglomerate that turned out such diverse products as sailboats, motor homes, law enforcement hardware, environmental control systems, and more. By the late summer of 1969, Bangor Punta had won. Piper Aircraft Corporation continued to produce airplanes but no longer operated as a family affair. W. T. Piper, Sr., died in 1970, and the Piper family eventually vacated their positions as company officers.

In the meantime, Piper's activities had expanded overseas, with projects in Latin America for the assembly of aircraft from components shipped from the United States. During 1974–75, Piper entered into a more comprehensive agreement with Brazil for the indigenous production of Piper designs under the direction of a government manufacturer, Embraer. Brazil wanted to develop its own technical, manufacturing, managerial, and marketing capabilities in general aviation as well as to control imports that drained Brazilian finances. As production of Piper aircraft got underway during 1975–76, the government imposed a 50 percent tax on imported planes, creating a protected market within the country. By 1983, Brazil was producing six single-engine types and two twin-engine types, based on Piper designs but marketed under specific Embraer designations and carrying distinctive Brazilian names like the "Carioca" model. In addition, Embraer produced an agricultural plane and developed its own line

The Dallas/Fort Worth Airport handled some 25 million passengers in 1982, making it the fifth busiest in the nation. The sprawling facility of 17,800 acres was planned to accommodate automobile traffic and to cope with increasingly larger aircraft carrying rising numbers of passengers. Within the airport complex itself are a 1400-room resort hotel, an automatic passenger transit system, a medical clinic with surgical rooms, a bank, restaurants, and other services characteristic of a self-contained city. Courtesy of Dallas/Fort Worth Airport.

of larger twin-engine designs, including the "Bandeirante," an 18-passenger commuter liner that sold well in the United States and in other foreign markets.

Cessna had begun overseas operations during the 1960s, establishing fabrication sites in Argentina and France. The French project eventually assumed greater autonomy, marketing planes in Europe under the designation Reims Aviation, taken from the city of its location. The Reims organization normally produced standard Cessna airplanes, although the European facility modified some designs for specialized needs in the European and African markets. In 1983, the Reims unit played a role in developing a new twin-engine design for utility and rugged bush-flying operations. For its larger twin-engine planes and jets that sold overseas, Cessna set up a marketing system based in the United States.

Beechcraft had been one of the companies to negotiate with Brazil in 1974, but then made a firm decision to keep all manufacturing at home, preferring to implement foreign sales through its own subsidiary for international marketing. Beech's brief experiment with international manufacturing had occurred during 1970–75, when the company imported the Hawker Siddeley 125 from England and sold the executive jet in the North American market. Like Cessna, Beech sometimes had undertaken the role of subcontractor. Beechcraft facilities in Wichita had begun turning out airframes for Bell Helicopter Textron in 1968, and were continuing to do so into the 1980s. The 1980s also included a major corporate change. Beech, like Piper, became a division of a larger corporate entity. With record fiscal sales exceeding $500 million, Beech wanted to avoid an unfriendly takeover and sought out its own partner, eventually coming to an agreement with Raytheon. The merger became official in 1980, providing Beech with ongoing management depth as Mrs. O. A. Beech turned over company affairs to a new generation of executives. Raytheon's principal activity in electronics gave Beech additional depth in technological capability and in product development, and Beech's sales in the civil sector added balance to Raytheon's government-related contracts.

Turbocharged power plants began appearing on high-performance, single-engine planes, often yielding performances that rivaled those of some twins on the market. In operating turbocharged singles at high altitudes, pilots needed to carry oxygen masks and adequate oxygen supplies in bottles even though the paraphernalia often proved cumbersome. For this reason, Cessna introduced the P-210 in 1977, a pressurized single that delivered high-altitude performance comparable to those of larger, more expensive aircraft. The versatility of the P-210 and similar high-performance aircraft also increased with the growing availability of compact, lightweight radar units that gave pilots a more detailed view of storms and air traffic along an intended flight path. Toward the end of the 1970s, Piper and Beechcraft introduced new types of primary trainers, with a NASA-inspired wing for increased aerodynamic efficiency and a "T-tail," said to enhance controllability. The T-tail arrangement, already a fixture in some modern airliners, appeared on additional models of singles and twins offered by several manufacturers. These aeronautical refinements and improved avionics brought general aviation designs to advanced levels comparable to those of modern airliners. In one respect, in fact, the general aviation industry stepped ahead of the airliners (and even the military) by producing the first plane equipped with "winglets," another NASA concept—small, vertical surfaces installed on the plane's wing tips. The first winglets on a production aircraft came from Gates Learjet, who christened the new version of their well-known executive jet the "Longhorn."

These elements of progress in the general aviation industry lacked the drama of the milestones of earlier years, like pressurization and gas-turbine power plants. Innovation proceeded but assumed a steadier profile. The major changes within the general aviation sector seemed to be in marketing. Although single-engine designs continued to sell well, the demand for twins and jets constituted a growing share of the market. This shifting pattern of sales reflected the continuing trend of business decentralization and the impact of deregulation within the airline industry. As airlines suspended service to many smaller communities and instead concentrated on heavily traveled routes instead, many firms decided to acquire a corporate aircraft to carry executives on business trips. During the early 1980s, although the production of general aviation aircraft declined,

Planes like the four-place Piper Archer II, with a cruising speed of 150 MPH and a range of 600 miles, allowed business pilots to cover a considerble territory, including smaller communities with limited or no scheduled passenger service. Advances in avionics also enhanced their utility in IFR (instrument flight rule) conditions. Courtesy of Piper Aircraft.

total sales increased to nearly $3 billion annually. The record billings came from the continuing demand for higher-priced turboprops and business jets. But the industry slumped in 1982–83 under the effects of a sluggish economy and high interest rates.

At the same time, several manufacturers pushed ahead with new products, hoping to take advantage of an improving economy by 1983–84. In addition to Cessna's large executive jet, the Citation III, Beechcraft, Mooney, and Piper announced new single-engine airplanes that featured pressurization. All had made preliminary flights by 1983, and Piper was ready to make deliveries. With wide, six-place cabins, all these aircraft could have arrangements for club seating, similar to twin-engine planes of similar performance. With pressurization and either a turboprop or a turbocharged engine, they could also duplicate the speed, operational ceiling, and comfort of many twins, while providing the lower price and operating cost of just one power plant.

Coming at a time of comparatively slow demand, the new jets, pressurized turboprop singles, and other designs revealed an aggressive, innovative drive in the general aviation sector. The general aviation industry, like the major airline manufacturers and military firms, also included a significant joint venture in the case of the Saab-Fairchild 340, which entered flight tests during 1982–83. Components were produced by Fairchild Swearingen, who had been manufacturing a line of similar twin-engine turboprops in the United States, and by Saab-Scania, a major military aerospace firm in Sweden, where final production of the 340 took place. The international collaboration has lowered reduced development costs for both companies and should give the plane an edge in certain international markets. Commuter airlines were the major market for the big twin-engine turboprop, but the Saab-Fairchild marketing team also aimed at the market for large executive aircraft.

For business travel, the value of a company plane in saving time could be observed in Europe as well as in the United States. In Europe, the Digital Equipment Corporation used its own plane to fly personnel from executive offices in Geneva, Switzerland, to production facilities in Galway, Ireland. Executives departed from Geneva in the morning, put in a day's work in Ireland, and returned that night. The same trip by commercial airline required three days. In the United States, changing ways of doing business also enhanced business flying at individual and corporate levels. Just after World War II, over a half of American factories were located in cities of 100,000 or more. By 1956, a third of all new factories were going up in smaller towns. Companies not only adopted general aviation to cope with decentralization but built plants, moved offices, and structured management organizational charts on the basis of general aviation operations. Of over 14,000 airports in America, only 425 had scheduled airline service in the late 1970s, and 70 percent of airline traffic originated from only 25 major city airports. For a decentralized business, general aviation therefore became essential. Gannett Newspapers, for example, owned papers in 55 towns and directed them with strong, centralized corporate policy that required heavy travel. From corporate headquarters in Rochester, New York, only 5 of the 55 towns could be reached by nonstop flights. Gannett estimated that its corporate planes allowed it to reduce its corporate staff by a third and eliminate a number of regional executives. Even the economists of major domestic airlines admitted that corporate aircraft could make "huge" time savings. General aviation became a major transportation network. Approximately 300 million passengers were flying between American cities annually by 1980, and one-third of them traveled in general aviation aircraft.

The numbers alone indicate general aviation's integral role in the national economy, but there were other subtle influences of significant quantitative and qualitative value. In emergency medical service, aircraft played a dramatic role. One hospital administrator in New York, assessing the sluggish and chaotic nature of highway traffic in modern cities, stressed the need for additional helicopters in medical emergencies. A service known as "Life Flight," supported by a Houston, Texas, hospital, offered convincing evidence of the value of such emergency helicopters. Begun in the summer of 1976, Life Flight used two helicopters to respond to 1,788 calls in 1978 alone. A third helicopter was added in 1979; in three years, Life Flight made over 4,000 emergency runs in its operational radius of 130 miles. Dr. David M. Boyd, an emergency care expert at the Department of Health, Education, and Welfare, cited the service as a model for other cities. He said that 110,000 people died annually from traffic accidents and other trauma cases across the nation, costing more than $62 billion annually; better rescue and treatment services like Life Flight could cut the death rate in

Employees of Conoco board the company's Grumman Gulfstream I corporate propjets for early-morning flights. From its headquarters in Ponca City, Oklahoma, Conoco runs its own "airline," a daily shuttle service embracing Tulsa, Houston, and Lake Charles, Louisiana. Many corporations have found that such operations are both economical and productive. Courtesy of Conoco, Inc.

half. Houston also added a specially equipped, pressurized Learjet, extending emergency air ambulance operations to Canada and South America. Besides saving lives, this kind of service facilitated regional rather than localized use of sophisticated medical technology, enabling smaller and less affluent hospitals to control costs and avoid needless duplication. Worldwide, similar services made thousands of medical flights annually.

In the urban environment, police found that helicopters made continual contributions to traffic control, crime prevention, and general law enforcement. Houston acquired 3 helicopters in 1970 and eventually built a fleet of 16 machines. Between 1970 and 1978, the helicopter patrol division responded to 104,683 calls for specific assistance, participated in 6,370 arrests, and recovered over 1,000 stolen cars. "In our police department," the division commander stated, "helicopters have become an indispensable tool for effective law enforcement and support for our police ground units."

In crowded urban environments, the lives of thousands of accident victims are saved every year as a result of the mobility offered by helicopters. The Bell 206 Longranger above is operated by a hospital in St. Louis. Courtesy of Bell Helicopter Textron.

In other areas, general aviation contributed to energy production and facilitated environmentally sound operations. Recent studies have emphasized the growing role of helicopters in off shore oil production and exploration where a 15-hour boat journey to an offshore rig could be reduced to 1 hour by helicopter. In logging, helicopters airlift timber from selectively harvested areas, eliminating the need for expensive logging roads that contribute to environmental degradation. In remote and wilderness areas, helicopters assisted in uncounted ways during construction jobs. They performed surveys for high tension power lines, flew in the towers, and even strung the lines themselves—all of which reduced costs and environmental problems created by road construction.

In an air age keenly conscious of energy and transport efficiency, the role of general aviation in modern society was not only useful but also energy efficient, consuming less than 7 percent of all the fuel used in civil aviation, and less than 1 percent of the fuel consumed in all forms of transportation. At the same time, some activities prompted sharp issues, as in the case of aerial spraying. The use of highly toxic herbicides and pesticides raised questions about unharmful plant and animal life so often affected by powerful chemicals. For humans, there were also disturbing short-term and long-term physiological effects related to aerial spraying. But the speed and general efficiency of aerial spraying may be crucial to protect indispensable crops required for a growing world population.

Agricultural operations demand skilled flying and a knowledge of toxic chemicals. Planes like the Cessna AGtruck are equipped for dusting or spraying. The reinforced cockpit is designed for safety as well as visibility. Courtesy of Cessna Aircraft.

AEROSPACE WEAPONS

In view of combat experience in Vietnam, the modern air-war environment seemed to depend more and more on electronic capability. Indeed, upgrading the B-52's electronic capabilities and electronic weaponry (plus other refinements) appeared to be a feasible and economic way to extend the plane's useful life into the 1980s, especially after cancellation of the B-1 manned bomber in 1977. Between 1954 and 1962, some 744 production B-52s were built, and about 400 were still operational as the USAF entered the 1980s. Almost half of this strategic bombing force was composed of B-52G and B52H models, with a redesigned wing that featured integral fuel tankage and with improved defensive capabilities and other refinements. New electronic equipment included highly sophisticated gear to improve flight capability at low levels, where "ground clutter" helped to hide attacking planes from enemy radar. During the early 1970s, both models received further modifications, allowing them to carry up to twenty short-range attack missiles (SRAM)—8 in the bomb bay and 6 under each wing. The SRAM itself, propelled by a solid-propellant rocket engine, was designed to fly over supersonic speeds and carry a nuclear warhead for strikes at enemy surface-to-air missile (SAM) sites. Depending on its altitude, the SRAM's range varied from 35 to 100 miles, and its inertial guidance system not only made it impossible to jam but also permitted the choice of high- or low-altitude attack and included options for dog-leg courses to further confuse enemy defenses.

Electronic wizardry extended to modified EF-111A aircraft, which found gaps in enemy air defense systems; the F-4E "Wild Weasel," which detected, identified, and destroyed enemy radar; and the E-3A (a modified Boeing 707) AWACS (airborne warning and control system), which conducted aerial surveillance. Easily identifiable by the large, dishlike radome supported by a pylon atop its fuselage, the AWACS was designed to fly literally everywhere as an electronic surveillance system capable of detecting any kind of manned or unmanned air vehicle and of supporting tactical as well as air defense operations. In addition to these and other aircraft used for reconnaissance, intelligence, and surveillance, the USAF also employed the E-4 (modified Boeing 747) as a flying command post in case of national emergency; one such aerial command post remained in the air at all times.

The USAF also presided over an awesome array of missiles. The Titan II, a two-stage, liquid-propellant ICBM, became operational in 1963 and remained in service into the 1980s, with six squadrons of nine missiles each. Service improvements to the Titan II incorporated equipment to hamper detection and destruction by enemy defensive missiles. Newer ICBMs in service included the three-stage, solid-propellant Minuteman II and III versions, which together numbered about 1,000 missiles by 1980. The Minuteman III featured a MIRV (multiple independently targetable reentry vehicle) capability to put warheads on three separate targets. The Department of Defense also continued development of a new missile to carry up to ten MIRVs, the MX, along with proposals to make it less vulnerable to attack from nuclear-tipped enemy missiles. Original schemes for deployment in a maze of underground tunnels, to be constructed in the Southwest, created strong opposition because of costs (in the tens of billions of dollars) and environmental impact. The MX progressed into the flight-testing phase during 1982–83, even though Congress had not made a final decision about funding or the deployment issue until late in 1983.

Following cancellation of the B-1 manned bomber, the air force planned intensive development of the air-launched cruise missile (ALCM) as an alternative. Since the aging B-52 bomber force would be increasingly vulnerable in deep penetration missions over enemy territory, the best solution seemed to be a long-range missile released from a safe position farther from the target, where defensive rockets and other antiaircraft defenses were less of a threat. Thus, the logic of the ALCM of the early 1980s. Boeing and General Dynamics fielded competitive ALCM candidates; Boeing won the lucrative contract following a "fly-off" competition in 1980. Although many specifications remain classified, the ALCM concept embodied a turbofan-powered aircraft of high subsonic speed but capable of sustained flight over a comparatively long distance to the target.

The winged missile incorporated sophisticated terrain-avoidance systems for low-level flight, which, with its small radar "signature," made it a hard target to track and destroy. The ALCM's inertial guidance system enabled precision attacks against surface targets, and its (approximately) 12-foot wingspan and 21-foot fuselage provided room for all the necessary electronics and nuclear warhead. Since the ALCM's wing and tail surfaces were retractable, a B-52 could carry about 12 missiles externally and 8 more internally on a rotary dispenser. Despite the B-52's advanced age, this multiple ALCM threat promised to make the B-52 a highly potent aerial weapon some three decades after it had entered service. The administration of President Ronald Reagan revived the B-1 program in 1981, and the air force planned to equip it with ALCMs as well.

The location of potential targets for advanced, high-speed missiles posed problems of a different sort. In terms of extremely long-range, high-altitude, manned reconnaissance aircraft, the Department of Defense continued to utilize updated variants of the U-2. Although the U-2's long, tapering, high-aspect ratio wing was unusual, its unique looks and performance were surpassed by the SR-71, unofficially dubbed the "Blackbird." Carrying a two-man crew, the first Blackbirds entered service early in 1966, although details remained highly classified for many years. In 1976, however, the USAF took advantage of the plane's incredible performance to set several new speed and altitude records, flying the futuristic-looking craft to speeds over 2,000 MPH and reaching an altitude of over 85,000 feet. The remarkable SR-71 continued to make highly classified missions with correspondingly classified surveillance gear. Some sources stated that its mission profile included the capacity to make a "specialized surveillance" of up to 60,000 square miles of territory in just one hour. The Blackbirds flown by air force pilots thus provided the means for impressive flexibility and rapid action in response to fast-breaking events for which the Department of Defense wanted information.

But the SR-71, U-2, and other manned aircraft could not offer continuous electronic and photographic surveillance of a world full of potential trouble spots. For this purpose, the Pentagon relied on an array of military satellites, and the USAF maintained a stable of launch vehicles, including Titan III missiles, modified and uprated with a variety of strap-on solid boosters as well as various upper-stage boosters and payloads. The payloads themselves were designed and built for specific jobs, ranging from weather satellites to navigational satellites, communications satellites, photo-reconnaissance spacecraft, and early warning satellites, each often designated as "classified."

For all this otherworldly hardware, the immediate concerns of limited war, conventional tactical issues, and air superiority remained. One of the major problems facing NATO forces in Europe concerned the massive superiority of the USSR and its allies in tanks and other armored vehicles. The solution adopted by the United States and its NATO allies involved more flexible methods of air attack rather than match Soviet-bloc armor on a pound-for-pound basis. Various combinations of NATO partners developed promising designs, like the MRCA (multirole combat aircraft) "Tornado", a swing-wing design for air defense as well as strike roles. On the other hand, the SEPECAT Jaguar, an Anglo-French venture, was specifically developed as a strike fighter; additional similar aircraft were produced by other West European

The Lockheed SR-71 Blackbird was built for reconnaissance at exceptionally high altitudes and speeds. Its advanced aerodynamic design and structural features influenced other high-performance military aircraft. Courtesy of Lockheed Aircraft.

manufacturers. But one of the most interesting designs came from the USAF; the Fairchild A-10 "Thunderbolt II" (so named because one of Fairchild's precursors, Republic Aircraft, produced the famous World War II P-47 Thunderbolt). Conceived as a particularly versatile close air-support plane, designers provided the A-10 with multiple underwing pylons to mount unusually varied military stores totaling up to 16,000 pounds. These might include free-fall bombs, various rockets, napalm, or other combinations of tactical weapons. In addition, the A-10 carried a 30-mm multibarrel gun with selective firing rates between 2,100 and 4,200 rounds per minute, a feature intended to weaken and then penetrate the armor plating of enemy vehicles. Indeed, the A-10 was seen as a potent tank-killer, and its externally mounted twin engines, twin tail, high maneuverability (even when heavily loaded for combat), and other design characteristics were planned to give the aircraft maximum survivability from enemy ground fire during close air-support missions.

Nevertheless, in order for the A-10 and other aircraft, like helicopters, to operate effectively in close air-support missions, control of the air above the combat zone remained essential. For this requirement, the USAF equipped itself with a new generation of air superiority fighters. The McDonnell Douglas F-15 Eagle, first flown in 1972, epitomized the new breed. A single-seat aircraft with a relatively large wing area for extra maneuverability at high altitudes, the plane boasted a top speed of Mach 2.5. The Eagle mounted two after-burning turbofan engines, produced by Pratt & Whitney and designated as the F100-PW-100 power plant. Engine development preceded the Eagle by a considerable period, since the engine was initiated as an advanced, "new-generation" design and generated a thrust equal to nine times its own weight, an unusually high ratio for the time. The combined thrust of a pair of F100 engines topped the F-15's loaded weight (approximately of 40,000 pounds) by a large margin, giving the plane an exceptional climb performance for interception and combat maneuvers. Admittedly slower than the MiG-25 Foxbat, its principal Soviet counterpart, the Eagle was designed to be more agile than any other potential aerial antagonist. The F-15 armament consisted of a 20-mm multibarrel cannon, with provision for more than a half dozen air-to-air missiles. Fully equipped as an all-weather fighter, the plane carried an advanced radar system to detect and track small, high-speed objects (enemy fighters and/or enemy air-to-air rockets) and to fight from 70,000 feet down to treetop level. The F-15 pilot also relied on a "head-up display" system. The head-up display, integrated with the F-15's radar, consisted of a transparent screen mounted atop the instrument panel and directly in the pilot's line of sight ahead of the plane. The system projected a symbolized picture of enemy aircraft as well as pertinent flight information and weapons data, eliminating the need for pilots to scan cockpit instruments while at the same tracking targets during combat.

Significant segments of the Eagle's components were interchangeable with a smaller, lighter fighter, the F-16. Produced by General Dynamics, the F-16 used a single F100 engine, putting the aircraft in the Mach 2 range, and carried the same 20-mm cannon as the F-15. The lightweight F-16 incorporated several new technological features, such as composites to decrease structural weight, a special canopy shape for high visibility, and a cockpit designed to improve the pilot's tolerance for high G-forces in combat maneuvers. The cockpit configuration included a "fly-by-wire" system, which used electrical signals actuating mechanical linkages to move control surfaces (another feature to cope with high G-loads on the pilot during combat) as well as carefully blended wing-body aerodynamics, special refinements for low drag and additional maneuverability, and other improvements. The commonality of engine and armament with the F-15 promised significant economies in acquisition and operating cost. The F-16 also received considerable press attention because it was selected by several NATO countries, and sales were estimated as high as $15 billion through the 1990s.

In 1975, Belgium, Denmark, Norway, and the Netherlands settled on the F-16 to replace their aging fighter squadrons. Agreement on an airplane common to all air forces was significant because the choice meant greater standardization of NATO forces. The negotiations made international headlines since French and Swedish manufacturers, as well as Northrop and General Dynamics, were keen to win the multibillion-dollar deal, heralded as the arms contract of the century. General Dy-

The General Dynamics F-16 was designed as a low-cost air superiority fighter. Its designers made considerable use of special composite materials and lightweight alloys. The plane shown here is finished in low-visibility paint and insignia. On order from some 16 foreign nations, the F-16 exemplifies the importance of exports and arms sales in the U.S. aerospace sector. Courtesy of General Dynamics.

namic's successful bid was due partly to the company's complex manufacturing and financial arrangements, which gave the four European partners coproduction rights. The Europeans not only produced a major proportion of their own planes but also shared in the sales to the USAF as well as to other foreign customers. The European firms also gained advantages in the acquisition of advanced aerospace production procedures. By 1983, the F-16 had been chosen to equip the fighter squadrons of five additional countries.

The air combat environment put a premium on specialized types, but escalating costs and maintenance considerations led designers to keep multirole aircraft in mind. Although designed for air superiority tasks, the F-15 and F-16 incorporated features for certain ground-support duties as well. Naval requirements placed even greater reliance on multirole aircraft because of limited space aboard aircraft carriers. Faced with this constraint, the navy bucked considerable opposition from Congress and ordered production of the Grumman F-14 Tomcat, which made its first flight in 1970. The F-14 was designed to serve as fighter escort, for fleet defense, and for ground attack, requiring considerable versatility in performance and armament and dictating a two-man crew of pilot and weapons officer. Equipped with a pair of high-thrust turbofan engines, the F-14 featured a variable-sweep wing programmed to move automatically in response to performance requirements in its various combat roles. Although the Tomcat was capable of speeds in the Mach 2 range, its variable-sweep wing gave it a landing speed even lower than many of the planes it replaced. Pilots admired its high rate of climb and impressive maneuverability at all altitudes, although the complex airplane required particularly rigorous maintenance schedules in order to keep its combat status up to par.

The navy, like the air force, also recognized the need for a simpler, less expensive lightweight fighter and supported development of the F-18, a joint design project involving Northrop and McDonnell Douglas. Navy and marine procurement officers attentively followed F-18 progress, which, unfortunately showed severe cost overruns. The plane was designed to perform as a fighter as well as a fighter-bomber, a dual capability desired for the limited number of planes that could be carried aboard aircraft carriers. Like many complex new aircraft, the F-18 experienced its share of teething troubles during two years of flight tests, 1978–80. Production commenced in 1980, and by 1983, difficulties were being resolved. Following competitive evaluations, additional orders came from Spain, Canada, and Australia, whose limited air forces also required a versatile fighter/attack aircraft. Versatility was also a strong virtue of the Harrier attack jet, a remarkable warplane capable of vertical takeoff and landing in front-line combat zones. Flown by the U.S. Marines, the Harrier was one of the few foreign postwar military aircraft to be manufactured under license in the United States, a task undertaken by McDonnell Douglas. The Harrier had originally been conceived and developed by the British firm of Hawker. An advanced version, the AV-8B, with improved performance and weapons systems, entered regular U.S. service during the early 1980s. Supporting all three services (army, navy, and air force) were helicopters, trainers, and transports of remarkable diversity.

The army continued to develop helicopters as a potent battlefield weapon. In forward combat zones, the army planned to use several types of choppers for carrying cargo airlifting troops, and attacking enemy armor and strong points. Specialized attack helicopters, like the Hughes AH-64, typified the careful design and reliance on sophisticated electronic systems that characterized aerial weaponry of the 1980s. Carrying a two-man crew, the AH-64 fuselage was designed to withstand hits from most ground fire. The main rotor blades were constructed with stainless spars and made extensive use of glass fiber and other composite materials. The helicopter was powered by a pair of gas-turbine engines; to confuse heat-seeking enemy missiles, engineers devised special engine exhaust outlets in which the hot exhaust gasses would be masked by a secondary flow of cooler air. Complex electronic systems were developed by several other contractors and were tagged with inevitable military acronyms: PNVS (pilot's night vision system) and TADS (target acquisition and designation sight), jointly developed by Martin-Marietta and Northrop; IHADSS (integrated helmet and display sight system) from Honeywell. Further, the weapons-sighting equipment in the nose of the awkward-looking helicopter incorporated infrared radar plus laser equipment to track and mark the target.

A bow-on shot of the USS Nimitz, a nuclear-powered carrier, shows the angled decks and multiple catapults permitting concurrent launch and recovery of aircraft. A pair of F-14s have been positioned at the forward launch catapults. Having a complement of approximately 6,000 personnal, such carriers operate about 85 aircraft, including all-weather fighters and attack planes, anti-sub aircraft and others for specialized electronic warfare missions. Most of the aircraft can deliver either conventional or nuclear weapons. As sea-going bastions, carriers have often been deployed to trouble spots around the globe as a symbol of American national security interests. Courtesy of U.S. Navy.

The lethal AH-64 packed a variety of weapons to make it a killer above the battlefield. These included a rapid-fire 30-mm gun, mounted on a turret under the fuselage; hardpoints of the stubby wings carried various combinations of rockets, ranging from 76 smaller rockets to 16 Hellfire guided missiles. It was necessary for pilots of predecessors to the new attack helicopters to point the aircraft toward the target before firing guided missiles. Now, in the AH-64, the pilot only had to turn his head. Looking through a device like a monocle, attached to his helmet, the pilot would see the target. As he turned his head to keep the target in sight, sensors in the aircraft would monitor the helmet's movement and automatically lock on the target. On the command of the pilot, a laser-guided missile would be launched. Kill rates by such weaponry are expected to be very high. However, since ground troops of modern armies will have similar antiaircraft weapons, the losses for aircraft in future combat are also expected to be high.

FOLKLORE, FANTASY, AND ARTIFACTS

By the 1980s, the phenomenon of aviation and space flight had become ingrained within the sociocultural framework of American life. The perception of aerospace patterns shifted during various decades, but the concept of flight had unquestionably become a permanent fixture within the life styles and consciousness of the American public. Literature, cartoons, motion pictures, fine art, and music incorporated various aspects of flight. Because many older examples of flight themes in books, comic strips, films, and so on have not only survived but are also experienced by contemporary audiences, they serve as cultural evidence of the present as well as the past. Older works have thus become a source of current folklore and fantasy and represent identifiable artifacts that illuminate the evolution of flight.

Among many older Americans, one example is the humorous doggerel concerning the hapless "Darius Green and His Flying Machine," a popular story poem of the late-nineteenth century. The hero of J. T. Trowbridge's poem failed in the attempt to build a successful flying machine "with thimble and thread and wax and hammer and buckles and screws, and all such things as geniuses use." Even after the successful flight by the Wrights, generations of grammar-school students chuckled at Darius's ill-advised flying experiments. Once the public accepted the Wright brothers' feat, airplane stories became a common literary device. Much of the fictional literature involving airplanes seems to have appeared in juvenile titles, and the authors have been long forgotten. One exception is Sinclair Lewis, who in three weeks dashed off an adventure story called *Hike and the Aeroplane,* published in 1912 under the pseudonym of "Tom Graham." An adult novel, *The Trail of the Hawk,* appeared three years later, under Lewis's own name. The protagonist, Carl Ericson, found excitement, fame, and fortune as a flier and emerged as a hero of the machine age. The book won modest critical approval, but never sold well. Nevertheless, it represented one of the earliest efforts by a serious novelist to treat aviation as a contemporary phenomenon and to view it in the context of rapid technological change.

A comprehensive analysis of the kind of aeronautical literature that sold well, serving as a sort of barometer of public interest, is a formidable task. Most of the titles that follow achieved the status of bestsellers, in part because they were chosen by the Book-of-the-Month Club (BOMC). Selection by the club not only conferred a degree of prestige but also denoted books that its editorial board considered newsworthy. Thus, the aviation and space titles offered by the club say something about the tenor of the times over several decades.

Up to World War II, successful aviation books were most likely to be nonfictional. In the 1920s, Charles Lindbergh's chronicle of his trans-Atlantic flight, *We*, became a bestseller, although the first to win critical acclaim was the French author Antoine de Saint-Exupéry. His mystical, romantic memoirs of the pioneering French air-mail lines appeared as *Night Flight* (1932), followed by *Wind, Sand, and Stars* (1939), both chosen as main selections by the BOMC. The latter book marked a change in mood, since it included Saint-Exupéry's first-hand experiences of air warfare during the Spanish Civil War of 1936–39. During the opening phases of World War II, the savage bombing of European cities, coupled with the long-range capability of large bombers, generated a wave of concern in the United States, and three other BOMC selections reflected wartime themes. Alexander P. de Seversky, founder of Republic Aviation Corporation, a major firm of the era, wrote *Victory through Air Power* (1942), a detailed critique of the air war (prior to American entry) and advice on programs needed to prepare the United States for modern aerial conflict. Maps and text dramatized America's proximity to overseas combat zones in the air-age world. Ted W. Lawson's book, *Thirty Seconds over Tokyo*, recapitulated the Doolittle raid, which had boosted American morale during the dark days of retreat in the Pacific. The choice of the latter two books by the BOMC can be interpreted as a mark of recognition for the growing significance of aviation, not only in the war but as a major force in global politics. The stark realities of air power's nuclear component were acknowledged in the selection of John Hersey's *Hiroshima,* distributed by the BOMC as a "pro bono publico" in 1946.

During the postwar era, the club offered selections on both aviation and space. In aviation, Lindbergh's *Spirit of St. Louis* (1953) and Guy Murchie's *Song of the Sky* (1954) represented, respectively, something of a nostalgic view of the aeronautical past and a romantic view of modern long-distance flying. In *The High and the Mighty* (1953), Ernest K. Gann recounted a tense episode as a crippled airliner struggled to remain airborne. The story line displayed a somewhat standardized formula, as various passengers reacted in essentially stereotypical fashion to their predicament, but the rising number of airline passengers undoubtedly made the book's aviation setting both believable and timely. The airliner, in this instance, landed safely. In 1961, Gann's own memoir, *Fate Is the Hunter,* elucidated many of the themes found in his novels. Readers were often startled to realize that so many of Gann's fictional episodes were so closely related to personal experiences. In Gann's occupation as a pilot, fate was never to be taken lightly; it had hunted in the wake of his flying career at every takeoff. The transition to the arcane world of astronautics found expression in Arthur C. Clarke's book *The Exploration of Space* (1952), a primer for the layperson as well as an informed analysis of current and projected programs. The murky world of military scenarios and aerospace weaponry formed the core of a cautionary novel by Eugene Burdick and Harvey Wheeler, *Fail-Safe* (1962). Written by two political scientists who were familiar with Pentagon politics, the plot chillingly detailed a routine response to a presumed Soviet violation of North American airspace. Unfortunately, an electronic malfunction did not recall a group of SAC bombers at the "fail-safe" point. When Moscow was obliterated by SAC's nuclear weapons, the U.S. president ordered an American bomber crew to destroy New York as a sacrifice in order to avert war.

There were, of course, dozens of other examples of the literary treatment of aviation and space, including fiction, nonfiction, and poetry. In the latter category, long poetical works celebrated heroes of flight and significant stages of aeronautical progress: *With Wings as Eagles* (1940) by William Rose Benet, and *The Airmen* (1941) by Selden Rodman. At one time or another, poets whose efforts gained currency include Hart Crane, "Space Conquerors"; Randall Jarrell, "The Death of the Ball Turret Gunner"; Stephen Spender, "The Landscape near an Aerodrome"; and William Butler Yeats, "An Irish Airman Forsees His Death." Similarly, novels, short stories, and essays with an aeronautical motif have been written by John Dos Passos, William Faulkner, and James Thurber. These and others have been anthologized in works specifically devoted to aeronautics.

The literature of space exploration likewise continued to multiply. For years, science fiction addicts had devoured the works of early writers like Jules Verne and H. G. Wells; later practitioners of space-age science fiction included Ray Bradbury, Robert Heinlein, Arthur C. Clarke, and many more. But nonfiction works also caught the attention of critics and made a mark on the bestseller lists. Norman Mailer, *Of a Fire on the Moon* (1970) contained the writer's pungent impressions of the American space program and the flight of *Apollo 11*. In *The Right Stuff* (1979), Tom Wolfe dissected the astronauts' psyches and careers in a racy and fascinating bestseller that was also a BOMC selection.

The outpouring of poems, short stories, and books dealing with aviation was paralleled in music. Aeronautical themes in music date to the era of eighteenth-century ballooning and to the "birdman" era just before World War I, when the romance of early flight inspired many aerial tunes. One of the earliest pieces by a serious composer came from the mercurial George Antheil, who delighted in his reputation as the "bad boy of music" during the 1920s and 1930s. *Ballet Mécanique,* which had its American premier in 1927, had notations in the score for airplane motors and propellers; Antheil revised it in the 1950s to include jet engines. World War II spawned a new wave of popular military tunes, and serious composers also made efforts to capture the spirit of flight. "The Airman's Hymn," by Ralph Vaughn Williams, appeared in 1942, but the most ambitious work was *The Airborne,* a symphony composed by Marc Blitzstein during 1943–44. Commissioned by the Eighth Air Force, *The Airborne* refers not to parachute troops but to the human epic of flight from mythology to air warfare. Blitzstein did not finish the symphony's orchestration until after the war, and the work was finally premiered in 1946 in New York City. The performance attracted considerable attention as a major musical event: Leonard Bernstein conducted the New York City Symphony; Robert Shaw directed the male chorale; Orson Welles intoned spoken parts. Reviews from critics like Irving Kolodin, Olin Downes, and Virgil Thomson were favorable, and the audience responded enthusiastically. Although *The Airborne* was eventually recorded, neither it nor Antheil's pieces, which included two versions of The "Airplane Sonata" (1925 and 1931), were frequently performed works. They merit notice, however, as evidence of intellectual interest in flight as a facet of modern life.

In visual terms, comic strips presented another kind of evidence in the study of popular American culture. The 1930s spawned a series of aeronautical comic strips, partly due to the depression era's penchant for escapism and partly due to the rapid aeronautical advances of the period. A motley crew of aerial heroes came and went in the 1930s, including space-flight pioneers like *Buck Rogers* and *Flash Gordon*. One cartoon story featured an aerial heroine, *Flying Jenny,* who winged through the comic strips of the late thirties and early forties. Although some of the cartoons survived several decades of rapid social and technological change, the advent of television seems to have precipitated the end of popular action and adventure strips. In the meantime, aviation cartoons fed the curiosity of a generation of aeronautical enthusiasts.

Zack Mosley, creator of Smilin' Jack, filled the strip's panels with a variety of aircraft as well as a cast of characters whose names became familiar to a host of readers: Smilin' Jack himself, a mustachioed epitome of the dashing male aviator; a paunchy comic-relief character named Fat Stuff; and an enigmatic buddy, Downwind, who attracted multitudes of adoring females and whose full face remained tantalizingly obscured for the life of the strip. Mosley perfected the standard formula of the hero besting sky pirates, vicious storms, crippled aircraft, and scheming villains. At the same time, he passed along considerable aeronautical lore and subtle commentary on the practicality of aviation, primarily in the civil sector.

Military aviation had its own set of heroes. Three cartoon series stand out: "Terry and the Pirates," "Steve Canyon," and "Buzz Sawyer." Terry Lee, created by Milton Caniff, was something of a desultory adventurer in China until he received flight training early in the war and joined the famous Flying Tigers. Terry's adventures thus paralleled the wartime headlines, creating an immediacy and graphic appeal that was hard to beat. Terry eventually gravitated into the regular U.S. Army Air Force; the U.S. Navy found a counterpoint cartoon hero in Roy Crane's Buzz Sawyer. The wartime strips, drawn with great accuracy, mixed adventure with patriotism and attained an admirable depth of personality in their presentation.

In 1946, Caniff turned over "Terry and the Pirates" to a new illustrator, George Wunder, and began a new story, "Steve Canyon." During the postwar era of cold war confrontations, Steve Canyon and Terry Lee took on the duties of propagandists for the U.S. Air Force, while Buzz Sawyer did the same for the U.S. Navy. Caniff, in particular, maintained close ties with the air force's public information officers in the Pentagon. One of them, Maj. William J. Lookadoo, appeared in Caniff's strip as another air force major, "Luke Adew," both majors being slightly stocky types with an Arkansas drawl. During congressional sessions, when committees were reducing air force budgets, Caniff and Wunder regularly produced episodes in which their heroes not so subtly argued for restored funding and more air power.

"Sky King" *began as a radio show in 1947. In the 1950s it became a television series, carrying on a tradition of air adventure drama dating from the radio broadcasts of the 1930s. Sky King, Kirby Grant, and his niece, Penny, Gloria Winters, pose here beside their Cessna 310,* Songbird III. *Courtesy of Cessna Aircraft Company.*

Pentagon officials overtly supported these messages and assiduously courted the cartoonists whose strips were read by millions every day. "Steve Canyon," for example, ran in newspapers published in over 600 American cities, plus 17 countries overseas. When Buzz Sawyer's creator, Roy Crane, decided he was running short on story ideas, and considered taking his character out of the U.S. Navy, the naval brass regarded it as a genuine crisis. The admiral in charge of the navy's office of information put out a global request for story ideas, and Crane also got a lengthy VIP tour of the Pacific fleet. Crane was inundated with letters, arguing that the strip gave the service prestige and sustained public appreciation of the navy's role. For reasons like these, the postwar comic strips should be remembered, since they continued into the 1970s and made an impact in terms of propaganda as well as entertainment. Although such military aviation strips have disappeared as regular fare in major newspapers, a vestige of the genre lingers on. In Charles Schulz's *Peanuts*, the canine insouciance of Snoopy as an intrepid World War I pilot dueling the Red Baron captures a lasting example of nostalgia.

The visual impact of comic strips was magnified by the dimensions of the motion picture screen. During the 1920s, film makers learned the unique choreography required of aerial sequences, as demonstrated in several classic films like *Wings* (1927), *Hell's Angels* (1930), and *Dawn Patrol* (1930). These productions also fostered the clichés of numerous aviation films that followed—the devil-may-care aviator and inveterate womanizer who becomes the cool and incredibly skilled hero once he is in the air. Fortunately, a few films in the 1930s attempted to portray something about significant contributions of technology to aviation progress and to suggest the impact of aviation on travel and international relations. Box-office stars in these films also drew sizable audiences who could not help picking up useful insights about aviation from these technological subplots: *Night Flight* (1933) starred Clark Gable and analyzed navigational techniques; *Ceiling Zero* (1935) had James Cagney perfecting deicing equipment; *Test Pilot* (1938) pitted Gable and Spencer Tracy against the hazards of stratospheric flying; *China Clipper* (1938) featured Humphrey Bogart and Pat O'Brien in the aerial conquest of the Pacific Ocean.

The Second World War inaugurated films that were packed with patriotism and heroic sacrifice. Film critic Judith Crist called them "propaganda melodramas, leaving us with the conviction that every bomber crew was required to include a hero, a heavy, and a comedian from Brooklyn." For the most part, the titles are self-explanatory: *Aerial Gunner; Bombardier; Dive Bomber; Eagle Squadron; A Guy Named Joe;* and so on—mostly grade "B" films, and mostly forgettable. Some of the wartime documentaries were exceptions. Film clips of actual combat gave them a gritty reality; talented directing and skilled editing made them memorable. The talent associated with the series called *Why We Fight* explains the impact as well as the importance of other outstanding wartime documentaries. Direction for *Why We Fight* was by Frank Capra and Anatole Litvak; writing and research staffers included Litvak and William Shirer; musical scores came from Dmitri Tiomkin and Alfred Newman. Released between 1942 and 1945, the series emphasized major theaters of the war, like Western Europe, the Soviet Union, and China, although one film, *The Battle of Britain* (1943), spotlighted air power in the war. Two other films that deserve mention were directed by William Wyler. *The Memphis Belle* recorded the final mission of a B-17 and its crew over Germany. One of the cameramen was killed while filming the flight, bringing home the risks of documentary projects. A second film, *Thunderbolt* (1945), portrayed the effectiveness of P-47 fighter-bombers during Allied offensives in Italy. One statement in the film, "The airman never sees the face of the people, but only the face of the country," exemplified the impersonal nature of combat dominated by machines. Some of the best World War II motion pictures, probing psychological pressures and the burdens of crisis, appeared in the postwar era. *Command Decision* (1948) starred Clark Gable, Van Johnson, Walter Pidgeon, and John Hodiak. Based on a book by Biernie Lay, the dialogue avoided the clichés of most Hollywood aerial epics. *Twelve O'clock High* (1959), also scripted by Lay, had a talented cast led by Gregory Peck and Dean Jagger. Peck played a commander posted to a demoralized B-17 bomber group in England; in the course of instilling a new spirit and repeatedly ordering crews through the deadly barriers of Germany's aerial defenses, Peck himself suffers a physical and mental collapse. The film became one of the best Hollywood productions that treated the strain and terrors of relentless combat missions.

Other postwar productions, built around contemporaneous themes, included *Big Lift* (1950), about the Berlin Airlift; *Strategic Air Command* (1955), on the global reach of USAF's intercontinental bombers; *The Bridges at Toko-Ri* (1954), on the Korean War.

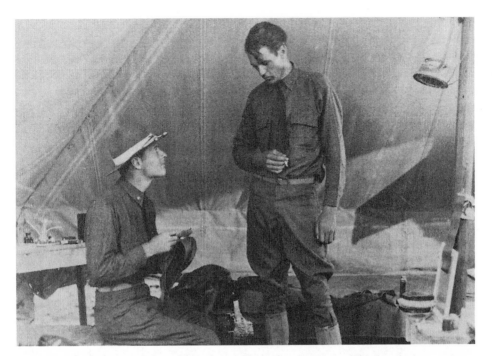

Aviation dramas captured the public's fancy during the twenties and thirties. Action sequences in Wings, *the World War I drama directed by William Wellman, had the authentic stamp of Wellman's own experiences in the Lafayette Flying Corps. The film was chosen best picture at the inaugural Academy Award ceremonies in 1928. Gary Cooper (right) received good notices for his supporting role in the movie, which featured Clara Bow and also included Hedda Hopper. Courtesy of U.S. Air Force.*

There were some colorful "period-piece" films as well, such as *Those Magnificent Men and Their Flying Machines* (1966), an entertaining film about pre–World War I aerial contests whose flying replicas of prewar biplanes recaptured the thrills of a bygone era. The lighthearted theme of this film represented the contrast in images generated by aviation. On the darker side, Stanley Kubrick's macabre comedy, *Dr. Strangelove* (1964), repeated a familiar theme of the 1960s in which miscalculations between the United States and the USSR lead to a nuclear exchange. In this case, doomsday air force generals and crazed military advisers fail to stop a die-hard, flag-waving bomber crew from attacking their prearranged targets. Francis Ford Coppola's *Apocalypse Now* (1979), a bitter indictment of the Vietnam war, uses helicopters in one telling episode to demonstrate the devastating, uninhibited aerial onslaughts of the conflict. Flying in over an azure sea, white surf, and tropical beaches, the attacking choppers form a malevolent wave of dark shapes. They are a presentiment of death, and the hollow beat of helicopter blades is repeated in a sinister counterpoint that frequently intrudes into the film.

The evolution of space exploration in the postwar era found expression in various science-fiction thrillers as well as adventure films that tried to portray space travel in a convincing way. Most of these efforts achieved little success, either at the box office or in technical verisimilitude. It was left to television to produce a series that was not only convincing but that became enshrined in the temple of popular culture. "Star Trek," which appeared in the fall of 1966, managed to make the flight deck of the starship *USS Enterprise* into a believable piece of space-age fantasy. Although the series ran for three years, the show's biggest success occurred after it departed the prime-time slot and went into reruns during the 1970s and 1980s. "Star Trek" developed a fanatical cult following, and thousands of "Trekkies" still attend national conventions to swap memorabilia and television trivia.

In the meantime, motion pictures had begun to catch up, epitomized by *2001: A Space Odyssey,* which premiered in 1968. Produced and directed by Stanley Kubrick, the film adroitly combined striking visual effects with knowledgeable science fiction using a screenplay developed by Kubrick and Arthur C. Clarke from one of Clarke's own stories. Briefly, the plot concerned an expedition to Jupiter to probe the mysteries of a slablike black monolith; the enigmatic conclusion left open several avenues for interpretation. The film took full advantage of the Cinerama wide-screen process, along with stereophonic sound for the score, which incorporated themes from classical music. The production appeared just as Americans were beginning to relate to the phenomenon of space flight and the otherworldly images that NASA's manned missions made real in the late 1960s. The film's visual imagery had little competition until the arrival of the spectacular *Star Wars* in 1977, written and directed by George Lucas.

Moviegoers who may have been perplexed by the ambiguous ending of *2001* had no such doubts about *Star Wars,* a space fantasy of heroic adventure in which a courageous few triumph over an army of malevolent villains. But the straightforward plot alone did not explain the phenomenal box office success of *Star Wars.* Its visual impact was stunning. Using the most advanced techniques of photography, graphics, model-making, animation, and set design, the Lucas production team achieved new levels of visual fidelity in action sequences. It was no accident that the swirling aerial duels between the X-wing fighters of the heroes and the TIE-fighters of Darth Vader and his evil minions were reminiscent of World War II engagements. Lucas had watched many filmed dogfights of that era in plotting his own space-age combat sequences. More-over, *Star Wars* had romance, upbeat philosophy, an appealing population of extra-terrestrial inhabitants, and robots with charmingly human foibles to appeal to the moviegoing audience. Two sequels were also hugely successful.

In addition to the commercial spin-offs of toys, calendars, and clothing, *Star Wars* provided cultural side-benefits. Composer John Williams provided an original score of impressive orchestral creativity. Millions in the film's audience—adults as well as adolescents—were subliminally caught up in the rich tapestry of the symphonic medium and left the theater with positive impressions of orchestral compositions. Finally, the film helped generate a strong wave of interest in topics like space travel and other galaxies, a trend that boosted the success of other space films as well as interest in television productions like Carl Sagan's "Cosmos." Presented as a 13-part series in 1980, "Cosmos" was developed for public television networks and became one of the

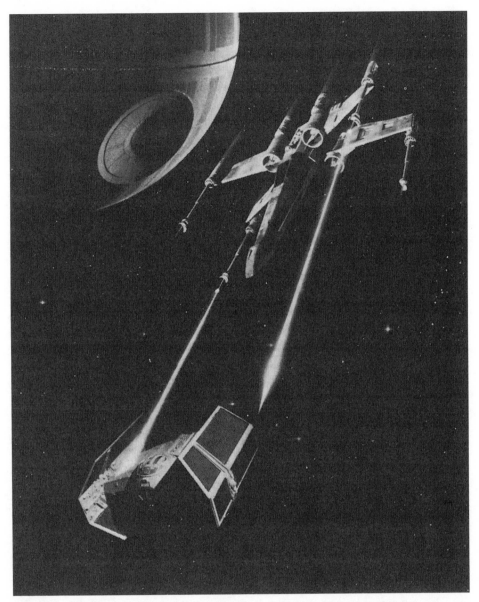

Star Wars *and its sequels have won both popular and critical acclaim during the 1980s. The films successfully blended fantasy and science fiction brought to life by skillful animation and photography as TIE-fighters (villains) and X-wing fighters (heroes) shot it out in galactic dog-fights. Courtesy of Lucasfilm, Ltd.*

most successful shows in the history of public television. As the title suggests, the content dealt with cosmology, offering viewers a fascinating review of astronomical lore like black holes, time travel, possible life in remote galaxies, and so on. With astronomer Sagan as commander of an imaginary space ship soaring through nebulae and other phenomena, it was difficult not to recall similar images from the genre of space fantasy films.

The visual imagery of films and television depicted the awe and fascination of space flight as well as the awesome destructive power of space-age weaponry. This paradox was clearly evident in earlier treatment of flight themes, as seen in examples of painting during the late 1930s. For the new administration building at Newark Airport, Arshile Gorky painted a series of murals with modernistic aeronautical images in 1936–37. Bold and colorful, the murals depicted a positive view of flight and of aviation technology. On the other hand, one of Gorky's contemporaries created a canvas that evoked the terror of aerial warfare. Pablo Picasso's *Guernica* (1938), painted in the aftermath of a bombing raid on a small town during the Spanish Civil War, continues to be an internationally recognized statement of protest. As World War II unfolded, combat artists on both sides depicted the role of aircraft in heroic terms, as technologically impressive military machines, or as the harbingers of death and destruction.

For many observers, modern aeronautics has presented a positive image, symbolizing the best of American ingenuity and technology. Although the concept of streamlining seems to have entered the realm of industrial design by way of aerodynamically shaped railway engines, the airplane came to represent the best of the streamline style in terms of lightness, economy of design, and use of modern materials like aluminum. As social critic Lewis Mumford noted in *Technics and Civilization* (1934), aeronautical design set the standards for refined and exact engineering. For Mumford, aircraft also represented a desirable aesthetical logic. In 1938, Mumford did the scripting for *The City,* one of the outstanding documentary films produced by Pare Lorenz during the New Deal era. Toward the end of the film, as narrative and photography urged the need for better urban planning and emphasized that a better life was possible, the camera focused on a DC-3 as it taxied into position, accelerated, and gloriously soared into the sky. Above the crescendo of Aaron Copeland's score, the narrator instructs the audience, "Science takes flight at last—for human goals." The plane represented an attainable goal for modern urban environments, "moulded to human wants as planes are shaped for speed."

By contrast, the destruction caused by air power during World War II was a sobering experience. The combination of air power and nuclear weaponry in the cold war era, discussed earlier, intensified apprehensions. Nonetheless, the wartime and postwar eras also experienced air-age travel as a potent and useful cultural force. During 1942, Wendell Willkie, the respected Republican political leader, made a globe-girdling flight of 31,000 miles. His best-selling and widely discussed book about the experience, *One World* (1943), noted the role of modern aviation. "If I had ever had any doubts that the world has become small and completely interdependent," he wrote, "this trip would have dispelled them altogether." Max Lerner's influential study, *America as a Civilization* (1957), referred to aviation in terms of its effect on the mobility of Americans as well as the democratization of vacation travel. He also awarded accolades for aviation's aesthetic influence. "In contrast to the auto, the American airplane illustrates the best in American design. Every element of it is shaped and built for speed, engineering, reliability, and safety. There are no frills or unnecessary gadgets," Lerner wrote. "Yet—or perhaps therefore—the beauty of the airplane has scarcely been surpassed in the history of industrial design." The airplane's place in the evolution of American design found affirmation in a roster of artifacts illustrated in an article, "The 100 Best-Designed Products," published by *Fortune* magazine in 1959. The list was compiled by the Illinois Institute of Technology, who had polled 100 individuals acknowledged as leaders in their fields of design. Two airplanes made the final list: the classic Douglas DC-3 of 1935 and the distinctive Beechcraft Bonanza of 1945. Americans continued to regard aircraft as an epitome of industrial quality. In 1972, *Fortune* named the Boeing 747 as the "machine of the year," citing the plane for its guidance system, on-board computer, structural design, and passenger comfort.

The technological symbolism of aviation had its droll episodes as well. As a token of top-drawer quality, Beechcraft products received press attention in 1960 when the company's planes were featured in the Nieman-Marcus Christmas catalog. "His and her" gifts had appeared before, but a merchandising executive of the exclusive department store in Dallas started a new trend in 1960 by spotlighting high-priced esoteric gifts. The catalog for that year included a double-page spread in full color that offered a pair of airplanes: "hers" was a trim Beechcraft Bonanza, "his" was a muscular Twin Beech. Both had matching paint schemes, with the appropriate gender lettered in gold script on the nose of each plane. The photograph was appropriately romantic in tone, showing the plane's reflection from pools of water on the airstrip; in the foreground, an elegant couple in formal evening wear projected an image of high style. Nieman-Marcus received a potential order from a rancher in west Texas: "I can't use both planes, for I already have one I use myself; but if you will break the pair, I'd like one for the little woman, who has been hankering for a plane of her own for a long time." The store naturally obliged, and added the lady's name to the plane's nose as specified—"Mamie Bell."

Evidence of the continuity of aeronautical folklore can also be found in the language of news stories and in the lingo of engineers. Such cultural symbols are as current as the Space Shuttle. Consider, for example, the gremlin. Reference to gremlins popped up in comments made in 1981 by John Yardley, NASA's associate administrator for space transportation systems. Explaining the rationale for extensive tests and flight trials of the shuttle, Yardley stressed the need to uncover unanticipated problems that might be lurking in the shuttle's unique flight profile. "This is entirely different," Yardley said. "A new engineering gremlin could crawl out of the woodwork, one nobody could have predicted." "Gremlins" in this context obviously referred to the pesky equipment failures that invariably cropped up in complex engineering projects. His casual use of the word implied that he expected others to understand its meaning. The term dated back to World War II, when it was usually associated with cartoon characters developed by Walt Disney. Actually, gremlins originated in the fertile imagination of British author Roald Dahl.

As an RAF fighter pilot in World War II, Dahl had first-hand experience with inexplicably balky aircraft. After receiving a medical discharge early in the war, Dahl

In 1960, trendy Nieman-Marcus launched its famous "His-and-Her" holiday gifts with the twin-engine Beechcraft G-18 (for him) and the trim Bonanza (for her), noting that "both can be bought in your choice of color, style, cabin arrangement, and any number of combinations of individual navigational equipment." For the ladies, a jacket in Russian white ermine with a bleached white fox border added "the right dash" for just $2,975.00. Courtesy of Nieman-Marcus.

was posted to the United States as an air attaché in the British Embassy. During this assignment, he began writing, and introduced gremlins in a short story for children. In wartime, Dahl wrote, infuriating little creatures called "gremlins" created the problems causing equipment to malfunction and planes to crash. Gremlins resembled tiny men; females were called "fifinellas"; offspring were "widgets." Some of Disney's whimsical illustrations accompanied the story in a magazine article and in a book; an animated film was planned but was never finished. But gremlins became well known in the U.S. Air Force as well as the RAF, and won Dahl an invitation to the White House because Mrs. Roosevelt's grandchildren had loved the story. The air force used Disney's gremlin figures in a series of instructional pamphlets supplied to training schools for aircraft mechanics. The term appeared in postwar dictionaries, although the etymological information was uniformly vague. In any case, the legendary gremlin has survived, bedeviling a new generation of hardware in the era of space exploration.

ASTRONAUTICS

Successful missions to distant planets, providing unusually dramatic visual images of undiscovered features, coupled with the equally impressive success of the Space

Shuttle sustained the sense of discovery and adventure in space exploration. The shuttle embodied the concept of a true space transportation system, capable of conducting a variety of missions in earth orbit. Like Apollo-Saturn missions, shuttle operations required several steps. A pair of solid-fuel booster rockets, along with the shuttle's trio of liquid oxygen/LH$_2$ main engines, lifted the shuttle off the launch pad and into the ascent trajectory. After the solid rocket boosters burned out, they were jettisoned and parachuted back to earth, to be recovered and refurbished for repeated use. The shuttle's main engines, supplied with fuel from a large external tank, continued to fire, driving the Orbiter close to the optimum altitude for its mission. Finally, the non-recoverable external fuel tank fell away and the Orbiter refined its correct orbital trajectory by firing its maneuvering engines. After carrying out its assigned mission, the Orbiter, with wings and control surfaces intact, returned to earth, landing on a conventional airstrip. Since the Orbiter could be used many times, the Space Shuttle could carry out space missions at a fraction of the cost of earlier systems, which were used only once and were not suitable for refurbishing. Because the Orbiter was designed to carry self-contained manned orbital workshops and laboratories for later retrieval, many nonastronauts could participate in research in space. The Orbiter could also carry satellites and other unmanned space probes into orbit and could retrieve satellites and scientific packages for repair and analysis. The flexibility and reusability of the shuttle made it an extremely attractive system for space missions in the years ahead.

Reflecting the trend toward multinational aerospace programs in commercial ventures, foreign agencies developed key elements of NASA's shuttle system. The electro-mechanical arm used to manipulate payloads in and out of the Orbiter's cargo bay was developed and built by a group of Canadian firms under the aegis of the National Research Council of Canada. An ingenious, 50-foot-long device, the robot arm had electrically operated joints at its "shoulder," "elbow," and "wrist," ending in movable clamps (acting like a hand, but officially called the "end effector") to grasp satellite payloads or other items. A large percentage of shuttle missions planned to carry the Spacelab, a laboratory module permitting shirtsleeve missions. A consortium of ten European nations under the European Space Agency developed and fabricated the first Spacelab, with subsequent versions to be commissioned by users. But launch of the shuttle ran into continuous, frustrating delays. A complex array of individually machined and fitted thermal tiles covering vulnerable sections of the Orbiter as it descended through the earth's atmosphere before landing broke and fell off the Orbiter under simulated operational conditions. The search for a solution to the tile problem consumed months of tests, and balky main engines on the Orbiter caused other delays. The original launch scheduled for 1978 slipped year by year into 1981.

In the meantime, a series of seemingly improbable tests explored the Orbiter's flight characteristics. Wind tunnel data usually gave engineers a clear idea of an aircraft's probable handling characteristics in the air. Still, one could never be sure until the "bird" actually flew. The Orbiter, designed to take off by rocket power and return to earth in a dead-stick landing presented some problems at the flight test stage. Short of an actual launch (totally impractical), how did one get any trustworthy flight data on the

The Space Shuttle inaugurated a new chapter in NASA's manned missions. Taken during the STS-7 flight in 1983 the photo shows the Orbiter with cargo bay doors open, exposing an array of scientific experiments. From the forward section of the cargo bay, the remote manipulator arm has been deployed to the left. The picture was snapped by a remote-control camera in a free-flying satellite just released from the shuttle. Courtesy of NASA.

unique Orbiter? NASA decided to carry the shuttle aloft in "piggy-back" fashion aboard a specifically modified 747. This solution also allowed the Orbiter to be carried back and forth between the contractor's maintenance facilities on the West Coast and the launch site in Florida. The first Orbiter, christened the *Enterprise* (after the space ship in the hit television show "Star Trek"), arrived at Edwards Air Force Base early in 1977 for flight trials.

A set of special jigs held the Orbiter above the ground while it was mated to the 747. Several test flights of the 747/*Enterprise* combination followed, giving crews of both craft confidence that this unlikely configuration would actually work for flight trials and for cross-country transportation of other Orbiters. Finally, on August 12, 1977, NASA tried the first free flight of the *Enterprise*. At about 28,000 feet, astronauts Fred Haise and Charles Fullerton punched the buttons to fire a series of explosive bolts, setting the *Enterprise* free. The 747 pilot quickly rolled the huge airliner into a diving turn to the left as Haise rolled the *Enterprise* into a right bank—evasive maneuvers to avoid possible collision with the 747's towering vertical fin. Fortunately, this flight and subsequent trial runs turned up no serious difficulties. Meanwhile, engineers solved other stubborn problems, clearing the way for the first orbital mission of the unique Space Shuttle in the spring of 1981.

Because of concern about tiles and engines, the first orbiter built for actual missions, the *Columbia,* languished at NASA's Kennedy Space Center on Cape Canaveral for nearly two years. Its flight crew, astronauts John Young and Robert Crippen, joked that they had spent so much additional time in simulators that they were "130% trained and ready to go." But John Yardley, who headed NASA's space transportation programs, insisted on repeated vehicle tests—not just for tile and engine problems, but to turn up any other last-minute shortcomings. It was necessary, he said, to scrutinize and triple-check for "things that people have never thought of simply because we have a configuration that's considerably different from the thirty years of history of launch vehicles." Finally, even Yardley was satisfied. Launch was scheduled for April 10, 1981.

Moments from ignition, however, mission controllers glumly had to scrub the launch due to a faulty computer. But two days later, the countdown matched a day of perfect weather at the Cape, and the shuttle soared away toward space, boosted by 7.0 million tons of thrust from its combined solid booster rockets and LH_2 engines. At an altitude of 130 nautical miles, the *Columbia* settled into orbit for a two-day mission. Some consternation occurred when a remote television camera aboard the *Columbia* showed gaps around the tail section, where some tiles had separated during the launch. Had other tiles, in critical areas along the Orbiter's vulnerable underside, also fallen out? The searing descent back into the earth's atmosphere carried more than the usual sense of drama. At a blinding speed of Mach 24, *Columbia* plunged back into the earth's upper atmosphere, where the intense heat of reentry began to build to over 3,000°F. Several minutes passed as the plummeting spacecraft developed a sheath of ionized gases that prevented radio communications. At 188,000 feet, a relieved mission control heard a matter-of-fact report from Crippen and Young that the *Columbia,* now slowed to Mach 10, was performing as planned. Unlike the Apollo spacecraft, which hurtled earthward like a ballistic missile until parachutes floated them somewhat unceremoniously to a soggy touchdown in the ocean, the *Columbia* swooped in for a landing like an aircraft—a true "spaceliner" symbolizing a new era in astronautical ventures.

The *Columbia* racked up another astronautical record in November of 1981, when it became the first spacecraft to return to orbit, proving the validity of NASA's program for a reusable vehicle. Although a faulty fuel cell caused mission controllers to cut the planned five-day flight to two, the two-man crew completed nearly all of the assigned scientific experiments, plus a successful demonstration of the robot arm. Additional flights racked up successive milestones in development of the Space Shuttle. During one of *Columbia*'s missions in March 1982, heavy rains at Edwards Air Force Base caused a switch to an alternate landing site, at White Sands Missile Range in New Mexico, demonstrating the progam's flexibility. After five flights, a new orbiter was launched. In April 1983, the *Challenger* began a successful series of missions, including a spacewalk on the seventh Space Shuttle flight. During June, the *Challenger* lifted off with the first full complement of five astronauts; mission commander, pilot, and three mission specialists. One of the mission specialists, Sally Ride, became the first American woman to travel in space. During a busy flight, the *Challenger*'s crew launched communications satellites for Canada and for Indonesia. Using the remote manipulator arm, the missions also included deployment and retrieval of a West German satellite that tested solar cells and carried out a series of experiments such as making metal alloys in space. Flights like these not only highlighted the ability of the Space Shuttle to carry out multiple tests and satellite launches during any one mission but also indicated the increasing international scope of NASA's space program. In addition, the Space Shuttle crews undergoing training included American ethnic and minority representatives as well as women. The eighth shuttle mission included Guion Bluford, Jr., the first black astronaut to make a space flight.

An important share of NASA and commercial satellite efforts involved direct earth applications. Fifteen years after the first Syncom satellite, more than 50 similar satellites had arrived in their respective orbits, vastly improving the world's communications systems. By the end of the 1970s, Intelsat, a global communications satellite network, numbered over 100 subscriber countries and handled television news, business data, and telephone calls. Even with inflation, international phone rates cost less than the first Intelsat service of 1965; Intelsat's television channels could reach over a billion people—one of every four people on earth. NASA research satellites demonstrated further applications to come: air traffic control far out at sea; lecture exchanges from classrooms thousands of miles apart; medical instructions and educational programs for remote outposts and hamlets around the world.

Weather observation and commodities information continued to be of great significance. Since 1966, weather satellites have photographed the entire earth on a daily basis and often appeared in scheduled television weather broadcasts. There is no way to estimate the thousands of lives and billions of dollars saved by this kind of coverage. No tropical storm escaped notice, and improved weather forecasting and storm warnings (plus information on the best times for planting, irrigation, fertilizing, and spraying) helped coastal and inland residents as well as farmers. During the mid-1970s, one NASA project perfected a satellite information system to project acreages and yields of wheat, integrating data from countries on several continents. Such information had inherent value in planning equitable food distribution and in market planning to avoid price dislocation resulting from crop shortages or bumper harvests.

Considering such commercial applications, a number of private firms entered the field. During 1982, one private company successfully launched its own vehicle, a solid-fuel rocket named *Conestoga 1*. The rocket, which carried a mock payload, was put together from surplus parts by Space Services Incorporated, who announced their intention of launching a series of low-cost vehicles to carry payloads into orbit. Possible cargoes included communications and resource management systems for customers in the petroleum industry.

Many types of groups realized benefits from data and pictures supplied during Skylab missions and from a new series of earth resources satellites. Users included

To maintain flying skills of its astronaut corps, NASA relies on Northrop and T-38 Talon jet trainers. Sally Ride, the first American woman to log a space flight, in 1983, was one of several women among U.S. astronauts. Courtesy of NASA.

foreign countries, federal agencies, state and local governments, private industry, and regional planning authorities. A partial list of applications cited by NASA included:

- mapping mountain snow cover to forecast spring runoff for irrigation and hydro-electric power
- detecting oil slicks at sea
- monitoring offshore dumping of sewage and industrial wastes
- taking inventory of timberlands and grasslands
- mapping changes in land use to assist in more intelligent urban planning
- mapping uncharted coastal shoals endangering shipping
- preparing and updating navigational charts and other types of maps

Costs of mapping from space compared favorably with other options. One state agency discovered that 14 satellite photos and associated labor-hours cost $75,000, as opposed to 20,000 aerial photos and total costs of some $1.7 million by conventional techniques.

Instrumented space missions progressed under many budget constraints, since Apollo had required immense allocations, and the Vietnam War, inflation, and the shuttle project all cut into available funds. Instrumented space programs persisted haltingly, sometimes kept alive only by reworking spacecraft and ancillary hardware left over from earlier efforts. Seen against such lean budgets, the record of successful missions between 1962 and 1976 stands out even more remarkably: exploration of the so-called inner planets (Venus, Mars, and Mercury) and the outer plantes (like Jupiter and Saturn) as well as continuing exploration of the interplanetary environment with spacecraft like Explorer, Pioneer, and others. Perhaps because of the time and distance involved in its journey, one of the more intriguing missions of the era involved the *Pioneer 10* probe. Launched from Cape Kennedy on March 2, 1972, *Pioneer 10* headed for a rendezvous with the distant, eerie planet of Jupiter. After a voyage of nearly two years, *Pioneer 10* reached the vicinity of Jupiter early in December of 1973. Nervous scientists were relieved that it survived the deadly belts of radiation around the planet and was able to transmit hitherto unknown data on Jupiter's atmosphere, temperature, magnetic field, and other properties. *Pioneer 10* did not land on Jupiter but continued its journey, speeding past the planet with a boost from Jupiter's gravitational field.

When *Pioneer 10* first reached Jupiter, 620 million miles from Earth, the space probe had already traveled farther than any man-made object in history. Most satellites and space probes relied on solar cells to convert sunlight into electricity to operate the instrument packages on board. *Pioneer 10*'s journey toward the far reaches of our solar system carried it into regions where the sunlight is too dim for solar cells to function, so the spacecraft was the first NASA vehicle to draw power from a set of four small, nuclear-fuel generators. As the spacecraft continued past Jupiter, its speed decreased from about 80,000 MPH to a mere 25,000 MPH—a rate of travel that nonetheless will permit its escape entirely from our solar system. A small plaque attached to the spacecraft is engraved with human figures and a mathematical code to explain *Pioneer 10*'s origins and odyssey into other galaxies to intelligent life that might exist some-where in outer space. In the meantime *Pioneer 10* will continue to cruise through space, crossing the orbits of Saturn, Uranus, Neptune, and Pluto. *Pioneer 10* was scheduled to reach Pluto in 1987; afterwards the spacecraft will plunge into the uncharted reaches of interstellar space. If nothing ever stops it, it will simply coast for infinity, a wanderer among the stars.

While *Pioneer 10* was being developed, a more thorough series of instrumented planetary excursions to the vicinity of Mars commenced. A succession of Mariner flights during the 1960s and early 1970s orbited the planet, photographing and mapping its topographical features while acquiring intriguing information about the Martian atmosphere. After the Mariners came *Viking 1* and *Viking 2*, launched August 20 and September 9, 1975, respectively. Each spacecraft consisted of an orbiter and a lander. Relaying computer-based images back to earth, the orbiters helped to determine final landing sites, then continued to map as they relayed data from the lander to earth receiving stations. The unique landers functioned as miniature automated laboratories. Once on the Martian surface, the landers extended armlike probes, to scoop up soil and rocks to test for moisture, organic compounds, and signs of life. Cameras snapped memorable panoramic views of the austere Martian landscape and horizon. No sign of life turned up on Mars. Pioneer-Venus probes in 1979, though compiling a wealth of information on the chemical composition of the Venusian atmosphere down to its sweltering surface, also yielded data pointing to a lifeless planet.

The outer planets came under close scrutiny in the late seventies and early eighties with NASA's advanced Voyager series. *Voyager 1* and *Voyager 2*, launched in 1977, reached Jupiter in 1979 and crossed Saturn's path in 1980/81. With startling detail, the pair of Voyagers relayed pictures of complex Jovian weather patterns and surprising differences among Jovian satellites—active volcanoes on Io, ancient impact craters on Callisto, and varied surface features on others. As the Voyagers sailed past Saturn, space scientists found more surprises as the spacecraft transmitted spectacular pictures of the planet's hundreds of concentric rings (previously thought to be only six), in addition to other unknown Saturnian moons and new surface features. In 1986, eight years after departing Earth, *Voyager 2*'s flight path carried it past Uranus, where the spacecraft could take the first closeup pictures of that far-off planet and its fascinating moons before vaulting onward towards Neptune. Then, both Voyagers will continue to coast, outbound from Earth on endless journeys. Planetary excursions like these, along with continuing missions in the near-Earth and deep-space environments, add immeasurably to the appreciation and understanding of our galaxy; comparison of the geologic patterns and atmospheric characteristics of other planets adds immeasurably to the appreciation and understanding of our own.

9

Turmoil and Transition, 1983–2000

The defining events of the 1980s and early 1990s unfolded in the dramatic thaw of cold war antagonisms, the collapse of the Soviet Union, and a commitment of massive U.S. military power to combat in the Middle East. Other trends already in motion were inevitably influenced by the new international order. Within the American aerospace community, a market that depended heavily on military sales, the corporate upheaval was intense and the decline of business in numerous urban areas soon blighted many local economies. Eventually, an increasingly healthy civil market helped pick up the slack as global business and travel flourished. At the same time, the effects of deregulation, along with an accelerating trend toward foreign association in American operations, dramatically reshaped the airline industry. Foreign partnerships in manufacturing also expanded, reflecting major patterns in the world economy.

All of this triggered astonishing changes in the aerospace industry as well as the airline business. In the United States, a series of exceptional mergers left Lockheed in control of combat aircraft formerly produced by General Dynamics; the company also absorbed missile and electronics manufacturer Martin Marietta. The legendary firm of McDonnell Douglas disappeared into the corporate matrix of its legendary rival, Boeing. Moreover, European industries mounted an unprecedented series of transnational mergers to compete more effectively with American corporations in the global market. At the same time, multinational partnerships in manufacturing also expanded, sometimes linking trans-Atlantic rivals in specific development programs.

These occurrences opened new vistas of international relationships as well as national change. This also meant that American aerospace programs operated within the context of fundamentally altered circumstances that brought a challenging and uncertain entry into the twenty-first century.

327

GLOBAL REACH — GLOBAL POWER

When Ronald Reagan referred to the Soviet Union as the "evil empire," many Americans agreed. In 1983, tension between the two superpowers was tangible; the year in which Reagan proposed his Strategic Defense Initiative, an ultrasophisticated network of satellite systems intended to detect Soviet missiles aimed at America and destroy them in mid-trajectory if they were launched. Characteristic of many cold war programs, it presumed billions of dollars allocated by Congress and big contracts for innumerable aerospace manufacturers. A series of unexpected events made the Strategic Defense Initiative, as well as the entire national defense budget, appear to be uncertain. During 1985, Mikhail Gorbachev assumed leadership in the Soviet Union and promoted thoroughgoing reforms. The Berlin Wall came down in 1989 as Communist regimes in the former Soviet satellites collapsed. Within the next two years, international conferences formally announced the end of the Warsaw Pact and the termination of the cold war itself. The Soviet Union abolished Communism as its official ideology and then splintered into a number of independent governments, many of which reformed with Russia to create the Commonwealth of Independent States. Against this background, a slow process toward Soviet-American arms control, initiated some twenty years earlier, suddenly took form.

During the 1970s, the American Department of Defense had expended considerable funds in R&D for advanced missile systems with extremely high accuracy. During the early 1980s, these systems went into production and became operational with the U.S. Air Force, Army, and Navy around the globe. The new systems were very expensive, more destructive, and more frightening in an armed world of increasingly lethal nuclear weaponry. As the Soviets endeavored to maintain parity with the United States and vice versa, the two superpowers began to begrudge the huge investments that compromised economic and social progress within their countries. Additionally, there was always the possibility of an accident or a miscalculation that might trigger the launch of nuclear weapons by both sides. The unparalleled destructiveness of such an exchange prompted rational leaders to initiate serious programs toward nuclear disarmament.

More than two decades of arms limitation pacts culminated in a series of crucial agreements in the late 1980s and early 1990s. In December 1987, the U.S. and Soviet conferees agreed to dismantle medium- and shorter-ranged missiles based on land and to verify the agreement through on-the-spot inspection teams. This resulted in remarkable scenes of American observers touring Soviet military sites to watch them dismantle deadly missiles while Soviet counterparts did the same at high-security American installations.

After months of discussion and drafting arcane briefs on the arms limitation process, in July 1991 President Bush and Russian premier Gorbachev signed the Strategic Arms Reduction Talks treaty (START I). The United States agreed to reduce and reorganize its force of land-based, fixed-silo intercontinental ballistic missiles. Under START I, the United States was allocated 550 silos housing MX Peacekeepers and Minuteman III missiles. Nonetheless, the United States retained its force of 34 ballistic missile nuclear submarines (SSBN), which carried a total of 640 sub-launched missiles. The

Navy also planned to retain 600 Tomahawk sea-launched cruise missiles in its inventory, providing a weapon that could deliver conventional or nuclear warheads.

In October 1991, Gorbachev presented a unilateral package of cuts even deeper than those proposed by the United States. By 1993, when a comprehensive START II agreement was drafted, the Bush administration had taken many similar moves, such as taking intercontinental bombers off alert status and removing additional nuclear-capable forces from global bases. With START II, the Americans and Russians agreed to limit respective nuclear warheads to approximately 3,000 units, roughly a third of their arsenals over the past decade. The Clinton administration did not abolish SDI research, although the Star Wars budgets received heavy cuts and the program itself lost bureaucratic status within the Pentagon. The U.S. Senate ratified START II in 1996, followed by the Russian legislature in 2000. This transition of traditional cold war attitudes was paralleled in far-reaching changes pertaining to conventional forces as well.

In the decades that followed the disquieting end to the war in Vietnam, American military services carried out a prolonged self-assessment of strategy, tactics, and equipment. Some decisions and some equipment had proven effective; in many other cases, the grueling analyses from Pentagon studies, think tanks, and postgraduate staff colleges like the Air University at Maxwell Air Force Base, in Alabama, resulted in blunt, negative conclusions. Moreover, when events in the late 1980s led to joint air, land, and sea operations involving Grenada and Panama, postaction analysis was hardly flattering. Fortunately, the main problems could quickly be resolved. By the time an Iraqi invasion of Kuwait in 1990 led to a United Nations operation known as Desert Shield, followed by the combat of Desert Storm, it was apparent that America's military capability had moved far beyond the tumultuous years of Vietnam.

In the process of studying combat in Vietnam, military planners acknowledged the effective role of helicopter gunships and the use of helicopters to carry airborne assault teams, despite the high attrition rate of these aircraft. Airlift—both long distance and short-range winged transports—played a key part in logistics. The improved reliability of "smart" weapons for pinpoint attacks and the sophistication of close air support tactics also got high marks. A long catalog of negative issues included the tragic number of noncombatants who died throughout the countryside and urban areas as a result of heavy air attacks. Despite close air support successes, there was a need for improved communications, and the ability to strike under cover of darkness demanded sophisticated electronic systems, including satellite networks. While acknowledging the paramount role of President and Congress, Pentagon planners argued effectively to restrain both of them from micromanaging the process of target selection and daily operational missions in the combat zone—a type of management interference that many believed to have confused military functions in Southeast Asia and contributed to higher casualties. At executive and congressional levels, the process of determining military policy was subjected to reevaluation and reorganization in order to avoid mistakes of the Vietnam era. These changing relationships of international diplomacy, strategic weapons, inter-continental mobility, satellite capability, and combat readiness formed the basis for a White Paper released by the Air Force in 1990, entitled *The Air Force and*

U.S. National Security: Global Reach—Global Power. Clearly, the U.S. Air Force intended to play a major role in post–cold war politics.

While high-level theory evolved, the armed services stepped up plans for more realistic training in peace time, including the U.S. Navy's famous "Top Gun" school for fighter pilots and the similar U.S. Air Force "Red Flag" maneuvers. In these exercises, American combat pilots were confronted by highly skilled "aggressor" aviators using tactics characteristic of potential adversaries. There was special emphasis in conducting joint maneuvers with U.S. Army forces, and air crews acquired experience in the use of electronically guided munitions, antiradar weapons, and new equipment for night-time operations. Combat objectives and service coordination were codified in a series of Pentagon directives, most notably the document known as FM 100-5, issued in 1982, which formalized an integral military doctrine characterized as "AirLand Battle." While much of this was directed towards a Soviet threat and involved NATO operations in Europe, it proved to be adaptable to other situations, particularly in the cooperative actions of U.S. Air Force, Navy, and Army units.

The new look in military doctrine was accompanied by a new look in the application of military air power. Various units were put to work as part of the effort to stifle international drug traffic, and eventually deployed as a highly flexible force to assert American influence elsewhere in the world—a sort of "gun-boat diplomacy" that depended heavily on American air power. From the military perspective, analysis of these actions revealed both positive and negative trends.

One of the first applications occurred on the island of Grenada, a small Caribbean country off the north coast of Venezuela, where a pro-Communist regime had taken over in 1979. During 1983, in a move endorsed by several concerned Caribbean governments, U.S. air, land, and sea forces occupied the island, although the operation revealed serious gaps in communications, planning, intelligence, and execution.

Three years later, U.S. Navy and U.S. Air Force planes carried out punitive strikes in Libya, when that country's strongman leader, Muammar Khaddafi, was suspected of supporting a variety of international terrorist attacks. Targets in Libya included an airfield, barracks, and port facilities used to train terrorists. The combined operation achieved complete surprise, with Libyan radars immobilized by pinpoint strikes while many aircraft were destroyed on the ground. Although U.S. sources claimed a high level of accuracy for the laser-guided bombs used in the attack, later reports of several dozen civilians killed or wounded raised some questions. Nonetheless, the Libyan operation demonstrated a very effective degree of operational coordination and combat readiness. Antiradar missiles smothered SAM installations, and the guided munitions achieved a favorable rate of success. American losses were low, with only one plane and its two-man crew reported missing in action. In challenging an adversary equipped with an extensive, modern, and deadly defense network, the U.S. forces had performed effectively.

As part of President Reagan's stronger stance in the war against the drug trade, the Department of Defense stepped up its support of patrol actions through the use of numerous military planes fitted with sophisticated radar to detect low-flying drug runners. There was also increased international cooperation with Bolivia and other countries. The use of electronically equipped planes for antidrug patrol increased, and the United States tested blimps and stationary balloons, called aerostats, equipped with special radar and electronic sensors, for use in border surveillance and patrolling over the Gulf of Mexico.

One such antidrug operation occurred late in 1989 against Panama's dictatorial regime, led by Manual Noriega. Noriega was widely believed to be a key figure in drug operations, and his heavy-handed rule seemed increasingly vicious. A combined American assault took place in December, introducing the Apache helicopter gunship and F-117 stealth attack plane into combat. Air Force transports carried over 9,500 troops into Panama during a 36-hour period. Numerous civilian casualties were reported, and additional shortcomings in military coordination came to light. Still, the action went reasonably well, leaving a sense of confidence and capability that gave momentum to the stiffer challenges of Desert Shield and Desert Storm during 1990–91.

Defying the accepted American intelligence summaries concerning Iraq's intentions in the Middle East, Saddam Hussein suddenly invaded neighboring Kuwait in August, 1990. Saddam apparently wanted control of Kuwait's substantial oil reserves to help pay off heavy debts left over from years of warfare with Iran. Additionally, Saddam would gain a strategic coastline on the Persian Gulf and build his own stature as military leader in the Gulf region. American strategists, now fully alarmed, predicted dire consequences from Saddam's control of huge oil reserves, his brutal show of force while occupying Kuwait, and the evidence of nuclear weapons development by a renegade government. Many observers also perceived an immediate danger to Saudi Arabia, with its own immense oil reserves and obvious significance to the economic health of Europe and America. Various diplomatic initiatives eventually resulted in a series of UN resolutions against the Iraqi actions in Kuwait, which were then shrugged off by Saddam.

Meanwhile, under UN auspices, the United States took the lead in building up coalition forces in the Persian Gulf, hoping this show of force would encourage Saddam to pull out. When he did not, the UN coalition forces prepared for combat. The rapid response of American military power demonstrated the revised doctrine and training of the previous two decades. On August 16, the Air Force received orders to send combat planes to the gulf; in less than 24 hours, the First Tactical Air Wing had two dozen of its F-15 fighters en route to Saudi Arabia, along with airborne tankers and transports loaded with supplies. The U.S. Navy ordered two carrier groups into the gulf, and the U.S. Army's 82d Airborne troops flew into Saudi Arabia not long after the F-15 fighters had arrived.

These forces were only the vanguard. As the U.S. Navy added carrier groups and orchestrated the indispensable sealift of bulk military stores, airlift began delivery of 8,000 troops per day into Saudi Arabia. The Air Force carried about three-quarters of the cargo and one-third of the personnel during Desert Shield/Desert Storm flight operations. The other airlift assets came from the Civil Reserve Air Fleet (CRAF), representing airliners from the United States that had been designated for emergency service with the Air Force. So many planes filled the air lanes converging on the gulf that one pilot commented that he could have walked across the Mediterranean Sea on their wings.

The huge C-5 transports airlifted combat troops and crucial field equipment, such as this heavy truck and auxiliary power unit. After landing on an airstrip in Saudi Arabia during Desert Shield, the nose of this C-5 is tilted upward, obscuring the cockpit, as equipment rolls off. Courtesy of U.S. Air Force.

The build-up of American and coalition forces failed to sway Saddam Hussein, who ignored the UN resolution to evacuate Kuwait by January 15. Two days later, U.S. Army helicopters, U.S. Air Force planes, and U.S. Navy missiles streaked across Iraq's border as Desert Shield became Desert Storm. The helicopter force (which included Air Force navigational choppers) used the Army's Apache helicopter gunships to wipe out early warning radar sites. At the same time, F-117 stealth fighters made undetected forays over Baghdad, the Iraqi capital, and other targets within the country. Launched by Navy vessels in the gulf, low-flying Tomahawk cruise missiles also threaded their way into Baghdad and other areas in coordinated attacks against air defense networks, communications complexes, and command centers. When a CNN television broadcast in Baghdad abruptly went off the air, American planners knew that a laser-guided bomb released by an F-117 had been successful, wiping out the telecommunications center in Baghdad which served as CNN's broadcast link. With other civil and military communications centers destroyed and air defense systems effectively blinded, the Iraqis had little chance of halting the waves of coalition planes that saturated Iraqi airspace. Television audiences in America sat mesmerized as news programs replayed

Electronically guided, Tomahawk cruise missiles proved to be effective weapons against targets inside Iraq. This launch came from the nuclear-powered cruiser USS Mississippi, *on station in the Red Sea during Desert Storm. Courtesy of U.S. Navy.*

sequences of electronically guided weapons that streaked down chimneys and through doorways to destroy key targets. Losses of coalition planes inevitably occurred, but air power certainly provided an awesome show of destructive capability during the following weeks.

Saddam's own air force proved ineffective as swarms of UN planes and helicopters wiped out tons of supplies, thousands of tanks, factories, and scores of other military objectives. From bases as far away as the United States, B-52 Stratofortesses pummeled Iraqi ground forces huddled in exposed positions across the bleak desert. After six weeks of an unprecedented aerial blitz, the ground attack by coalition troops plunged into Kuwait and simultaneously outflanked Iraqi armies along the Saudi border. In 100 hours of headlong advance, the enemy forces were cut off and Iraqi forces, in an attempted retreat, were brutally defeated by air attacks. A cease-fire came on February 28, four days after the ground attack had begun.

It was a "high-tech" war, integrating satellites for surveillance and communications, high-flying electronic command posts like the AWACS for aerial operations and JSTARS (joint surveillance and target attack radar system) for ground operations, fiery anti-missile salvos by the highly publicized Patriot batteries, cloak-and-dagger missions by the F-117 stealth airplane, and an array of other systems engaged in electronic feats of derring-do. Few weapons fascinated the public as much as the Patriot's dramatic missile engagements and the eerie success of the sinister stealth aircraft.

Still, the vast majority of U.S. and coalition aircraft in combat during the Gulf War represented conventional design and construction. For American operations, newer planes such as the Air Force's F-15 and F-16, as well as the Navy's F/A-18, were joined by a variety of other designs, some of them having an elderly design genealogy. On the Navy side, the Grumman A-6 Intruder, a two-seat carrier-based attack plane, entered service in 1963; the Air Force's F-111 attack plane, built by General Dynamics, became operational in 1965. These older designs had been extensively modified and updated over the intervening years, giving them both versatility and lethal force. For old and new planes alike, their effectiveness was greatly enhanced by a variegated catalog of military jargon that identified a mind-boggling array of electronic gadgetry. There was FLIR (forward-looking infrared system), TF/TA (terrain following/terrain avoidance radar), and other arcane systems that combined FLIR with TF/TA and additional oddments to make Pave Low II, installed on night-flying air force helicopters employed as pathfinders for army helicopter gunships. Fast jet fighters like the relatively small, single-seat F-16 and others used similar systems, including LANTIRN (low altitude navigation targeting infrared for night) for low-altitude high-speed night attacks on unsuspecting troop formations and fixed targets. More than any prior conflict, opposing forces were harassed by around-the-clock, pinpoint attacks in darkness as well as daylight. Most of the ordnance dropped during the war was comparatively unsophisticated, but electronically upgraded war planes delivered these "dumb bombs" with impressive accuracy.

To many observers, the aftermath of Desert Shield/Desert Storm implied a much different era of warfare, in which electronic weapons systems of awesome precision would provide the leverage for victory. More refined analysis of the aerial campaign suggested

Developed in great secrecy and tested from remote desert bases in the American Southwest, the Lockheed F-117 stealth combat plane incorporated advanced aerodynamics and computerized control systems. Its angular surfaces, shaped to frustrate radar detection, helped make it a star performer during the Gulf War. Courtesy of Lockheed Corporation.

that early estimates had been too optimistic. Still, the results were impressive, and the gulf experience helped shape air power planning in the decades ahead.

RETHINKING AIR POWER

Air power in the forseeable future would function in a markedly different political environment. Shaped for superpower confrontations involving the Soviet Union, the end of the cold war era forced a different way of thinking about potential conflicts and the aerial hardware required for different national strategies. The reduction of powerful missiles was paralleled in the reduction of budgets, bombers, fighters, military bases, and personnel. In the case of the Air Force, personnel strength of over 600,000 in the mid-1980s was targeted for under 430,000 by 1994 and numbered 367,000 in 1998. Air force planners, accustomed to budgets that customarily topped more than $110 billion annually during the 1980s, now had to rein in expenditures and work with tens of millions fewer dollars. The budget came in at $83.8 billion in 1993 and totaled $81 billion for 2000. The fleet of active duty aircraft dropped from 9,494 in 1986 to 7,640 by 1992 and numbered 6,111 by 2000. Looking ahead, many military analysts predicted increasing reliance on a new class of pilotless aircraft generally classified as *unmanned aerial vehicles,* or UAVs. Aircraft like these date back to the world wars, when they were known as *drones*—small aircraft guided by radio or a programmed internal guidance system and usually used for target practice. By the time of the Gulf War and the Balkan wars of the 1990s, UAVs had become far more capable, carrying sophisticated surveillance equipment and often guided by a human operator on the ground, who relied on electronic images from equipment aboard the aircraft. Entering the twenty-first century, the armed services possessed some 500 UAV units of varying size and sophistication, including smaller, propeller-driven versions and larger types that used jet engines and advanced aerodynamics for long-range surveillance at very high altitudes. The Northrop Grumman Global Hawk, for example, could fly up to 1,200 nautical miles, climb to 60,000 feet, spend up to 24 hours in orbit at that altitude, and return home in the process of a 32-hour mission.

The leaner budgets meant careful scrutiny of all new aircraft programs. They had been initiated before the collapse of the Soviet Union, and Pentagon analysts began to reconsider roles of military aircraft, for example, in the new post–cold war environment. For one thing, it seemed clear that reduced numbers would suffice, leading to proposals that presumed a stretch-out of production units so that an industrial base could be maintained in the interest of national security. Another likelihood concerned significant modifications and equipment upgrades for aircraft already in service as a way to extend their operational lives and maintain a reasonable force structure while smaller numbers of new planes entered the active inventory. Meanwhile, a number of programs moved ahead.

Several new aircraft in the American inventory originated as foreign projects. When the U.S. Navy needed a new tactical jet trainer, it finally settled on a navalized version of the British Aerospace Hawk. It was designated the T-45 Goshawk for American service. Designers retained a Rolls Royce power plant, added advanced avionics, and rebuilt much of the plane's structure to absorb the punishment of carrier operations undertaken by fledgling fighter pilots. Originally supplied by McDonnell Douglas, the merger in 1997 transferred production to Boeing. The Boeing Company additionally absorbed McDonnell's manufacture of the British-designed Harrier 2 attack plane used by the U.S. Marine Corps; the Harrier also relied on a Rolls Royce jet engine.

A new military trainer not only exhibited an overseas heritage but also embodied the armed services' growing commitment to joint collaboration in equipment as well as in operations. The Joint Primary Aircraft Training System (JPATS) bound the U.S. Navy, Marines, and Air Force together in a quest for the same plane. The goal was to save operational and development expenses and also capture international sales as a way to expand production and bring down unit costs. The Department of Defense openly solicited proposals from abroad, while stipulating that 70 percent of the plane must originate in the United States. The winner—Raytheon (Beech) Aircraft's T-6 Texan 2 turboprop tandem trainer—owed its heritage to the Swiss firm of Pilatus, which had sold its own earlier versions to several foreign air arms.

All the services were interested in long-range airlift, a role to be fulfilled by the Douglas C-17. Plagued by cost overruns and management snags, the C-17 featured several new technologies giving it striking short-field performance in forward areas, along with the ability to unload its cargo with exceptional speed. With its production budget still under a cloud, the first deliveries to operational squadrons began in the summer of 1993. In service, the C-17 proved to be an extremely effective airlifter for all of America's armed forces. It was not produced as part of a major international program, but the United States again hoped for overseas sales. In 2000, Britain's RAF finally signed leases for several C-17 transports, pending development of a suitable European-built transport. This arrangement underscored the latent atmosphere of competition in American versus European leadership in aerospace technology.

Also in 1993, the Northrop B-2 stealth bomber began its deployment at Whiteman Air Force Base near Knob Noster, Missouri, about 75 miles west of Kansas City. The B-2, with only a two-man crew, depended on its low radar observability and sophisticated electronics to avoid detection, frustrate enemy defenses, and deliver 40,000 pounds of precision guided bombs with unprecedented accuracy. It had a nonrefueled range of some 6,000 miles. Built as a flying wing, it spanned 172 feet, and each of 21 planned bombers cost $1.3 billion. Its complex construction of unusual materials, size, high maintenance requirements, and cost per plane made it highly controversial. Critics wondered whether some of its sensitive antiradar coatings and materials used in fabrication could actually stand up to combat operations. Moreover, its complex support infrastructure seemed to make its deployment overseas problematic at best.

A more unusual aircraft was the Bell V-22, planned for joint use by the army, navy, and marines to carry troops and supplies. With large horizontal rotors and engine pods at either end of its stubby wings, the V-22 could take off vertically, as a helicopter does, then tilt the rotor/engine pods forward so that the propellers, now vertical to the flight path, could move the aircraft at much higher speeds than conventional helicopters.

Meanwhile, the air force planned its next-generation fighter, intended as a multimission air superiority combat plane. A design developed jointly by Lockheed and Boeing, the F-22 Raptor, won the contract. Able to fly at about Mach 2, the F-22 did not represent a quantum leap in speed. But in overall performance, the plane set new standards. In addition to advanced stealth technology, the F-22 incorporated a thrust-vectoring system that yielded phenomenal agility. Its twin Pratt & Whitney engines possessed very high efficiency, permitting long-range supersonic cruising without wasting precious fuel. Its avionics suite allowed pilots to cope with multiple threats (such as enemy radar and SAMS) and still pursue their original mission; radar and antiaircraft rockets could detect and destroy hostile planes at extreme ranges before enemy pilots even knew an F-22 was in the neighborhood. But at prices estimated at $85 million to $100 million per plane (depending on production numbers), the F-22 stirred congressional budget committees into a righteous froth, beginning with the prototype's flight in 1990 and continuing through the preproduction version flight in 1997.

The air force wanted full production to crank up in 2000, with the first squadrons to be operational in about 2005. By that time, the air force warned, the average F-15 would be 26 years old, and the multinational Eurofighter, the French Rafale, and the Swedish Gripen—all newer planes and technologically advantaged—would be in service with European air forces. Moreover, if export sales expanded beyond western Europe, it was not unlikely that these European designs might sometime be in the armed services of a hostile air force. In which case, the F-15 would be outclassed. The air force argued that

F-22 costs would come down, given an adequate production run. At the same time, the plane's qualities gave air force squadrons the capabilities of a much larger force of aircraft. The Raptor promised not only clear air-to-air superiority, but also superb air-to-ground capabilities with advanced, precision-guided munitions. But the air force's preferred fighter for the new millennium, the F-22, continued to be plagued by the major issue of cost.

Issues of cost swirled around another new military plane designed for low-level attacks, plus the ability to hold its own in air-to-air combat. The U.S. Navy needed a plane capable of operations from carrier decks; the marines required short takeoff and vertical landing ability (STOVL) from rough airstrips near combat zones. The new plane also needed good stealth characteristics. Because budget constraints simply did not stretch far enough for separate air force, navy, and marine requirements, the three services blended their requirements into one aircraft, the Joint Strike Fighter (JSF). The British RAF needed a similar STOVL plane, and an RAF project team joined the Americans, resulting in a somewhat unique four-way design evolution and program management.

Two major firms competed for the contract: Boeing with the X-32 and Lockheed Martin with the X-35. Outwardly, the chief configuration differences seemed to be Boeing's chin-mounted air intake in contrast to Lockheed's side-mounted intakes. Both planes had to meet general expectations for at least Mach 1 speeds, coupled with excellent low-speed characteristics (for carrier deck landings as well as for crisp tactical maneuvers in close air support) and be adaptable for STOVL variants required by the marines and by the British. The new plane had to be stealthy, possess sophisticated electronics suites for multifaceted attacks as well as defense, and be affordable. Even at a projected cost of $28 to $35 million per copy, the JSF program was driven at every level to keep in line. The JSF effort became a lightning rod for attacks from skeptics, who pointed out that planes designed from the start for both land and sea-borne operations did not fare well. The McDonnell F-4 was something of an anomaly but had been designed and built as a carrier-borne plane from the start; only later was it adapted to USAF requirements. All the same, supporters of the JSF argued that much had been learned from earlier programs; that a multiservice, multirole JSF was the only budgetary alternative for future service requirements; that the prototype competition required contractors to keep a lid on expenses; that it could be done.

Additional considerations sharpened the JSF debate. For decades since World War II, the United States had served as the traditional supplier of advanced aircraft to NATO and other allies. But the new F-22 Raptor seemed far too expensive for foreign export and its technology much too advanced to share elsewhere. New, air superiority fighters like the Eurofighter seemed destined to replace U.S. planes as front-line equipment in air forces overseas. The JSF, on the other hand, was affordable. As a sophisticated, next-generation combat plane, it was bound to satisfy the tactical requirements of several international air forces. The United States could supply an attractive avionics suite while holding equipment with more advanced capabilities at home. More importantly, the JSF offered the U.S. aerospace industry a way to cut into the promising market for combat planes in the twenty-first century. Such arguments symbolized the enhanced capabilities of European aerospace and its successful competition in a sector that had long been the bailiwick of the United States. In fact, the European challenge to the American aerospace sector ran the gamut: transports, combat planes, helicopters, and missiles. By the end of the 1990s, Europe also challenged airline operations, as well as the production of airliners.

At the same time, America and its NATO allies demonstrated that they could still collaborate to meet an international emergency. In post–cold war Yugoslavia, age-old ethnic tensions contributed to the country's breakup. During the 1980s and 1990s, the province of Kosovo, along the country's southern border, became a hotbed of agitation spurred on by the Kosovo Liberation Army (KLA). Against a background of failed peace talks during 1998–99, Yugoslav president Slobodan Milosevic launched a series of punitive offensives. A flood of refugees recounted tales of "ethnic cleansing" by Yugoslav Serbs against Muslims, who represented the majority ethnic group in Kosovo; additional reports of widespread attacks against civilians in Kosovo made the situation a hot issue in the United Nations as well as within the North Atlantic Treaty Organization. When peace talks under international auspices collapsed, NATO forces inaugurated an aerial offensive against Serbian military units in Kosovo as well as military targets throughout Yugoslavia. The aim was to put pressure on the Milosevic government to withdraw its military units from Kosovo and cooperate with international peacekeeping forces. The air attacks began on March 24, 1999, and ended some 78 days later, on June 10, 1999, when Yugoslav forces announced their withdrawal from Kosovo.

All told, nearly 1,000 aircraft eventually took part in the aerial offensive, with U.S. Air Force assets numbering around 700. Among other NATO forces, Germany authorized one of its first armed forays since World War II, dispatching a squadron of 15 Tornado fighter bombers. Other countries with strong commitments included Italy, Britain, Turkey, Holland, and France. Additional aircraft arrived from Canada, Belgium, Portugal, Denmark, and Spain. These assets all played a significant role. The French contingent included an aircraft carrier that dispatched combat jets as it cruised the Adriatic Sea. In addition to an aircraft carrier and other planes from RAF bases in Germany, the British sent in a missile submarine, whose Tomahawk cruise missiles hit Yugoslav targets. Early in the campaign a Dutch F-16 claimed the first aerial victory by shooting down a Yugoslav Mig-29 fighter. Most of the land-based NATO aircraft operated from bases in Italy, and the commander of the combined Air Operations Center was an Italian air force general.

But U.S. forces carried out the majority of air strikes—and generally received the most criticism. In the first week, an F-117 stealth bomber went down, apparently struck by missiles, reportedly because of some shrewd guesswork and some luck by the Yugoslavs. The pilot was rescued; a shift in USAF tactics resolved the threat. In May, an American attack on Belgrade accidentally struck the Chinese embassy, resulting in several casualties, an abject apology from the United States, and tense relations with China. However, "smart" bombs and missiles rarely missed their targets, soon reducing Yugoslavia's electrical grids, petroleum refineries, communications, and key railway and highway bridges to a state of rubble. Large, B-2 stealth bombers flew round-trip trans-Atlantic missions from Whiteman AFB in Knob Noster Missouri, carrying highly accurate precision-guided munitions. USAF commanders expressed high praise for the B-2 bomber crews, who took out many high-priority objectives. About six B-2s from Whiteman flew 1 percent of all bombing missions but accounted for 11 percent of all targets hit. Air force officers thought that the B-2's performance in transoceanic strikes convincingly demonstrated the plane's abilities and answered critics of the bomber's utility.

Although NATO forces aggressively hit Yugoslav tanks, troop concentrations, and other targets in Kosovo, it was the air campaign against Yugoslavia's infrastructure that seemed to most affect Milosevic's decision to pull his forces out of Kosovo in June. In the aftermath of the NATO coalition's actions, considerable debate arose concerning claims about the number of tanks and other military targets destroyed. There were charges and countercharges involving the number of civilian casualties in the wake of NATO air strikes. Moreover, an uneasy peace settled over Kosovo, where several thousand coalition troops tried to manage the return of refugees, assist in delivering relief supplies, and wrestle with sporadic violence stemming from ancient ethnic conflicts. Nonetheless, the Yugoslav military forces were gone, rampant civilian massacres stopped, and attention was given to reconstruction rather than destruction. For the United States and NATO strategists, larger questions remained. Given the reality of regional armed conflicts that created humanitarian crises, what sort of guidelines ought to be considered in exercising armed force in the name of humanitarian intervention? American policy makers continued to debate appropriate responses.

The new era included major changes in military administrative structures. Within the Air Force, some reordering of assets had already occurred late in 1982, when a Space Command began to manage satellites and certain other activities. During 1993, Space Command also acquired day-to-day control of all the Air Force's ICBM installations, weapons, and personnel. Shifting such missile assets out of their traditional home in the Strategic Air Command (SAC) marked an even more dramatic event—the disappearance of SAC itself. As Donald Rice, Secretary of the Force, reported at the end of 1992, "Over the past year, the Air Force implemented the most significant reorganization since we became a separate service in 1947." Generally, the goal was to meet discrete regional crises, not face a continuous Soviet threat. Historic entities such as the Strategic Air Command, Tactical Air Command, and Military Airlift Command were folded into two new organizations, the Air Combat Command and the Air Mobility Command.

Attending these notable developments was an increased use of air power in humanitarian roles. Over the decades since 1945, air power had served admirably to carry relief supplies and evacuate victims from various natural or political emergencies around the world. As the world's sole surviving superpower, the extensive assets of the U.S. Air Force seemed increasingly in demand, including the airlift of emergency supplies to the former Soviet Union itself during the bleak winter of 1992. In the course of one particular week that year, the Air Force supported efforts to control forest fires in California and Idaho, flew emergency supplies into war-torn Yugoslavia, carried food and medicine to starving Somalis in Africa, and provided assistance to peacekeeping forces in Angola. During the 1990s, while military planes continued to play an important humanitarian role in world affairs, they also filled an expanding role in the war against drugs. Units of the U.S. Air Force, Navy, Army, and Customs conducted aerial surveillance throughout Central America and the Caribbean and across the northern coasts and jungles of South America. Aircraft used ranged from U.S. Army helicopters and small, twin-engine reconnaissance planes to navy patrol planes, U.S. Customs observation aircraft, and large aircraft like modified C-130 Hercules transports. Fighter aircraft were sometimes used for high-speed observation missions to track suspected drug shipments by air. When the Panama Canal reverted to Panamanian sovereignty in 1999, the United States lost its military air bases there, requiring deployment to new locations at civil fields in Ecuador and the Dutch islands of Aruba and Curacao off the coast of Venezuela. Congress appropriated $15 million (out of $42 million requested by President Clinton) to upgrade the new stations. From these sites, American service personnel continued to operate aircraft as one way to oppose international drug traffic.

The reorganization of commands, reduced funds, expanded duties in relief and antidrug operations, and imperiled aircraft production lines dramatized military air power's uncertain search for clarity in roles and missions in the post-Soviet era. If the aeronautical community in America expected a reassuring stability in civil aviation, they soon found a disconcerting environment there as well.

THE ADVERSITIES OF AIR TRAVEL

Although the airlines enjoyed the era's generally favorable statistics for travel and profits, a series of tragic events cast a pall over an industry that prided itself on security and safety. As an act of political terror, aerial hijacking created serious problems for airlines the world over. During the 1960s, a number of incidents occurred in which U.S. airliners were forced to deviate from normal routes, usually to carry the skyjackers to a different destination, such as Cuba, as a form of political protest. Electronic metal detectors and intensified security curtailed this sort of action, although lax security elsewhere in the world eventually led to a new wave of terrorism, often directed against the United States for its perceived role in Middle Eastern political issues. During the 1980s, terrorists escalated the level of violence with the deliberate murder of some hostages and, eventually, the use of explosives to destroy planes and all humans aboard.

Americans tensely followed the violent odyssey of TWA Flight 847, which had been hijacked on June 18, 1985, minutes after departing Athens for Rome. There were 145 passengers aboard, including 104 Americans, most of them returning home after a trip to the Holy Land. Waving grenades and guns, two young men, later identified as Lebanese Shiites, forced the pilots to divert to war-torn Beirut. Nearly two chaotic weeks later—after a terrifying odyssey that included the callous murder of a young U.S. Navy man—negotiations finally ended the incident. There were many acts of valor during the ordeal, but considerable praise went to the flight crew, who struggled with fatigue, incoherent terrorists in the cockpit, and dangerous confusion during each

Artist's rendering of Denver International Airport, opened in 1995 and designed to accommodate increased international travel. The site includes five full-service runways, with expansion plans for twelve. The facility initially served more than 30 million passengers annually. At 327 feet, the FAA control tower at DIA is the tallest in North America. Courtesy of DIA.

landing at the chaotic Beirut airport. In the aftermath of the TWA hijacking, American officials cautioned travelers about lax security measures at many world airports. Stringent measures for inspection of luggage were instituted at international airports, along with a general tightening of security. But the vulnerability of airline travelers to terrorism was underscored by a rash of similar episodes.

Sporadic acts of terrorism continued, even though tightened security procedures appeared to have averted numerous plots. Nonetheless, air travel in a world where political tensions remained at high levels could still lead to tragedy. Even in a supposedly reliable world of electronic verification, shocking mistakes happened. The United States experienced such a mistake during the summer of 1988, when Navy warships were patrolling the Persian Gulf as part of a multinational plan to protect oil tankers and other vessels threatened by hostilities between Iraq and Iran. The U.S.S. *Vincennes,* an advanced American cruiser packed with the latest in electronic warfare systems, mistakenly identified an Iranian jet airliner as a hostile combat plane and launched anti-aircraft missiles. The missiles destroyed the jet airliner in midair, killing 290 people. Most Arab leaders, understandably appalled by the tragedy, nonetheless recognized it as a grievous error and acknowledged the remorse expressed by the United States government. At the same time, a number of militant Arab groups promised retaliation, citing the U.S. air raid on Libya in 1986.

In December 1988, a Pan Am 747 flight, having originated in Germany, took off from London for the trip back to the United States. Over Lockerbie, Scotland, the plane suddenly vanished from air traffic control radar screens. Reports of a crash soon reached authorities; none of the 270 people aboard survived. Analysis of the wreckage confirmed that a bomb had destroyed the plane in flight, and an international investigation eventually cited a terrorist group based in the Middle East. Improved X-ray security devices were joined by new radiation machines to detect possible bombs in luggage and cargo containers.

Events such as these inevitably caused a drop in air travel, but recovery followed. Air transportation had become integral to leisure and business travel. Within the United States in particular, the strong growth of air travel seemed inexorable. The rising tide of air travelers seemed to be overwhelming the system, straining the capacity of air terminals and saturating the air space around landing fields. Many experts pointed to the inadequacy of existing airports, noting that the last major airfield constructed in the United States had been at Dallas-Fort Worth in 1974. New airports raised contentious issues of environmental effects from the mass of converging autos as well as planes. Bigger airliners seemed to drive up land requirements in geometric progression. In Denver, Colorado, boosters for a new airport spent years countering these objections until voters acquiesced in 1989. Opened in 1995, Denver International Airport was funded at $4.9 billion (skeptics predicted total costs up to $15 billion) and covered 53 square miles—more land than Dallas-Forth Worth and Chicago's O'Hare combined. Cities generally preferred to improve existing facilities, although that often seemed to be a losing battle.

Even though many airports had expansion or modernization plans in progress, the number of public-use airports actually declined by 129 between 1983 and 1985, and dozens of airports were forecast to exceed their annual capacity by the year 2000. The pressure on airport facilities and movements within national air space had a serious impact on the traveling public. The daily delays totaled 2,100 hours, the equivalent of grounding 200 aircraft every day, and the cost to passengers, the airlines, and the economy was estimated to be $3.2 billion per year.

At least part of the problem, according to some sources, was the growing reliance on the "hub-and-spoke" (or, as some cynics put it, "hub-and-choke") system, which funneled travelers into major airports. Prior to deregulation, airlines were often required to offer direct service between smaller cities. Frequently, this meant assigning an airliner with far more seats than necessary, so load factors on such flights were low. By eliminating many of these "thinner" routes, airlines could divert more travelers into major terminals and then coordinate schedules and aircraft more efficiently. Stated simply, the hub-and-spoke network represented a system that carried all air traffic from smaller cities through a central hub, where travelers were then sorted out and dispersed onward to their eventual destinations. The airlines squeezed more favorable load factors from this system, but passengers often grumbled about additional stops, crowded facilities at the hub airport, and the additional time consumed in waiting for connecting flights. At peak departure and arrival times, traffic control and terminal facilities became overloaded.

Most arrivals were big jets of the major airlines, accompanied by a growing fleet of commuter planes offering service from smaller cities to which the larger airlines had suspended service in the wake of deregulation. In partnership with a major carrier, a commuter agreed to organize its schedule to complement the national airline's operations at one of the national's hubs. Resort Air, for example, operated as Trans World Express, with flights from several different cities converting on St. Louis, where passengers could make convenient transfers to or from Trans World Airline's big jets. Such arrangements were useful to the commuter, which could use its partner's ticket counters, advertisement campaigns, corporate colors, and a variation of the partner's name. All this boosted the activity of regional airline operations and their role in the U.S. air traffic network. During the 1980s, the introduction of larger turboprop aircraft with pressurized passenger cabins proved highly popular with travelers. During the 1990s, the introduction of a new generation of small, twin-engine jet airliners brought another surge in traffic. The new planes also represented the growth of the aerospace industry abroad, primarily Bomdardier in Canada (Canadair Regional Jet, or CRJ series) and Embraer in Brazil (ERJ series).

The growth of airline travel flattened out during the early 1990s, affected by rising ticket prices (triggered by higher fuel costs resulting from the Gulf War), a sluggish American economy, and a severe downturn in the Asian economy. Late in the decade, when global finances and American business conditions improved, airlines enjoyed a new surge in air travel, even though success brought a different set of problems. As traffic began to recover, customer complaints to the Transportation Department began to multiply. From 1998 to 1999, they doubled. Congress began to query airline executives, pressing them to explain stories of nightmarish flights reported by constituents. One IBM executive, trying to get from New York City to Charlotte, North Carolina, managed to get through city traffic and reach the airport on time for his flight, only to fritter away two hours before boarding and then suffer through another five hours on the taxiway because of bad weather. Many airline patrons remained skeptical about foul weather as an explanation for delays and believed that simple mismanagement by the airlines lay at the bottom of the problem. Between 1995 and 2000, delays rose 50 percent, flight cancellations rose 68 percent, and at major airports, reports of jets sitting on the taxiways for an hour or more went up 130 percent.

In truth, the incidence of nasty weather in the last few years of the 1990s was actually worse—creating havoc all over the country as its ripples affected arrivals and departures at airports hundreds of miles away. The hub-and-spoke system, funneling travelers through major airport complexes, often seemed vulnerable to weather that compromised one or more such major centers. Added to weather problems was the sheer number of travelers aboard airliners. The strong economy boosted air travel about 5 percent from 1999 into 2000, with airliners carrying a record 665 million passengers. Most planes flew at about 76 percent of load factor, which meant healthy profits for the carrier on each flight. For passengers, everything just seemed more cramped and less enjoyable. A researcher at the Brookings Institution concluded that passenger complaints paralleled the rise in numbers of air travelers. In tight quarters, irritation seemed to rise exponentially. "I don't think the public would be so angry if there weren't so many people on the plane with them," he reasoned.

The Air Transport Association increased its efforts to make air travel more enjoyable, and individual airlines launched a variety of programs to provide friendlier service, reduce mind-numbing waiting lines, and appear less officious when enforcing rules about such things as carry-on luggage. And some airlines, such as American, ran full page ads in national magazines to tout their unilateral decision to make air travel more enjoyable by widening the space between seats.

Passengers in the 1980s had to contend with other frustrations, such as smoking and fluctuating fares. Although airlines intentionally seated smokers at the rear of passenger cabins for many years, this arrangement came under attack from health and consumer groups. During 1988, passengers with a nicotine habit ran into new problems, since smoking was banned on all domestic flights of two hours or less. Smokers aboard most U.S. flights of more than two hours' duration were still allowed to smoke in a reserved section. Beginning in 1990, smoking was entirely banned on all flights of fewer than six hours; a rule that affected virtually every domestic route, except those to Alaska and Hawaii. Groups like the Chicago Lung Association helped ease the shock when the law went into effect by supplying "emergency kits"—sugarless gum, mints, mouthwash, and advice to chew on toothpicks.

As the airlines struggled to soothe smokers, cope with long lines at ticket counters, and reimburse travelers who had filed a class action suit about collusion pertaining to airline fares, they also had to grapple with the massive and continuing impact of deregulation.

THE AIRLINE UPHEAVAL

The saga of Braniff and its demise during the early 1980s only marked the beginning of a tumultuous era for the airline industry. There seemed to be an unusual number of brash upstarts in the airline business, totally unlike such early figures as Tom Braniff, Pat Patterson (United), and C. R. Smith (American), who literally grew up with the airlines. Other pioneering operations—Pan American, Continental, and Eastern—faced bewildering dilemmas during the 1980s. The stormy aftermath of deregulation claimed many airlines and their leaders. For the survivors, life would never be the same, and individuals who now guided the industry often seemed to come from a different mold. The fortunes of Continental Airlines and Eastern Airlines exemplified this trend.

Under the shrewd, swashbuckling direction of Robert F. Six, Continental had become a major competitor in the U.S. market, with extensive routes across the Pacific as well. But competition in the new deregulated environment created serious problems across Continental's principal markets in the West and Southwest. When Six finally retired in 1982, at age 75, the airline was already under attack by one of the most aggressive entrepreneurs of the new era—Frank Lorenzo. In the minds of many airline analysts, Lorenzo personified the new breed of manager who had emerged during the 1980s. He was an innovator who turned discounted fares and relentless cost cutting into permanent realities throughout the industry. He also pioneered the leveraged takeovers of rival operations triggering the consolidations that swept away some of the industry's legendary airlines. In the process, he became one of the most controversial figures in postwar airline history.

Lorenzo grew up in New York City, majored in economics at Columbia University, and received an M.B.A. degree from Harvard in 1963. He worked for TWA and Eastern Airlines before organizing an investment group that bought a floundering airline in the southwest known as Texas International. He slashed costs, outfought the pilots' union, and—even before official deregulation in 1978—marketed deeply discounted "Peanut Fares" to several major cities across the nation. He became an exponent of bold financial offensives to seize control of an airline from other operators and then abrupt sell out to another contender for a hefty profit. By the end of 1981, after a bruising battle, Lorenzo had won his major objective, merging Texas International with Continental, whose headquarters he promptly transported from high-priced Los Angeles to budget-priced Houston. Lorenzo sold off some of Continental's valuable planes, laid off 15 percent of its employees, and then shut down the airline under Chapter 11 of the bankruptcy code. While still operating, he cut service, slashed salaries, and reintroduced Continental as a discount carrier. All this was followed by abortive efforts to buy out TWA and Frontier, after which he sold off his shares at considerable profit. With a bulging war chest, Lorenzo stalked Eastern Airlines, a company with its own sad history of errant management and arrogant labor organizations. During 1986, Lorenzo engineered a takeover that made Texas Air, his airline holding company, the largest of its type in the nation.

Lorenzo's tenure as the corporate head of Eastern became a spate of labor-management brawls, a series of losing court battles, and ongoing sales of Eastern's principal assets. Lorenzo suddenly withdrew from the field in the summer of 1990. He sold his interest in Continental Airlines Holding (formerly Texas Air Corporation) to Scandinavian Air System, which gave that foreign carrier a 16.8 percent share. Lorenzo himself walked away with an estimated $40 million, having signed an agreement to stay out of the airline business for seven years. While experiencing financial pressure, executives at Continental managed to keep the airline alive and worked toward resolution of its bankrupt situation. Ironically, its comparatively low operating costs, the legacy of Lorenzo's style, helped Continental keep its planes in the air. Eastern was not as lucky. Bereft of useful assets and battered by the high fuel costs triggered by the Persian Gulf Crisis of 1990, Eastern totally collapsed in January 1991, bringing an end to one of the nation's pioneers of air travel. The *New York Times* summed up Eastern's demise as "by far the largest casualty of the pressures brought by the deregulation of the industry 13 years ago."

The deregulation act of 1978 also led to expansion by the established airlines. Even though the short-term moves of major players such as United Airlines did not begin to keep pace with hustlers like Braniff, the momentum toward expansion seemed inexorable. Moreover, the bullish market of air travel led to large investments in ancillary ventures. United's new service into Mexico, beginning in 1980, seemed modest enough, but the parent company also began an unusually aggressive expansion into other areas. During the early 1980s, United assumed leadership in the Apollo computerized reservation system. The company trumpeted its goal of becoming a totally integrated travel giant, acquiring Westin International Hotels and the Hertz rental corporation. Although Pan Am had developed a foreign hotel chain some years earlier, the United scheme brought together domestic and foreign operations for air travel, lodging, car rental, and inclusive computerized reservations in an impressive travel package. These dramatic steps were only the beginning of rapidly expanded international service, significantly underscoring the new orientation of an airline that had been domestically focused for decades.

True, the company had been a fixture in the West Coast/Hawaii market for years, but Hawaii was U.S. territory nonetheless. From Hawaii, United offered service to Japan as early as 1983. Then, the company acquired big, long-range Boeing 747-400 airliners in 1989 and put them to work on the route to Australia. Next, it launched itself into decidedly exotic new markets. Australian service was quickly followed by schedules to Hong Kong, and United rapidly became the principal trans-Pacific carrier. Europe came soon after: direct flights from Chicago and Dulles International to Frankfurt, Germany, commenced in 1990. The decay of Pan Am offered new options to aggressive United, who bought Pan Am's Heathrow operations in 1991, creating its first European hub. These operations at once gave United a half-dozen key trans-Atlantic airways and seven intra-European routes. Late in the year, United picked off Pan Am's premier operation in Latin America, emerging as a global giant even as Pan Am degenerated into a shattered hulk.

By 1992, however, United faced the same sobering statistics that stalked the global airline industry, forcing the parent company to sell its travel subsidiaries. Political emergencies, fuel costs, interest rates, and a stuttering international economy bit deeply into total revenues. Like everyone else, United canceled orders and options and deferred delivery of some 400 planes on order. Nonetheless, the United fleet at the end of 1992 numbered more than 500 aircraft, compared to 300 in the early 1980s. Its worldwide operations in 1991 alone were still awesome: service to over 220 cities extended to 19 foreign countries. Each day the company listed approximately 1,926 flights to destinations in the United States and 50 more to airports elsewhere in the world. It had 79,000 people on its payroll, including 8,000 pilots and 15,000 flight attendants.

United survived; others faltered. Hard on the heels of Eastern's demise, Pan Am Corporation filed bankruptcy in 1991, issuing a symbolic death knell for America's traditional overseas airline. With its ethos of intercontinental pioneering across faraway seas to exotic foreign ports of call, Pan Am exuded the romance and appeal of a bygone era of air travel. Although Pan Am eventually acquired some domestic routes, stiff competition from both U.S. and foreign air lines, plus the growing pressures of fuel costs and recession factors, finally pushed the once-proud international giant to the wall. Over the years, Pan Am sold off considerable assets, including the 1990 deal in which United Airlines took Pan Am's London hub along with route authority between there and four major U.S. gateway cities. Finally, the airline had no option but to sell the remaining choice routes. Months of spirited negotiations involved TWA, United, and Delta. Delta, the normally conservative operator from Atlanta, finally emerged as the victor after offering a deal worth upwards of $1.4 billion. Delta picked up rights to dozens of Pan Am routes all over Europe, Africa, the Middle East, and as far away as India. Although Pan American World Airways sold the bulk of its route system to Delta, Pan Am hung onto its Caribbean and South American markets, hoping to survive as a smaller operation. The former international giant survived on contingency funds from Delta until a reorganization plan was in place, but Delta eventually declared its unwillingness to continue the payments, and the last vestiges of Pan Am collapsed in December 1991, bringing a final end to the global legend.

Pan Am was not alone. Early in 1992, TWA declared Chapter 11 bankruptcy and struggled to continue operations. Like TWA, airlines of the United States—and of the world—experienced a rough year in 1991, shadowed by the Gulf War, high fuel prices, and a global recession. For the first time, the International Civil Aviation Organization reported a worldwide downturn for passenger and cargo traffic. The impact of this trend was evident from the growing number of airliners arriving for storage at facilities in the Mojave Desert in California, where the dry climate and mild weather accommodated row after row of furloughed planes. Airline companies also deferred purchase of new transports and postponed delivery dates—actions that added to the woes of aerospace manufacturers.

The growing troubles of the industry led to reports that the competitive structure of the U.S. airline industry was becoming compromised. By 1991, the ''Big Three'' of the industry—American, United, and Delta—controlled over half the market. The rapidly shifting alignment of U.S. flag carriers made the North Atlantic a major battleground in the struggle for international market share. The North Atlantic market was the world's largest in the early 1990s, accounting for some 28.7 million scheduled passengers per year. By 1992, American and Delta each expected to gain just over 10 percent of the market, and United expected to account for nearly 8 percent. Two years earlier none of these three airlines had competed as major players in the Atlantic market.

The increasing clout of the ''Big Three'' in both domestic and foreign markets put pressure on the surviving major U.S. airlines, leading them to negotiate special arrangements with foreign flag lines. The hope was to generate more travel through domestic hubs in the United States, accommodating passengers outbound to foreign destinations or inbound from cities abroad. In contrast to the selective marketing agreements of earlier years, the new alliances involved partial ownership of American operators by airlines from other countries. This implied comprehensive coordination of schedules, equipment, and marketing campaigns. For traditionally independent airlines in the United States, it meant a revolutionary change in business style and created international debate.

Leading the way in this unprecedented international investment were airlines from the so-called second tier of carriers, such as Northwest, Continental, and USAir. In 1989, Northwest signed a deal with KLM, giving the Dutch carrier a significant equity stake in the American company. Another European carrier, SAS (Scandinavian Airlines System), became the largest single shareholder in the finances of Continental Airlines during the summer of 1990, when Frank Lorenzo announced he was stepping down as Continental's head. One of the most controversial airline deals of the early 1990s involved troubled USAir and British Airways. Many analysts believed the arrangement represented a harbinger of future airline operations. The chieftains of American Airlines, Delta, United, and other major operators in the United States viewed the event with raised eyebrows.

Studies of the U.S. airline market noted that some two-thirds of Americans headed for Europe began their travels from a city or town somewhere east of the Mississippi River. Theoretically, USAir ought to have been tapping a big share of that market, but the company simply did not have the resources to shoulder its way into trans-Atlantic competition. An agreement with British Airways would feed passengers into USAir's gateway cities, where they would have access to the British international network. Conversely, passengers arriving in the United States via British Airways would continue inland aboard a USAir flight.

And so, in 1993, British Airways and USAir signed a modified pact in which the British holding company invested $300 million in its American partner. Arrivals, departures, and ticketing for international passengers were integrated, offering a powerfully convenient travel product. There were also discussions about cooperating in the purchases of fuel and in servicing contracts in which the advantage of size would offer bargaining leverage. By the end of 1997, however, the deal had soured, as British Airways opposed the American carrier's desire to expand its access to London's Heathrow Airport. USAir also changed its name to US Airways in a move to reshape its image as a global carrier rather than a niche operation within the United States. But economic troubles followed, a legacy of the airline's rapid expansion and acquisition activity during the early years of deregulation. In the spring of 2000, United and US Airways announced a proposed merger, setting the stage for a prolonged period of legal scrutiny for potential antitrust violations.

By the turn of the century, relationships among various airlines resembled a kaleidoscopic maze. KLM kept its share of Northwest, which held a sizable stake in Continental; other arrangements came and went as KLM flirted with mergers involving European airlines and Continental dallied with Delta. But one new scheme took hold with remarkable speed—the *alliance system*. Trends toward internationally integrated airline systems continued, often involving a financial stake. Eventually, the prospect of a full merger came up again when British Airways and American Airlines unveiled just such an arrangement. Government opposition—from both sides of the Atlantic—averted this blockbuster proposal, but in the meantime, the airline community came up with an even more expansive scheme. This was the alliance system. The new plan involved not just two partners, but a string of partners representing different global regions. Travelers could accumulate bonus miles to be used on any airline within the system; flights to numerous out-of-the-way destinations could be booked as a seamless journey requiring only a single ticket within the system. For travelers having access to special airline lounges in passenger terminals, such privileges within the alliance could easily number into the dozens, located in nearly any corner of the world.

Frustrated in their merger attempt, American Airlines and British Airways launched the Oneworld Alliance in 1999 and quickly added several affiliates: Canadian Airlines, Cathay Pacific, Qantas, Finnair, Iberia, Aer Lingus, and LanChile. Others joined later.

Early the following year, Oneworld announced the formation of its own central management company to carry on customer services, marketing initiatives, information technology, and financial activities. Returns to individual alliance airlines varied, but early reports indicated increased passengers worth tens of millions of dollars in added revenues. In many ways, the alliances reflected advantages of actual mergers but skirted potential trouble spots with labor groups, pilot seniority, and so on.

Rival alliances soon appeared. United and Lufthansa became the principals in launching Star Alliance ("The Airline Network for Earth"), which soon embraced Ansett Australia, Air New Zealand, Air Canada, Varig (Venezuela), SAS (Scandinavian Airways System), and Thai Airways. Other carriers scrambled to organize themselves, but not always with success. What was to have been the Wings Alliance (Northwest, Continental, and European carriers like KLM and Alitalia) foundered on hopes by the European carriers to adopt revenue sharing. While talks continued, other systems began to coalesce. Left out in the cold, Air France began seeking substantial partners, especially from the United States. Delta Air Lines responded. The two signed a comprehensive, bilateral pact in mid-1999, and executives from both sides emphasized goals to expand into a multicarrier, global network. It took several months to agree on a name—SkyTeam—and the pact did not reach its multicarrier goal until the summer of 2000, when Aeromexico and Korean Air joined the alliance. Subsequent negotiations pointed toward potential agreements with additional carriers in Europe, northern Africa, and South America. In yet another symbol of shifting post–cold war arrangements, Aeroflot Russian Airlines (the remnants of the former Soviet state airline, Aeroflot) voiced its intent join SkyTeam in the future.

Not all airlines in the United States wanted to dance to an international tune. Southwest Airlines, based in Dallas, Texas, continued to fly the old-fashioned way, by attracting passengers through inexpensive fares, on-time arrivals and departures, and sometimes quirky service. Led by its shrewd, ebullient chief executive, Ed Kelleher, Southwest shed its image of female cabin attendants in hot pants while sticking to its no-frills service of no dinners, no box lunches, but lots of free peanuts. Cabin attendants still livened up flights with occasionally outrageous commentary over the plane's intercom system. The airline continued to rely on only one type of plane—the Boeing 737—keeping its maintenance costs and spares inventory at minimum levels. Southwest also expanded out of its regional territory to become a national carrier by carefully selecting new destinations and sometimes bypassing a market's major airport in favor of an outlying airfield with less air traffic and lower landing fees. Southwest's revenues continued to climb, its employees remained loyal, and its passengers continued subscribing to bargain fares.

READJUSTMENTS IN THE AEROSPACE INDUSTRY

Significant shifts in the world aerospace manufacturing industry became markedly stronger during the mid-1980s. The traditional leadership by suppliers from the United

States suffered under stiff competition from foreign producers. One market niche, for smaller turboprop transports used by commuter airlines for short-haul routes, became dominated by manufacturers based in Canada, Latin America, and Europe. In the United States, manufacturers of big airliners continued to enhance existing jet transport designs with additional versions tailored to consumer needs. The Boeing 737-300, which added 21 more seats to the 120 seats of the 737-200 series, enjoyed strong sales; McDonnell Douglas signed contracts for its MD-87, a smaller, 130-passenger version of its MD-80 series. Both American manufacturers also introduced improved models of their jumbo jets. McDonnell Douglas's new MD-11 trijet featured winglets and other design features to reduce weight and drag. Boeing unveiled the 747-400, also with winglets and advanced cockpit systems, designed to carry as many as 624 passengers.

The strong demand enjoyed by U.S. manufacturers of large jet transports reached a peak during the mid-to-late 1980s, when changing global politics—accelerated by the break-up of the Soviet Union—and foreign competitors changed the nature of the business. American manufacturers like Boeing and McDonnell Douglas dominated the U.S. domestic market and carved a commanding niche in the world market as well. Between 1980 and 1986, U.S. manufacturers accounted for 80 percent of the total value of commercial transports sold in the non-Communist bloc. The European manufacturing consortium, Airbus Industrie, accounted for 15 percent of the business, and the remaining 5 percent was shared by Fokker (Netherlands) and British Aerospace.

With deliveries of a new narrow-body commercial jet airliner in 1988, the A320, Airbus Industrie underscored its intent to win a larger share of worldwide sales. The organization firmed up plans to develop new wide-bodied aircraft, the A330 series, giving it a varied product line that included a long range, four-engine jet, the A340, that would compete with some versions of the 747. With the ability to contend against U.S. manufacturers in virtually every category of narrow-body and wide-body aircraft, Airbus Industrie hoped to control as much as one-third of the market for airliners having more than 100 seats.

As the decade of the 1990s began, a healthy U.S. airline and aerospace industry faced a number of new challenges as a result of political changes in Europe. The sudden thaw in cold war antagonisms meant less reliance on defense contracts for both U.S. and European manufacturers. In America, aerospace companies announced major personnel reductions. In Europe, the pending elimination of trade barriers in 1991–92 was a factor in restructuring Europe's aerospace industry and airline operations, all of which sharpened the challenge for American industry. Moreover, the Iraqi crisis during the summer of 1990 brought higher oil prices, which triggered an increase in airline fares, checked the increase in air travel, and caused many airlines to cancel or defer plane orders due to huge operating deficits. Despite a healthy backlog, McDonnell Douglas and other American aerospace firms faced cutbacks caused by declining civil and military orders.

News of McDonnell's cuts—plus those announced by Hughes, Boeing, Northrop, and others—created dramatic headlines and triggered some decline in corporate stock prices. As the dust cleared, the overall position of the aerospace industry, though serious enough, did not appear to be catastrophic, as some news stories implied. Many companies recruited heavily during the 1950s and 1960s, as both the cold war defense build-up and the space program gained momentum, and many of these employees had now reached retirement age. Early retirements and natural attrition made personnel reduction projections much less threatening. Moreover, diversification into nondefense ventures promised to create new opportunities for aerospace employees. In cooperation with state agencies, the aerospace industry in California also launched a reemployment program in nonaerospace sectors.

Among the major aerospace contractors, Boeing Corporation remained bullish about the future, although the company expected downturns in its various defense operations. Meanwhile, sales of its commercial transports, like the 737, 757, and 767 twinjets, as well as the mammoth four-engine 747, continued to generate substantial profits. In addition, the company announced a trend-setting new airliner, the Boeing 777.

In the decades after World War II, engineering drawings for an entire aircraft, down to nuts, bolts, and rivets, had arrived on the production floor as huge paper drawings and thick handbooks. The increasing size and complexity of these production documents led to increasing use of computerized systems to reduced the avalanche of paper. By the 1970s, computer-aided design and computer-aided manufacturing, or CAD/CAM, had brought revolutionary changes to the design and manufacturing sequence. CAD/CAM equipment permitted engineers to generate different components as three-dimensional images and change them at will early in the design process. At the same time, production engineers could begin to plan appropriate jigs and tooling for the production phase. The high costs of developing new airliners meant that many sub-contractors were also risk-sharing partners, contributing R&D capabilities and production capacity. If the airliner sold well, subcontractors recovered their investments. If not, they shared the same financial downside as the prime manufacturer. In planning the Boeing 777, a Japanese consortium held a shared-risk stake representing 20 percent of the airframe.

In 1991, when Boeing announced plans for its new twin-jet, wide-body 777 airliner, the company also implemented a design/build concept, in which mock-up aircraft and prototype tooling were essentially eliminated. The 777 evolved as a "paperless" airplane. Designers, manufacturing engineers, subcontractors, and airline maintenance personnel were all tied together in a comprehensive computerized network. Within the first year, the system had pinpointed 2,500 interface problems. Using CAD/CAM and the design/build approach saved tens of millions of dollars and helped keep the 777 on schedule. It entered airline service in 1995.

As the respective American and European transports began to take shape on assembly lines, the marketing competition between Airbus Industrie and U.S. manufacturers began to assume crisis proportions. At issue were the alleged subsidies that the Airbus consortium had used to develop six different commercial airliners. By 1991, the consortium accounted for some 30 percent of all new airliner orders in competition with McDonnell Douglas and Boeing. Although Boeing still led the world in sales of airliners, McDonnell Douglas had dropped to third place. The United States claimed that the Airbus consortium had the benefit of European government subsidies of 60–90 percent (over $13 billion) of program costs, allowing the sale of Airbus jets at reduced prices in the world market. European industry countered that charge by pointing to benefits that U.S. manufacturers derived from NASA and military research; American manufacturers contended that such assistance was negligible in comparison to the massive funds contributed by Airbus member nations. The United States and its European competitors held many months of talks over the subsidy issue, but a mutually satisfactory solution seemed illusive.

There was little doubt about the significant role played by aerospace in the American industrial base. In 1991, the U.S. aerospace industry's trade balance of $31 billion marked a record year and made aerospace the nation's leading exporter of manufactured goods. The aerospace business, with 1.6 million workers, represented 6 percent of all workers in manufacturing. But the figures highlighted a decline of 13 percent since 1989, with a further drop of 7 percent predicted by the end of 1992. Moving into the mid-1990s, the industry looked far different than it had a decade earlier, with more changes likely. Some of the stalwart veterans, like Grumman, had reduced personnel by as much as 30 percent, and the company's historic role as manufacturer seemed close to ending, with only a handful of navy jets for electronic warfare in production. Grumman seemed financially solid, having transformed itself into a company that specialized in electronic systems and development. Nonetheless, a merger in 1994 created the Northrop Grumman corporation. Fairchild had gone out of business as an airframe manufacturer and, like Grumman, survived in the electronics field. But Fairchild was no longer an American company, having become a subsidiary of the French firm Matra, underscoring the growing trend toward industrial globalization. Fairchild later joined a German organization to build a regional jet.

Other sales and mergers dramatically altered the long-established roster of major aerospace companies. General Dynamics, a principal aerospace manufacturer, sold off major aerospace subsidiaries, retaining its space and ballistic missile business along with divisions for producing battle tanks for the army and submarines for the navy. Its tactical rocket and cruise missile division went to Hughes; its Cessna division, to Textron. In a stunning move in 1993, its Fort Worth operations, producer of the internationally marketed F-16, went to Lockheed. General Dynamics argued that it wanted to focus only on "core business," where it could operate unchallenged as number one in that field. In the process, the sale of its various lucrative divisions generated big profits and pushed its stock sharply up. Martin Marietta emerged as the major electronics powerhouse, having purchased the military electronics division of General Electric in 1993. Then, in a surprising move, Lockheed and Martin Marietta announced a merger in 1994, creating an airframe, missile, and electronics powerhouse second only to Boeing. While subsequent mergers of various U.S. firms occurred, none made more news than Boeing's buyout of its longtime rival, McDonnell Douglas. The latter had failed to win several contracts for major military projects; accounting and management systems still seemed to lag in comparison to those of other aerospace industry leaders, and its role as the second-largest builder of airliners had fallen to the aggressive Airbus Industrie consortium from Europe. Announced in 1996, the official merger took place the following year, and one of the most historic names in aviation/aerospace history disappeared as a corporate entity.

There were parallel developments on the other side of the Atlantic. For many European defense experts, the reliance on American combat planes represented a disagreeable dependence. As a result, various multinational coalitions evolved to develop European warplanes, like the Anglo-French Jaguar in the 1960s and the Panavia Tornado, which involved Britain, Germany, and Italy in the 1970s. During the 1980s, the most ambitious plan for a combat design involved Britain, France, and Germany in an effort to develop a new, extremely advanced fighter plane. Political arguments and budgetary disputes caused the French to withdraw and produce their own product, the Dassault Rafale, scheduled to enter service early in the 2000s. After further problems, the British and Germans enlisted Italy and Spain in the European Fighter Airplane (EFA) program. Also known as the Eurofighter, the plane made its first flights during 1995. The partners agreed on a production of about 600 aircraft, and it was scheduled to enter service not long after the Rafale.

Despite the long development period, efforts such as the EFA program demonstrated growing expertise by the Europeans in terms of conducting very sophisticated multinational aerospace projects for both military and civilian service. At the same time, there were significant political and economic changes in Europe, represented by the expansion of the European Union, a European Parliament, and a uniform currency on the continent, the Euro. In this context, additional aeronautical collaboration transpired as European industry took steps to challenge large American corporations on the world scene.

After the wave of mergers involving American aerospace firms, similar restructuring began in Europe, partly due to leaner defense budgets and partly as a means of creating corporate entities big enough to compete against American counterparts. In the United Kingdom, a merger of British Aerospace and Marconi Electronic Systems created a new entity, BAE Systems, with 115,000 employees in nine countries and 1998 revenues of $19.7 billion. Even larger organizations were in the works. Early in 2000, the European Aeronautic Defense and Space Company (EADS) took shape. Some sources referred to it as "Europe's most complex cross-border industrial undertaking." The EADS group joined principal firms from France, Germany, and Spain, with French and German companies each controlling about 30 percent of the new organization. EADS reported an approximate size of 96,000 employees with annual sales of $23 billion. Even though BAE Systems employed more personnel, the sales of EADS made it the world's third largest aerospace company, trailing Boeing ($65 billion) and Lockheed Martin ($27 billion).

The new EADS group owned 80 percent of the Airbus organization, with the remaining 20 percent held by BAE Systems. The EADS merger coincided with moves to remold the Airbus program into a single corporate entity doing business on its own rather than functioning as a consortium of government-controlled units. As a single corporate entity, Airbus would also be better poised to contend with its American rival, Boeing, in the high-stakes competition for leadership in the market for jet airliners. In fact, Airbus claimed to have taken the lead during 1999, when it booked orders for 420 aircraft, compared to 386 for Boeing. However, American analysts pointed out that Boeing actually delivered 620 airliners during the year, considered a better benchmark

In Washington, technicians complete work on new Boeing 747-400 airliners in final assembly. Foreign contracts account for the lion's share of the airliner business, so success in sales overseas constitutes a vital market for American manufacturers. Courtesy of Boeing Company.

of performance. Some controversies still swirled around the system Airbus used for reporting real numbers relating to backlog orders and actual deliveries.

At the start of the new century, Airbus declared its intention to produce a new jumbo airliner, which press releases referred to as the A3XX, pending a formal designation. This was a huge, double-deck plane intended to overtake the Boeing 747 as the leading jumbo-sized airliner and to surpass Boeing as the world leader in civil airline transports. With some configurations seating up to 650 passengers, it could carry up to 200 more travelers than could a standard 747. Preliminary illustrations of the A3XX portrayed such amenities as a gym, an airborne shopping area, and private berths with showers. Boeing managers maintained that their European rival had overestimated the market for such a plane and contended that a stretched version of the 747 represented a better match for probable air traffic scenarios of the future. A major sales contest of international proportions loomed. In any case, Airbus had clearly made significant gains in the competition to sell airline transports.

By the end of the 1990s, European aerospace manufacturers had crossed swords with American firms in virtually every arena of the aviation and space business. Because the European consortium Arianespace already reported about half of the launches of commercial satellites, the European aerospace industry challenged America in that sector as well. Moreover, when NATO partners like the Greek government announced orders for several dozen Eurofighters and the Norwegian government short-listed the same plane in competition with the Lockheed Martin F-16, these decisions marked a significant shift in the traditional domination of the export market for jet fighters. Also, the Swedish manufacturer, Saab, entered into a joint marketing effort with BAE Systems and successfully won a contract to supply its new multirole Gripen fighter to the South African Air Force. The French carved out additional sections of the fighter market with sales of upgraded versions of the Dassault Mirage 2000 to the air forces of Quatar, the United Arab Emirates, and Taiwan, long a traditional customer for American warplanes. For helicopters, missiles, and a wide range of avionics, the same trends gained momentum. As it entered the twenty-first century, the American aerospace industry faced demonstrably greater challenges in the international marketplace.

GENERAL AVIATION: DARK CLOUDS AND SILVER LININGS

News stories about record flights cheered the general aviation sector, although production figures fell short of expectations during the 1980s and 1990s. At the same time, general aviation operations continued to demonstrate its value in contemporary civilization, especially in the area of business flying. By the end of the 1990s, federal legislation had breathed new life into the manufacturers of light planes. People in the industry savored press coverage concerning an incongruous airplane called the Voyager, conceived by the gifted designer Bert Rutan. With its long, trim wings, twin tail booms, and canard surfaces connecting the booms to the plane's nose, it was a startling contraption in the air. Designed for maximum lift, the plane also made extensive use of lightweight composite materials in its construction. In the winter of 1987, the unique Voyager captured world headlines in a nonstop flight around the world without refueling. Another nonstop record fell in 1999, when Swiss and British fliers navigated the "Breitling 3" balloon in the first nonstop flight around the world by a lighter-than-air craft. After launch from the Swiss Alps, they covered 29,000 miles in just under 20 days, coming down in the Egyptian desert southwest of Cairo.

Flying activity in the general aviation community remained at high levels, with business flying on the increase in various regions of the United States. This development occurred in certain areas to which airlines had reduced service during the era of deregulation. Also, many business travelers chafed at inconvenient schedules resulting from the hub-and-spoke system. Recreational aviation remained popular, enlivened by hot-air ballooning, hang gliding, and ultralight planes with small engines.

The general aviation manufacturers, though not so adversely affected by the cold war thaw, found that market changes and other factors presented serious problems. For much of the 1980s, high interest rates depressed sales, which fell below 1,000 deliveries during 1992. The industry's trade association spent much time dealing with the issue of high liability expenses. Manufacturers argued that they should not be held accountable for an airplane after it had been operated by several different owners, who may or may not have observed maintenance requirements or flown the plane within recommended guidelines. After several years of failed efforts to change federal legislation, Beech, Piper, and Cessna all but eliminated small, single-engine planes from their product lines as unprofitable and concentrated on expensive twin-engine turboprops and corporate jets.

In one sense, general aviation manufacturers became victims of their own success, because earlier years of high production had created a large pool of affordable used planes in good condition—a factor that eventually cut into the market for new aircraft. The general aviation fleet, which remained stable at approximately 215,000 airplanes into the 1990s, continued to support the lively activities of pleasure flying as well as fundamental business travel. Because general aviation capabilities remained integral to contemporary society, manufacturers stayed in business by finding a niche. Beech courted business and corporate sales with big turboprops, jets, and an updated version of its single-engine Bonanza. Similarly, Cessna prospered with business planes, although it also built a single-engine plane, the Caravan I, designed for utility jobs and light cargo. Other manufactures maintained a slow but steady business by specializing in agricultural or single-engine personal planes.

Still, most of these firms experienced considerable change. Beech enjoyed stability under Raytheon's umbrella, but in the early 1990s Cessna wound up as part of General Dynamics, only to be spun off to Textron. An investment group in France controlled Mooney; Learjet passed through a series of corporate owners until the Canadian aerospace manufacturer, Bombardier, purchased the firm in 1990. The Piper Saga continued, replete with downsizing, Chapter 11 filings, court-ordered reprieves, white knights, overseas bids, valiant employee efforts, and a spunky last stand in 1993 to make deliveries on several dozen new orders—an illustration of customer loyalty that exemplified much about Piper's strong reputation over many decades. The persistent Mooney organization reestablished control of its own destinies in 1991 and continued to build its sporty, high-performance single-engine designs. Piper's precarious fortunes finally turned around in 1995, when major creditors and investment bankers worked out a restructured entity officially called The New Piper Aircraft Incorporated. In 1999, General Dynamics reentered the aviation business with the purchase of Gulfstream Aerospace Corporation, builders of luxury executive jets like the Gulfstream V, priced at $38 million. The spacious G-V could whisk corporate travelers nonstop from New York to Tokyo or Los Angeles to London.

The general aviation community had received a huge boost in 1994, when Congress passed legislation that stipulated limited liability for the light plane industry. A "statute of repose" of no more than 18 years established the time limit during which the manufacturer would be liable for a product and covered the existing piston-engine fleet as well as other aircraft. In particular, the effects of the legislation rejuvenated sales of single-engine, piston-powered light planes and gave a jump-start to renewed production. In 1999, Cessna delivered nearly 900 aircraft in this category, ending the year with all five of its piston-engine models in simultaneous production.

When the National Business Aviation Association (NBAA) met in 1999 for its 52nd annual convention, a bullish atmosphere prevailed. The NBAA reported the strongest market in its history, and officials noted that acceptance of aviation as a useful means of conducting business had never been higher. The NBAA membership alone numbered over 6,000 companies. All told, more than 9,000 firms operated turbine-powered aircraft.

As more and more companies decided to become plane owners, the future for manufacturers of corporate aircraft looked brighter than ever. Forecasters predicted sales of 6,800 new planes worth $89 billion by 2010. Over a thousand entry-level business aircraft were expected to be sold, representing smaller jets averaging four to six passenger seats plus pilot and copilot. The remaining sales represented various types of corporate jets having larger passenger cabins and more amenities. An elite group of executive jets at the top of the scale included modified airline transports, expected to number about 170 planes. Boeing, for example, produced two versions of its 737 transport, dubbed the BBJ, or Boeing Business Jet, priced at $40 million plus. Capable of hauling around 140 passengers as a commercial airliner, the BBJ series conversions offered ample room for corporate luxury, including meeting rooms, private suites, and spacious seating for as many—or as few—as the buyer might want. Some corporate versions featured a private bedroom and shower. By the end of 1995, Boeing reported 56 BBJ orders, with over three dozen in service. Traditional builders of large-cabin, long-range executive jets, such as Gulfstream Aerospace and Bombardier, reported a strong demand as well. Such planes could carry corporate executives on long, trans-oceanic jaunts, including nonstop flights across the Atlantic.

One of the more unique designs available to private operators was the Bell/Agusta 609 civil tilt-rotor, based on Bell's V-22 design for the armed services. Bell's partnership with Agusta produced the more compact 609 version, scheduled for service early in the 2000s. The industrial team reported 77 orders from 42 customers in 18 different countries around the world.

Given the cost of many corporate jets, from one million dollars or so for a small jet to ten or twenty times that for cabin-class planes, many buyers chose an option known as fractional ownership. This allowed an individual or company to acquire shares in an aircraft at sharply reduced fees for overhaul, maintenance, and other costs, paying operational expenses only during the time they flew in it. An important goal was to cut the high cost of convenient charter flights. For companies unsure that a corporate jet would really pay its way or for users who needed jet transportation only on a seasonal basis or at widely spaced intervals, fractional ownership solved a variety of needs.

When the idea got started in the mid-1980s, only a handful of planes operated as fractional jets, and only 32 with 140 owners fell into that category in 1994. By 1997, fractional ownership operations had become big business. Executive Jet, the industry leader, operated a NetJets program, which listed over 260 aircraft with several dozen more on order. Across the United States, the fleet of fractionally owned jets included some 350 planes and over 1,500 shareholders. Fractional owners cited basic reasons for ownership: executive planes flew to more airports; multiple stops could be tailored to an executive's schedule; with associates on board, business meetings proceeded while en route. Quite simply, participants reported that fractional ownership boosted productivity while dramatically reducing the high costs of corporate jet ownership. Owners were individuals, like movie star Arnold Schwarzenegger, as well as corporations, like General Electric Company, Gillette Company, and a range of other businesses.

Although manufacturers and business flying usually received the most news coverage in the general aviation sector, the general aviation community continued to include a broad spectrum of utility flying. During the summer of 2000, several stories referred to an epidemic of forest fires across western American—the worst fire season in more than 50 years. Hard-fought campaigns against these awesome fires included specialized aircraft counted in the general aviation fleet and represented some of the most dramatic chapters in general aviation.

California—because of its size, forested regions, and meteorological contrasts—relied heavily on aircraft to help contain wildland fires, especially in the heavily timbered areas in its northern reaches. Bureaucratically, the responsibility for handling these threats fell to the California Department of Forestry and Fire Protection, or CDF. During California's dangerous dry season, usually the eight months of April through November, the CDF dealt with approximately 7,500 wildland fires, the majority accidentally triggered by human beings. With the U.S. Forest Service, which oversees national forests and federal lands, the CDF watched over 51,000 square miles, about one-third of the state. The CDF alone commanded 3,500 firefighters, 600 fire stations, 1,000 vehicles, and 50 aircraft, including fixed-wing fire bombers and helicopters. In severe emergencies the CDF could call on a special command team to tap additional statewide personal and equipment, including aircraft leased from commercial firefighting companies.

In nearly every case of wildland firefighting, aerospace technology played an important role. Using the satellite-based Global Positioning System, CDF receivers helped fix the precise location and size of fires; heat-sensing aircraft probed through heavy smoke to trace the fire's perimeter; squadrons of planes and helicopters remained on call to swoop in with fire-retardant liquids and hundreds of gallons of water. Still, as one veteran admitted, "wildland firefighting is part science, part witchcraft, and part sheer luck."

Characteristically, an aerial offensive delivered the first salvos against new fires. The CDF relied on flying "tankers" like the military surplus Grumman S-2, modified to haul some 800 gallons of fire retardant in large tanks inside its fuselage. The retardant, a mixture of water and chemicals, was dyed a bright red to aid pilots in gauging its fall and to mark the location for following aircraft. The retardant covered the drop zone with a recipe including a sticky substance, which also coated the vegetation, reducing its flammability, and hampered the fire's tendency to spread. Pilots homed in on their targets at about 100 feet above the tree tops, holding a course through thick smoke, violent updrafts, and orange flames. After the drop, planes banked sharply to head back for refills; the reloading circuit required about 20 minutes. To keep the flames at bay, planes followed each other very closely, and slower helicopters usually operated on the other side of a fire at the same time. The separate streams of aircraft rolled in and out, sometimes only a few hundred feet apart and as close as a quarter of a mile on each flank of the fire. Circling above them, an air commander scouted for effective drop routes that ran in favorable directions downhill; pilots wanted to avoid flying into rising terrain if an aircraft suffered a loss of power or other malfunction. All of this often tended to create a sort of aerial melee. "It's not exactly FAA-regulated airspace," admitted one air controller. A 25-year veteran, he also grimly remembered 32 crashes.

SPACE: A STAR-CROSSED SAGA

When the Space Shuttles began flying with payloads in the 1980s, NASA's expendable rocket launches did not cease. Several rocket launches had already been scheduled, and NASA also intended to maintain this capability as a backup through the mid-1980s. NASA boosters orbited a variety of communications and environmental satellites as well as several spacecraft involving space science. Moreover, the audacious *Voyager 2* continued its richly rewarding "Grand Tour" of the outer planets. Shuttle launches may have received the lion's share of news coverage, but rocketed payloads continued to demonstrate their utility and value in space exploration.

Planetary probes continued to turn up surprising insights into the nature of the Solar System. Four and a half years after uncovering a wealth of new data on Saturn and its spectacular rings, *Voyager 2* approached Uranus in January 1986. By the time the intrepid Voyager completed its flyby, the spacecraft had revealed more information about the planet and its company of moons than observers had learned since its discovery by the English astronomer William Herschel more than 200 years earlier.

Meanwhile, the shuttles performed remarkably well through five years and 24 missions. The fleet completed several international missions, and such cooperation received keen attention during a 1983 mission in which *Columbia* carried the European Space Agency's (ESA) Spacelab into orbit.

The Orbiter *Discovery* enlarged the fleet in 1984, and *Atlantis* followed in 1985. The demographics of the Orbiter crews reflected growing diversity, encompassing more women, Canadians, Hispanics, Asians, assorted Europeans, a Saudi prince, a Senator (E. J. "Jake" Garn), and a Congressman (Bill Nelson). The various missions engaged astronauts in extended extravehicular activity, such as untethered excursions using manned maneuvering units. During a mission in 1984, an astronaut using one of these units assisted in the first capture of a disabled satellite, the Solar Maximum payload (Solar Max), followed by its repair and redeployment.

Continuing missions included more Spacelab flights, although the shuttle cargo bay carried a growing variety of other payloads, such as satellites, as well as American and international scientific experiments. One involved electrophoresis, in which an electric charge was used to separate biological materials; the goal in this case was the production of a medical hormone. Additional experiments emphasized vapor crystal growth, containerless processing, metallurgy, atmospheric physics, and space medicine, among other areas. The payload manifests for most missions were recognizably similar, listing satellites, experimental biomedical units, physics equipment, and so on. The manifest for STS-16 (51-D) in 1985 had a decidedly different quality: in addition to a pair of satellites, it listed a "Snoopy" top, a wind-up car, magnetic marbles, a popover mouse named "Rat Stuff," and several other toys, including a yo-yo. For die-hard yo-yo buffs, a NASA brochure solemnly reported that the "flight model is a yellow Duncan Imperial." The news media gave considerable attention to the whimsical nature of the Toys in Space Mission, although the purpose was educational. The toy experiments were videotaped, with the astronauts demonstrating each toy and providing a brief narrative of the scientific principles involved, including different behaviors in the space environment. The taped demonstrations became a favorite with educators— and the astronauts obviously delighted in this uncustomary mission assignment.

Despite occasional problems, shuttle flights had apparently become routine—an assumption that dramatically changed with *Challenger*'s mission on January 28, 1986. On the morning of the flight, a cold front had moved through Florida and the launch pad glistened with ice. It was still quite chilly when the crew settled into the shuttle just after 8:00 A.M. Many news reports remarked on the crew's diversity; seven Americans who seemed to personify the nation's heterogenous mix of gender, race, ethnicity, and age. The media focused most of its attention on Christa McAuliffe, who taught social studies at a high school in New Hampshire. She was aboard not only as a teacher but as an "ordinary citizen," since Space Shuttle missions had seemed to have become so dependable. Scheduled for a seven-day flight, the *Challenger* also carried a pair of satellites to be released in orbit.

NASA officials, leary of the icy state of the shuttle and launch pad, waited two extra hours before giving permission for launch. When the shuttle's three main engines ignited at 11:38 A.M., the temperature was still about 36°F—the coldest day ever for a shuttle lift-off. After a few seconds, the solid-fuel boosters also ignited, and the *Challenger* thundered majestically upward. Everything appeared to be working well for 73 seconds after lift-off. At 46,000 feet in a clear blue sky, the shuttle was virtually invisible to exhilarated spectators at Cape Canaveral, but the telephoto equipment of television cameras captured every moment of the fiery explosion that destroyed the *Challenger* and ended the lives of its crew. In the aftermath of the tragedy, stunned government and contractor personnel took action to recover remnants of the shuttle and to begin a painstaking search for answers.

In the meantime, President Reagan appointed a special commission to conduct a formal inquiry—the Rogers Commission, named after its chairman, former Secretary of State William P. Rogers. The Rogers Commission discovered that NASA had been worried about the booster joints for several months. The specific problem involved O-rings, circular synthetic rubber inserts that sealed the joints against volatile gases as the rocket booster burned. It was believed that the O-rings lost their efficiency as boosters were reused; their efficiency was even less in cold weather. The Rogers Commission report, released in the spring of 1986, included an unflattering assessment of NASA management, calling it "flawed," and recommended an overhaul to make sure managers from the space centers kept other top managers better informed. Other criticisms not only resulted in a careful redesign of the booster joints but also led to a long list of technical improvements to the shuttle.

Even though President Reagan authorized construction of a new shuttle for operations by 1991, the existing fleet of three vehicles remained inactive for more than a year and a half, severely disrupting the planned launch of civil and military payloads. For some scientific missions, desirable "launch windows" were simply lost, and other missions, rescheduled for sometime in the future, were severely compromised in terms of scientific value. In the case of the Space Shuttle program, NASA had not only stumbled but was left staggering. Although the flight of *Voyager 2* past Uranus and on toward Neptune represented a striking success, it was almost lost in the clamor triggered by the loss of *Challenger*. During the next several months, the agency's frustrations multiplied.

In 1986, Halley's Comet made its appearance again after an absence of 76 years. However, during Halley's dramatic swing across Earth's orbit, many American scientists lamented that no American spacecraft made a mission to meet it and make scientific measurements. Critics charged that excessive NASA expenditures on the shuttle had robbed America of the resources to take advantage of unusual opportunities such as the passage of Halley's Comet.

A series of subsequent launch failures, operational mishaps, and the brooding shadow of *Challenger* dulled the otherwise bright successes. Early in 1987, determined launch crews had successfully put two important payloads into orbit, including a weather satellite for Caribbean hurricane research and an international communications satellite over Indonesia. While debate over the nation's space program persisted, NASA continued its spacework.

Resumption of Space Shuttle missions (1988) for which special payloads were developed held the promise of a renaissance in astronomical science, especially in the case of the Hubble Space Telescope. Weighing 12½ tons and measuring 43 feet long, the Hubble Telescope, with its 94.5-inch mirror, was to be the largest scientific satellite built to date. The most alluring prospect of the Hubble Telescope's operations was the potential to search for clues from other solar systems and gather data about the origins of Earth's universe, perhaps resolving once and for all the debate between the "big bang" theory of the origin of the universe and the steady state concept.

Nor was the Hubble Space Telescope the only major effort in astronomy, astrophysics, or planetary research. NASA initiated a new family of orbiting observatories, often developed with foreign partners, to probe more deeply into the background of gamma rays, infrared emissions, celestial X-ray sources, ultraviolet radiation, and a catalog of other perplexing subjects. There were also several bold planetary voyages. The Magellan mission compiled a detailed map of the planet Venus; Ulysses (planned with ESA) was designed to explore virtually uncharted solar regions by flying around the poles of the Sun.

Another significant milestone occurred in January 1984, when President Reagan endorsed the Space Station Freedom program in his State of the Union message. Meanwhile, NASA and contractor space station studies proceeded through several versions before one design was chosen by NASA as the "baseline configuration." This structure, which emerged during 1987–88, was scaled down in size due to budgetary controversies and the reduced number of shuttle flights after the loss of *Challenger*. The revised baseline configuration called for a horizontal boom approximately 360 feet long, with pairs of solar panels at each end to generate power. At the center of the boom,

The International Space Station over Lake Balkash (Kazakhstan) during its occupation by the first trio of astronauts. The Unity module is at the top; then Zarya with solar panels; then Zvezda with another set of panels. At the bottom is the Soyuz vehicle, which carried the first crew aloft on October 31, 2000, and docked on November 2, 2000. Courtesy of NASA.

four pressurized modules, linked together, provided the focus of manned operations in a 220-mile orbit above the Earth. Given the cost and complexity of the project, the American space station initiative included an invitation to foreign partners to share in its planning and operation; refining the details of this partnership engaged negotiators from the United States, Canada, Japan, and the ESA over the next several years.

Even though Space Shuttle missions began again (including the Orbiter *Endeavor,* delivered in 1991 as the *Challenger*'s replacement), ambitious launches like that of the Magellan spacecraft continued, and technicians groomed the Hubble Space Telescope for what was planned to be a spectacular operation, NASA's reputation soured once more. True, some things seemed to perform as promised. Late in 1991, the Magellan spacecraft began to return startling data about Venus, such as the presence of active volcanoes. But this was not enough to compensate for persistent problems in other planetary missions and news stories about severe cost overruns with Space Station *Freedom,* which had already swallowed $8 billion and was still in the planning stage even though it was supposed to be operational in 1992. Various estimates put total costs at $30 to $50 billion and more, with a launch date sometime before the decade's end. Other nasty revelations concerned a weather satellite, two years behind schedule and 230 percent over budget. And the vaunted Hubble Space Telescope became a haunting, continuing, embarrassment.

When Hubble was finally launched in April 1990, it was already seven years late and $700 million over budget. Once in orbit, the telescope's first pictures had a fuzzy quality that did not surprise scientists, since adjustments and fine tuning were to be expected. After some two months of mounting frustrations and increasing tension, horrified specialists acknowledged the ghastly truth: a flawed mirror that would never produce the clarity required. "The Hubble," reported *Newsweek*, "which had promised to give man a crystal-clear view of the heavens, was seeing the cosmos as if through a cataract." As a Harvard astronomer bitterly remarked, "This is one of the worst things to happen to astronomy since the Pope strung up Galileo."

Imaginative computer programming enhanced some subsequent images, although the Hubble's anticipated performance not only remained dismal but grew progressively worse. NASA had planned all along to change or improve some of Hubble's equipment in orbit, but the catalog of major failures put the entire telescope in jeopardy. The agency planned a major repair expedition for late 1993, in which a team of seven astronauts aboard a shuttle would spend several days in a desperate series of repair jobs in orbit. "It's a tremendous gamble," admitted one astrophysicist. "If it succeeds, it will be a really impressive feat." The December mission appeared to achieve its goals.

During the mid-nineties, NASA seemed to hit its stride once again. A string of finales for space probes, dispatched several years earlier, yielded considerable scientific data. Scientific instruments aboard the Magellan, launched in 1989, accomplished a successful mission before disintegrating (as expected) in a descent into the Venusian atmosphere during 1994. Launched in 1989 with German participation, the Galileo spacecraft eased into Jupiter's orbit during 1995, directing a probe into the Jovian atmosphere and leaving the main spacecraft in orbit to send back revealing images and information about Jupiter and its fascinating moons. Ulysses, planned with several international groups coordinated through the European Space Agency, took three years to reach the sun's polar regions in 1993, followed by a return orbit in 2000. Again, reams of scientific data about the little-known solar poles promised to keep investigators busy for some years to come.

Four years after the last-ditch, but effective, manned mission to repair the ailing Hubble in 1993, a more thorough upgrade of its instrumentation took place during a manned mission. Subsequent images from the Hubble Space Telescope mesmerized scientists and the public alike, showing collisions of ancient galaxies and other spectacular, previously unknown events in the universe. These images made the front pages of newspapers and magazines around the globe. NASA's fortunes seemed to be on a roll.

There were disappointments. During 1993, the Mars Observer, the first U.S. mission to Mars in 17 years, went silent only a few days before its scheduled maneuver into Martian orbit. But probes to Mars continued, often focused on the tantalizing prospect of finding water—a key ingredient for life in our universe. A pair of payloads, Mars Pathfinder and Mars Global Surveyor, lifted off in 1997. The Surveyor performed its assigned tasks while in Mars orbit; Pathfinder obligingly touched down on the Fourth of July. At home for this national holiday, millions of Americans—along with several hundred million viewers elsewhere on Earth—watched the proceedings through the spacecraft's camera as it landed and debarked the Sojourner, a wheeled robot about four feet long and three feet high, weighing 25 pounds. Both units posted images from Mars to Earth's internet, some 141 miles distant; Sojourner's Martian meanderings made it a media phenomenon. The Surveyor/Pathfinder/Sojourner results continued to spice scientific analysis of Mars and the issue of water in its evolutionary history.

Making 1997 a banner year, the $3.3 billion Cassini mission successfully blasted off on its way to Saturn, a journey of one billion miles requiring seven years to complete. Cassini's planning had begun in 1982, engaging an international team of scientists and engineers. In addition to analysis of Saturn and its rings, Cassini packed along the Huygens probe, the European Space Agency's first planetary lander. One NASA official dubbed Cassini "the most ambitious and complex deep space mission ever launched." Rather than Cassini, most people probably remembered Senator John Glenn's return to space. At age 77, Glenn joined the Space Shuttle crew of Mission 95 (October–November 1998) as the oldest person to fly in space. Glenn became the subject of biomedical experiments to determine why space missions cause muscle loss and sleeplessness—research that might also provide insights into similar features of the aging process. The mission carried out numerous other research tasks, from bio-medicine to astronomy and arcane facets of space science. But ex-astronaut Glenn clearly stole the spotlight. As the first American to go into orbit in 1962, Glenn's reputation as an early space hero quelled much of the cynicism about his flight. In any case, NASA harvested reams of press coverage, and Glenn's return to space was hailed as a positive boost to the image of millions of American senior citizens. In the meantime, other projects and spacecraft probed various corners of earthly space as well as the universe at large. Unfortunately, their successful results seemed to get lost in the uproar following a pair of failed space missions.

In September 1999, the Mars Climate Orbiter burnt up during attempts to achieve a proper orbit; NASA and its prime contractor, Lockheed, had gotten mixed up over the conversion of pounds to metric units in navigational calculations. Barely three months later, a much-heralded Mars Polar Lander designed to prospect for Martian water apparently made a descent to the planet's surface but failed to respond to repeated attempts by NASA's Jet Propulsion Laboratory to make contact with it. The Lander, as one news writer reported, "failed to phone home." It took the better part of four months to discover a reasonable cause: a software glitch that signaled the descent engine to shut down prematurely, causing the craft to crash into the hard Martian surface. NASA Administrator Dan Goldin shouldered the blame, admitting that his policy of "faster, better, cheaper" for space exploration hardware had stretched the agency too thin. Personnel lacked both time and money to test hardware properly; harried senior managers failed to exercise appropriate oversight. Chastened, NASA took corrective action to resolve these shortcomings.

NASA also weathered increased skepticism and budgetary harassment from Con-gress about the halting progress of the Space Station. Financial realities in both Rus-sian and American manned space exploration helped pave the way for Russian collab-oration on what was billed as the International Space Station (ISS). After weeks of discussions, Russia became a partner in 1992. With this agreement, Russia's own aging space station, Mir (translated as "Peace"), launched in 1986, was to receive NASA crew members, who would verify docking techniques and check out other operational procedures involving the Space Shuttle Orbiter. This preliminary linking of the Mir station and the NASA Orbiter was important, since Russia agreed to fabricate several large components with docking for the ISS. The Russian space program also received several hundred million dollars to help subsidize space hardware and training. Between 1995 and 1999, seven NASA astronauts served aboard Mir, providing invaluable training and operational experience for ISS missions to come.

The Russian-built hardware experienced chronic slippages and schedule hang-ups until the Zarya ("Dawn") module went into orbit on a Russian booster in November 1998. Early in December, a Space Shuttle from Kennedy Space Center carried the American-built Unity module into orbit, where an international crew docked it with Zarya and checked out the systems of the first elements of the ISS. Russia continued to struggle with the schedule for the key Service Module, Zvezda ("Star"), designed to carry essential life support systems and house the first ISS crew. But it lagged nearly two years behind everything else, plagued by budgetary shortages within the Russian political and financial morass. Finally, Zvezda made it into orbit in July 2000, and ISS managers cheerfully made preparations for the first three ISS crew members (two from the Russian program and one American) to occupy the Space Station before the year was out. Plans called for the trio to arrive aboard a Soyuz module launched by a Russian booster whose lineage had originated as a ballistic missile during the Soviet era—another event symbolic of the political changes set in motion by the end of the cold war.

Aeronautical research began to receive more attention. During testimony given to a congressional subcommittee in 1993, NASA officials acknowledged that the market for civil aircraft continued to grow while the United States industry continued to lose its share of the business. In 1969, the United States built 91 percent of the world's civil transports but claimed a diminished share, of 67 percent, in 1993. With approximately one million workers in an industry that sold over $90 billion in aircraft and related products, the need for shoring up against foreign competitors gained credence. Accordingly, NASA announced an aggressive program in spring 1993 to maintain high-speed research for advanced civil transports. At the same time, a broad program in subsonic flight research received strong support for such apparently mundane but marketable aspects as environmental effects, composite structures, short-haul aircraft design, and traffic control involving both ground and airborne movements. Development of tilt-rotor vehicles like the V-22 continued, and funding for more sophisticated wind tunnels was mandated to match many of the newer facilities built in Europe since 1945. NASA also earmarked special funds for advanced computers used in aeronautical design and for support of university research. Moreover, NASA continued to play a critical role in experimental high-speed research, with yet another generation of X-planes in the skies over Dryden Flight Research Center in California. In partnership with the armed services and contractors, the agency studied thrust-vectored maneuvers for jet fighters, a variety of electronic control systems, exotic aerodynamic studies, new techniques and materials for fabrication of structures, and remotely piloted vehicles. The fallout from these technological studies found its way into Boeing's new 777 transport, the Bell V-22, and new military planes for all three services. NASA also continued its studies of high-speed flight with an eye to a potential supersonic transport (SST) for the future. When engineers bemoaned the lack of a full-size airplane to compare actual flight data with wind tunnel tests and computer simulations, the possibility of renting a retired Russian SST became alluring. In the wake of the cold war thaw, a NASA inquiry to Russian authorities yielded a mutually advantageous contract. After traveling to Russia, American and Russian engineers refurbished a Tupolev TU-144 SST to carry extensive instrumentation for joint research flights. In 1996–98, a TU-144 with a NASA logo and appropriate Russian markings completed over a dozen research missions with pilots from either country at the controls.

NASA's pioneering role in boosting nonmilitary satellites began to fade as a result of changes in domestic politics. Through the 1980s, civilian payloads—television satellites and communications satellites—arrived in space aboard rockets built and launched under NASA contract. New initiatives by the Reagan administration shifted these payloads to commercially marketed boosters in 1989. Launch operations became the province of commercial entities that operated from ex-government sites in the vicinity of Cape Kennedy. Alternative launch sites and boosters were marketed by European and other international organizations. Energized by competition and by recent achievements in lighter, compact, and more powerful electronics systems, new, space-oriented companies mushroomed during the 1990s. The field of communications became a hotbed of activity, particularly with the advent of portable, hand-held cell phones that relayed messages by means of satellites passing overhead.

The market for various commercial satellite services continued to heat up, despite the ill fortune that befell two early contenders in the field of satellite telephone ventures. Both wound up in bankruptcy court. One of them, Iridrum, had built its hopes on business executives criss-crossing the globe. Another, ICO Global communications, misread other sectors of the market. Backed by Loral, a major electronics conglomerate, a third contender, Globalstar, prepared to inaugurate service during 1999–2000, when the last dozen of its 52 low-Earth-orbit spacecraft reached their positions in space. Globalstar's enterprise, at a cost of $3.8 billion, intended to offer phone service in regions that were underserved or that had no service at all with cellular and land-line systems. The company expected to snare the business of millions of users in such regions beginning in North America and western Europe. In operation, the user placed a call from a mobile phone, with the signal going directly to a satellite, which in turn down linked to a regional gateway; the gateway routed the call through terrestrial lines. Calls between Globalstar phones went through the regional gateway, back up to a satellite, then down to a receiver. Although the system required coordination through terrestrial routes, Globalstar's plan permitted simpler satellites and lower operation costs. Industry observers expected competition from regional competitors scheduled to start up during 2000, but the potential worldwide market of 40–50 million customers allowed considerable room for several services.

Although the pace of major Russian spacecraft missions fell off, several reorganized Russian design bureaus worked out partnerships with American and European initiatives. Lockheed Martin created a subsidiary, International Launch Services, that collaborated with Russian companies to offer combinations of Russian stages and engines. The Atlas launch vehicle, a 1950s design built originally as an ICBM to strike Soviet targets, soldiered on with more powerful propulsion units supplied by the Russian firm of Energomash.

One of the most unique partnerships linked America, Norway, Russia, and Ukraine in plans to launch big boosters from a floating platform anchored near the Equator. This undertaking became marketed as a business dubbed Sea Launch. Boeing accounted for 40 percent of the project, manufacturing numerous components, holding responsibility for overall integration of the Sea Launch system, and providing overall management from Long Beach, California. A Norwegian firm took charge of modifying an offshore oil-drilling platform as the launch complex, as well as outfitting a command ship for transporting the launch vehicle segments. The Russians supplied the vehicle's third stage; a Ukrainian firm supplied a pair of lower stages. The launch site near the Equator, about 1,400 miles southeast of Hawaii, was chosen to give the rocket an extra "boost" from the earth's rotation, allowing it to carry heavier payloads such as communications satellites into orbit. Launch operations began successfully in 1999, with 18 missions booked through 2004.

The confluence of international cooperation, space technology, capitalism, consumerism, and the cash-strapped Russian effort sometimes had its droll moments. At the launch of the Zvezda module from the Russian space center in Baikonur, the booster vehicle sped upward with a singular logo emblazoned on its side: the familiar Pizza Hut insignia, complete with its rakish, red-roof trademark. The Pizza Hut folks paid $1.2 million to achieve their first symbolic delivery involving an address in space.

During the course of the twentieth century, the development of aviation technology into aerospace technology often paralleled significant trends in the evolution of science and engineering. There were similar parallels in the social and demographic profile of participants in the aviation and aerospace community. Even though the percentages of minority groups and women may not have reflected the population mix of the nation as a whole, change had undeniably occurred. Over time, it became more commonplace to find members of minority groups and women as pilots and astronauts. Moreover, such individuals became important figures in highly visible management roles.

After a career as a U.S. Air Force officer, Elwood Quesada, son of an Hispanic father and an Irish mother, became an aerospace industry executive and served as the first head of the Federal Aviation Agency (1959–61). In 1975, Gen. Daniel "Chappie" James Jr., a U.S. Air Force combat pilot in Korea and Vietnam, became the first African-American to hold the rank of four-star general, taking charge of the North American Air Defense Command. Women became more prominent as well. By the early 1990s, women in the U.S. Air Force included 100 navigators, 300 pilots, and 600 enlisted aircrew personnel. Women also played principal roles as aerospace executives. In 1993, Sheila Widnall took the oath of office as secretary of the Air Force. With a Ph.D. in aeronautical engineering, Widnall had been on the faculty of the Massachusetts Institute of Technology and had served on countless advisory boards and committees before her appointment as air force secretary; in 1997, she returned to the MIT faculty. The president of the Air Transport Association, Carol B. Hallett, assumed that position in 1995 after a political career in California and several managerial posts in the federal government. Along the way, she also accumulated more than 5,000 hours as a private pilot licensed to fly small aircraft through business jets. The high-profile attainments of such individuals clearly defined social change in the modern aerospace community.

Although space exploration and research continued to inspire assorted Cassandras, most people seemed to view astronautical ventures as part of society's continuing pursuit of knowledge and understanding. The historic voyage of three astronauts to the moon and back fulfilled one of humanity's oldest quests. That leap into space placed the world on the threshold of a new frontier, one whose promise may far exceed that opened by the Wrights' first faltering flight not so long ago. Both manned and unmanned launches have accumulated data that is still under analysis, having answered some questions, while raising new ones. In the mid-1990s, the dramatic shift in global politics suggested a stronger likelihood that manned and unmanned missions would reflect a stronger international bias than aerospace rivalries had done in the past.

Aviation experienced similar patterns of change. During the 1980s and 1990s, international collaboration became commonplace. American and European firms jointly produced jet engines; American airliners were built with a variety of components manufactured in Asia as well as Europe, and American components appeared in many aircraft produced abroad. At the same time, the growing sophistication of aerospace capabilities abroad meant increased competition in markets in which the United States had long enjoyed preeminence. In the case of air transport, the same paradox evolved, with aggressive international competition for the air travel market paralleled by global collaboration involving airlines from different countries. As the twenty-first century began, the story of flight in America promised to be as dynamic and fascinating in the future as it has been in the past.

Notes

CHAPTER 1. THE AWKWARD YEARS

Introductory Paragraphs

The historical miscellany is culled from Bernard Grun, *The Timetables of History: A Horizontal Linkage of People and Events* (New York: Simon & Schuster, 1979). Comments on the Wrights are taken from Tom D. Crouch, *A Dream of Wings: Americans and the Airplane, 1875–1905* (New York: Norton, 1981).

Aeronauts and Balloons

There are many surveys of the history of flight. The starting point for any study is still Charles H. Gibbs-Smith, *Aviation: An Historical Survey from Its Origins to the End of World War II* (London: Her Majesty's Stationery Office, 1970). Among the "coffee table" books, heavily illustrated but reliably written, see Frank Howard and Bill Gunston, *The Conquest of the Air* (New York: Random House, 1972), as well as John W. R. Taylor and Kenneth Munson, *History of Aviation* (New York: Crown, 1976). The vignette on Eilmer of Malmesbury is from Richard P. Hallion, *Test Pilots: The Frontiersmen of Flight* (Garden City, NY: Doubleday, 1981). An indefatigable scholar, Hallion not only visited the abbey but also investigated the Flying Monk, a local pub.

Standard surveys of early ballooning in America include Jeremiah Milbank, Jr., *The First Century of Flight in America* (Princeton: Princeton University Press, 1943), and Munson Baldwin, *With Brass and Gas: An Illustrated and Embellished Chronicle of Ballooning in Mid-Nineteenth-Century America* (Boston: Beacon, 1967). For a colorful survey of ballooning in Europe and the United States, see Donald Dale Jackson, *The Aeronauts* (Alexandria, Va.: Time-Life, 1980). An informed survey of American ballooning is Tom D. Crouch, "The Gasbag Era," *Aviation Quarterly* 3, no. 4 (1977):291–301.

Gliders and Airplanes

For an assessment of early pioneers like Cayley, Lilienthal, Chanute, and others, see the basic work by Charles H. Gibbs-Smith, *The Invention of the Aeroplane, 1799–1909* (London:

Her Majesty's Stationery Office, 1966). A colorful and heavily illustrated survey is Valerie Moolman, *The Road to Kitty Hawk* (Alexandria, Va.: Time-Life, 1980). The definitive analysis of early American efforts is Crouch, *A Dream of Wings*. Crouch gives due credit to Chanute's American ventures as well as to the efforts of Langley and many others at work on flying machines. Thus, while never detracting from the Wrights' historic achievement, Crouch puts their work in proper context in this richly detailed work. Crouch is my source for Manly's blasphemies (p. 290) and for Johnny Moore's excited report of the Wrights' success (p. 305). Indispensable for a detailed study of this era is Marvin W. McFarland, *The Papers of Wilbur and Orville Wright*, 2 vols. (New York: McGraw-Hill, 1953). A convenient and useful book by a contemporary is Fred Kelly's authorized biography, *The Wright Brothers* (New York: Harcourt, Brace, 1943).

The Skeptics and the Wrights

Much of this section, including the remarks by Wilbur and Orville Wright, is based on Roger E. Bilstein, "The Airplane, the Wrights, and the American Public," in *The Wright Brothers: Heirs of Prometheus*, ed. Richard P. Hallion, pp. 39–51 (Washington, D.C.: Smithsonian Institution Press, 1978). The comment on Adams is from Ernest Samuels, *Henry Adams: The Major Phase* (Cambridge, Mass.: Harvard University Press, Belknap Press, 1964). Mark Sullivan, *Our Times: The United States, 1900–1925*, 6 vols. (New York: Scribner's, 1927), discusses Bell and other contemporary topics in vol. 2. For samplings of turn-of-the-century news accounts and comments on Langley, see Sullivan, *Our Times*, pp. 562–67 (which includes the Bierce quote), and Harold U. Faulkner, *The Quest for Social Justice, 1898–1914* (New York: Macmillan, 1930). Orville Wright's remark on persistent public skepticism is from Sullivan, *Our Times*, p. 599.

The Birdman Era

Aspects of early schools and exhibition flying are discussed by Bilstein, "The Airplane, the Wrights, and the American Public." For comments by veterans of the "Birdman Era," consult the interviews of Thomas Milling and Beckwith Havens, housed at Columbia University, Oral History Collection (cited hereafter as COHC). The exuberance of early aeronautics and the era's flying meets is evident in Glenn H. Curtiss and Augustus Post, *The Curtiss Aviation Book* (New York: Stokes, 1912). Musical themes can be traced in the remarkable collection of Bella C. Landauer, housed in the collections of the National Air and Space Museum. An extensive sampling of juvenile books can be found in the museum's holdings, as well as the holdings of the Ross-Barrett Historical Aeronautics Collection in the Denver, Colorado, Public Library.

Recollections of early flying films and public stunts can be found in Guy Gilpatrick, *Guy Gilpatrick's Flying Stories* (New York: Dutton, 1946). Additional stories of Beachey and other colorful figures are recounted by Curtis Prendergast, *The First Aviators* (Alexandria, Va.: Time-Life, 1980). Typical flying techniques of the prewar era are covered in Curtiss and Post, *Aviation Book*.

The photograph of Arch Hoxsey appeared in the magazine *Aero* 1 (January 7, 1911):4.

The sensation of flying in early aircraft made a lasting impression on apprehensive passengers. The person who grasped the struts is cited in John B. Huber, "Psychology of Aviation," *Scientific American* 103 (October 29, 1910):338; Ida Tarbell's recollections appeared as "Flying: A Dream Come True," *American* 76 (November 1913):66. Contemporary clippings, poetry, and doggerel cited here are from "Aeronautical Archives (1783–1962), American

Institute of Aeronautics and Astronautics," Manuscripts Division, Library of Congress, Boxes 9 and 270. Comments from Havens, Coffyn, and Arnold are from COHC.

Commercial Trends

The first experiment in cargo-by-air is described by Roger E. Bilstein, "Putting Aircraft to Work: The First Air Freight," *Ohio History* 76 (Autumn 1967):247–58. For a discussion of early airlines, see R. E. G. Davies, *Airlines of the United States since 1914*, rev. ed. (Washington, D.C.: Smithsonian Institution Press, 1982). The summary of the St. Petersburg–Tampa venture is drawn from my unpublished M. A. thesis, "Public Attitude toward the Airplane in the United States, 1910–1925," Ohio State University, 1960. For a roster of hopeful manufacturers, see Welman Austin Shrader, *Fifty Years of Flight: A Chronicle of the Aviation Industry in America, 1903–1953* (Cleveland: Eaton, 1953). Aspects of early corporate evolution and mergers are outlined by Elsbeth E. Freudenthal, *The Aviation Business: From Kitty Hawk to Wall Street* (New York: Vanguard, 1940). On the origins of NACA, see the first-hand account by Jerome C. Hunsaker, "Forty Years of Aeronautical Research," *Smithsonian Report for 1955*, pp. 241–71; see also George W. Gray, *Frontiers of Flight: The Story of NACA Research* (New York: Knopf, 1948).

War in the Air, 1914–1918

The literature on military aviation is formidable, though uneven in scholarly quality. In developing this section, I relied on the basic histories and scholarly monographs that follow. For an excellent overview, see Robin Higham, *Air Power: A Concise History* (New York: St. Martin's, 1972). Early years of army service through World War I air included in Alfred Goldberg et al., *A History of the United States Air Force, 1907–1957* (Princeton: D. Van Nostrand, 1957), and similar coverage for the navy can be found in Archibald D. Turnbull and Clifford L. Lord, *History of United States Naval Aviation* (New Haven: Yale University Press, 1949). Aspects of production and planning are found in the penetrating monograph by Irving Brinton Holley, *Ideas and Weapons: Exploitation of the Aerial Weapon by the United States during World War I* (New Haven: Yale University Press, 1953). Realities of aerial warfare are convincingly detailed by James J. Hudson, *Hostile Skies: A Combat History of the American Air Service* (Syracuse, N.Y.: Syracuse University Press, 1968). A colorful and well-illustrated general survey of American and European highlights is Ezra Bowen, *Knights of the Air* (Alexandria, Va.: Time-Life, 1980). Orville Wright's comment on war appeared in a letter to Dr. Wallace C. Sabine, November 7, 1918, in McFarland, *The Papers of Wilbur and Orville Wright*, 2:1121.

CHAPTER 2. THE AVIATION BUSINESS

The Army's Wings

For an informed review of the evolution of air power, both land-based and ocean-based, see Robin Higham, *Air Power: A Concise History* (New York: St. Martin's, 1972). Irving Brinton Holley, *Ideas and Weapons: Exploitation of the Aerial Weapon by the United States during World War I* (New Haven: Yale University Press, 1953), includes a trenchant analysis of early postwar decisions. Alfred Goldberg et al., *A History of the United States Air Force, 1907–1957* (Princeton: D. Van Nostrand, 1957), is a basic survey with informative comments on the 1920s.

On the career and conflicts of Billy Mitchell, see Alfred F. Hurley, *Billy Mitchell: Crusader for Air Power* (Bloomington: Indiana University Press, 1975). A recent book that emphasizes personalities of the interwar years is Dewitt S. Copp, *A Few Great Captains: The Men and Events that Shaped the Development of U.S. Air Power* (Garden City, N.Y.: Doubleday, 1980). The valuable memoir by Henry H. ("Hap") Arnold, *Global Mission* (New York: Harper, 1949), includes many details of the 1920s.

Air Power at Sea

Although somewhat dated, a basic survey is still the study by Archibald D. Turnbull and Clifford L. Lord, *History of United States Naval Aviation* (New Haven: Yale University Press, 1949). For an excellent photographic coverage, see Martin Caidin, *Golden Wings: A Pictorial History of the United States Navy and Marine Corps in the Air* (New York: Arno, 1974). The role of dirigibles in naval thinking between the wars is incisively interpreted by Richard K. Smith, "The Airship, 1904–1976," in *Two Hundred Years of Flight in America: A Bicentennial Survey*, ed. Eugene M. Emme, pp. 67–89 (San Diego: Univelt, 1977).

Marine aviation is concisely reviewed and illustrated by Edward C. Johnson, *Marine Corps Aviation: The Early Years, 1912–1940* (Washington, D.C.: U.S. Government Printing Office, 1977). Peter Smith's study of the evolution of dive bombing, *The History of Dive Bombing* (Annapolis: Nautical and Aviation, 1982), gives due credit to RAF strikes in World War I. For an outstanding treatment of the political, tactical, and technological evolution of carriers in U.S. naval doctrine, see Charles M. Melhorn, *Two-Block Fox: The Rise of the Aircraft Carrier, 1911–1929* (Annapolis: Naval Institute Press, 1974).

Air Mail and Commercial Airlines

Henry Ladd Smith, *Airways: The History of Commercial Aviation in the United States* (New York: Knopf, 1942), is still an impressive work of careful scholarship. Page Shamburger, *Tracks across the Sky: The Story of the Pioneers of the U.S. Air Mail* (Philadelphia: Lippincott, 1964), is a popularly written study with much useful information. Donald Dale Jackson, *Flying the Mail* (Alexandria, Va.: Time-Life, 1982), includes Europe, but abounds with colorful vignettes and excellent illustrations of the American effort. For details of the socioeconomic impact of air mail in America, see Roger E. Bilstein, "Technology and Commerce: Aviation in the Conduct of American Business, 1918–1929," *Technology and Culture* 10 (July 1969):392–411.

Although opinionated, Benjamin B. Lipsner, *The Airmail: Jennies to Jets* (Chicago: Wilcox & Follett, 1951), represents the recollections of one of the founders of the air mail. Byron Moore's autobiographical reminiscences are titled *The First Five Million Miles* (New York: Harper & Brothers, 1955). Dean Smith's memoirs are, appropriately, *By the Seat of My Pants* (Boston: Little, Brown, 1961). The growth of aviator's unionization is the subject of George E. Hopkins, *The Airline Pilots: A Study in Elite Unionization* (Cambridge, Mass.: Harvard University Press, 1971).

On airlines of the era, see R. E. G. Davies, *Airlines of the United States since 1914* (New York: Rowman, 1972). Frank J. Taylor, *High Horizons: The United Air Lines Story* (New York: McGraw-Hill, 1962), is still a useful book. Oliver Allen, *The Airline Builders* (Alexandria, Va.: Time-Life, 1981), is one of the better volumes of Time-Life's Epic of Flight series. For a contemporary review of airline operations, including Aeromarine's efforts, see Thomas Hart Kennedy, *An Introduction to the Economics of Air Transportation* (New York: Macmillan,

1924). The significance of Aeromarine's example was noted in another pre–World War II textbook, Albert E. Blomquist, *Outline of Air Transport Practice* (New York: Pitman, 1941). Details and statistics of passenger and express activities are from Bilstein, "Technology and Commerce."

The Genesis of General Aviation

For a colorful story of the adventurous barnstormers, see Don Dwiggins, *The Barnstormers: Flying Daredevils of the Roaring Twenties* (New York: Grosset & Dunlap, 1968). General aviation services like those of Bob Johnson are discussed in Marion Templeton Place, *Tall Timber Pilots* (New York: Viking, 1953). The origins of business flying and relevant statistics can be found in Bilstein, "Technology and Commerce," and in Tom D. Crouch, "General Aviation: The Search for a Market, 1910–1976," in Emme, ed., *Two Hundred Years of Flight in America*.

Aircraft in Field Work

Eldon W. Downs and George Lemmer, "Origins of Aerial Crop Dusting," *Agricultural History* 34 (July 1965):123–35, covers the first experiments. Various applications of aerial photography are detailed in Roger E. Bilstein, *Flight Patterns: Trends of Aeronautical Development in the United States, 1918–1929* (Athens: University of Georgia Press, 1983). Aspects of archaeology are covered in Leo Deuel, *Flights into Yesterday: The Story of Aerial Archaeology* (New York: St. Martin's, 1969). On New Guinea, see Richard Archbold, "Unknown New Guinea," *National Geographic*, March 1941, pp. 315–44. For photographs, specifications and performance of typical general aviation aircraft used in business and miscellaneous utilitarian roles during the twenties, see Joseph P. Juptner, *U.S. Civil Aircraft*, 9 vols. (Fallbrook, Calif.: Aero, 1962–82); volumes 1 through 3 cover the 1920s.

Technological Trends

George W. Gray, *Frontiers of Flight: The Story of NACA Research* (New York: Knopf, 1948), is still useful. A more recent survey, with excellent illustrations, is David A. Anderton, *Sixty Years of Aeronautical Research, 1917–1977* (Washington, D.C.: U.S. Government Printing Office, 1977). Technological advances in engines, airframes, and allied equipment are impressively assessed by Ronald Miller and David Sawers, *The Technical Development of Modern Aviation* (New York: Praeger, 1970).

The striking success of the Guggenheim Fund's numerous benefits to aviation is skillfully told by Richard P. Hallion, *Legacy of Flight: The Guggenheim Contribution to American Aviation* (Seattle: University of Washington Press, 1977). The significance of Theodore von Karman's permanent arrival in the United States is the subject of Paul A. Hanle, *Bringing Aerodynamics to America* (Cambridge, Mass.: MIT Press, 1982). Douglas Robinson, *The Dangerous Sky: A History of Aviation Medicine* (Seattle: University of Washington Press, 1973), is an authoritative book written by a doctor who is also a certified flight surgeon. The early association of meteorology and aeronautics is noted in Donald R. Whitnah, *A History of the U.S. Weather Bureau* (Urbana: University of Illinois Press, 1961).

An Aviation Community

Production statistics and financial aspects are drawn from several sources, including Victor Selden Clark, *History of Manufacturers in the United States*, 3 vols. (New York: McGraw-Hill,

1929), 2:338; Welman Austin Shrader, *Fifty Years of Flight: A Chronicle of the Aviation Industry in America, 1903–1953* (Cleveland: Eaton, 1953); U.S. Department of Commerce, Bureau of the Census, *Historical Statistics of the United States, 1789–1945* (Washington, D.C.: U.S. Government Printing Office, 1949), pp. 412, 466; U.S. Bureau of Air Commerce, *Bulletin,* no. 1 (April 1, 1934), p. 9. Airline finance and mergers of the era are traced in H. Smith, *Airways;* a more thorough analysis of corporate maneuvers of both airlines and manufacturers is John B. Rae, *Climb to Greatness: The American Aircraft Industry, 1920–1960* (Cambridge, Mass.: MIT Press, 1968). Geographic factors in industrial location is the subject of William G. Cunningham, *The Aircraft Industry: A Study in Industrial Location* (Los Angeles: Morrison, 1951).

The impact of federal regulation, the improvement of navigational aids, and development of airports is thoroughly chronicled in Nick Komons, *Bonfires to Beacons: Federal Civil Aviation Policy under the Air Commerce Act, 1926–1938* (Washington, D.C.: U.S. Government Printing Office, 1978). For a contemporary view of airline and airport management, see Archibald Black, *Transport Aviation* (New York: Simmons-Boardman, 1926). Contemporary issues of law, insurance, education, and so on are each discussed by the decade's leading experts in International Civil Aviation Conference, vol. 1: *Papers;* vol. 2: *Proceedings* (Washington, D.C.: U.S. Government Printing Office, 1928).

Aviation and American Culture

Aeronautical images in music, books, comics, and international relations are treated in Bilstein, *Flight Patterns.* Charles E. Planck, *Women with Wings* (New York: Harper, 1942), though somewhat dated, still has much useful information. The often reprinted article by John William Ward, "The Meaning of Lindbergh's Flight," *American Quarterly* 10 (Spring 1958):3–16, is still a definitive statement. Among the several biographies of Lindbergh, Kenneth Davis, *The Hero: Charles A. Lindbergh and the American Dream* (Garden City, N.Y.: Doubleday, 1959), offers fascinating insights.

Scholarly notice of aviation as a promoter of international good will includes Robert Ferrell, *Peace in Their Time: The Origins of the Kellogg-Briand Pact* (New Haven: Yale University Press, 1952), and E. W. Burgess, "Communication," in *Social Changes in 1928,* ed. William F. Ogburn (Chicago: University of Chicago Press, 1928).

International Airways

Latin American activities are included in Henry Ladd Smith, *Airways Abroad: The Story of American World Air Routes* (Madison: University of Wisconsin Press, 1950), but the definitive study of U.S. aviation in Latin America is Wesley P. Newton, *The Perilous Sky: U.S. Aviation Diplomacy and Latin America, 1919–1931* (Miami: University of Miami Press, 1978). Matthew Josephson, *Empire of the Air: Juan Trippe and the Struggle for World Airways* (New York: Harcourt, Brace, 1944), is still the starting point for the study of Pan Am, to be supplemented with Robert Daley, *An American Saga: Juan Trippe and His Pan Am Empire* (New York: Random House, 1980).

The aerial geography article by A. P. Berejkoff, "Aerial Map of the World," appeared in *Aviation* 19 (August 24, 1925):208–9. Polar flights are viewed by Richard Montague, *Oceans, Poles, and Airmen* (New York: Random House, 1971), which includes a critical assessment of Byrd's flight to the North Pole in 1926.

CHAPTER 3. ADVENTURE, AIRWAYS, AND INNOVATION

The Adventurers

C. R. Roseberry, *The Challenging Skies: The Colorful Story of Aviation's Most Exciting Years, 1919–1939* (Garden City, N.Y.: Doubleday, 1966), chronicles the repeated assaults on time and distance records up to World War II. Roseberry also covers many milestones in commercial aviation between the wars. A recent survey that includes both European and American efforts in time and distance flights is David Nevin, *The Pathfinders* (Alexandria, Va.: Time-Life, 1980). The careful, professional approach of James H. ("Jimmy") Doolittle is conveyed in Lowell Thomas and Edward Jablonski, *Doolittle: A Biography* (Garden City, N.Y.: Doubleday, 1976).

The New Airliners

The most comprehensive survey of technical advances is Ronald Miller and David Sawers, *The Technical Development of Modern Aviation* (New York: Praeger, 1970). See also Peter W. Brooks, *The Modern Airliner: Its Origins and Development* (London: Putnam, 1961). A contemporary review is Edward P. Warner, *Technical Development and Its Effect on Air Transportation* (York, Pa.: Maple, 1938). Written by a pair of active professionals in their respective fields of engines and fuels, Robert Schlaifer and S. D. Heron, *Development of Aircraft Engines and Fuels* (Boston: Division of Research, Graduate School of Business, Harvard University, 1950), is extensive, detailed, and invaluable.

Harold Mansfield, *Vision: The Story of Boeing* (New York: Duell, Sloan & Pearce, 1966), is a bit too laudatory but very instructive. Frank J. Taylor, *High Horizons: The United Air Lines Story* (New York: McGraw-Hill, 1962), is an uncritical, although useful, study. David W. Lewis and Wesley P. Newton, *Delta: The History of an Airline* (Athens: University of Georgia Press, 1979), is an exemplary study by two professional historians. The DC-3 has been the subject of numerous popular studies; Douglas J. Ingells, *The Plane That Changed the World: A Biography of the DC-3* (Fallbrook, Calif.: Aero, 1966), is detailed and informed. The role of wind tunnel research in the DC-3's evolution is documented by Richard P. Hallion, *Legacy of Flight: The Guggenheim Contribution to American Aviation* (Seattle: University of Washington Press, 1977).

Carl Solberg, *Conquest of the Skies: A History of Commercial Aviation in America* (Boston: Little, Brown, 1979), written with journalistic flair, includes many interesting vignettes. Oliver E. Allen, *The Airline Builders* (Alexandria, Va.: Time-Life, 1981), emphasizing the period 1920–40, includes considerable material on foreign flag lines and Pan Am's international operations, as well as domestic routes. R. E. G. Davies, *A History of the World's Airlines* (New York: Oxford University Press, 1964), is encyclopedic and the best overall history of its type. My quotations from Masland come from one of his letters, reprinted as "The Ships Had Wings," *Boeing Magazine* 34 (December 1964):12–13.

Kenneth Munson, *Airliners Between the Wars, 1919–1939* (New York: Macmillan, 1972), conveniently summarizes the origins and performance of the era's principal transports. A slow but steady steam of Zeppelin books continues to appear, the most recent of which is Douglas Botting, *The Giant Airships* (Alexandria, Va.: Time-Life, 1980). See also Douglas Robinson, *Giants in the Sky: A History of the Rigid Airship* (Seattle: University of Washington Press, 1973).

Airways and Aviators

My account of Roosevelt's flight in 1932 is taken from Arthur M. Schlesinger, Jr., *The Crisis of the Old Order, 1919–1933* (Boston: Houghton Mifflin, 1957), pp. 296–313. The general outline and details of this section are drawn on my unpublished paper, "The Airline Pilot: Changing Profession in a Progressive Industry." My quotes from the Boeing pilot are those of Ray Little, one of the first mail pilots of the early 1920s. His recollections, "The Passing of Pioneer Days," appear in *Aces of the Air*, ed. Joseph L. French, pp. 168–72 (Springfield, Mass.: McLoughlin, 1930). On the CAA, airway modernization, and the Cutting crash, see Donald R. Whitnah, *Safer Skyways: Federal Control of Aviation, 1926–1966* (Ames: Iowa State University Press, 1966), and Nick Komons, *Bonfires to Beacons: Federal Civil Aviation Policy under the Air Commerce Act, 1926–1938* (Washington, D.C.: U.S. Government Printing Office, 1978). The definitive treatment of the Air Line Pilots Association is George E. Hopkins, *The Airline Pilots: A Study in Elite Unionization* (Cambridge, Mass.: Harvard University Press, 1971).

The value of the barnstorming experience and the flexibility required of early fliers is noted in an article, "From Barnstorming to Air Liners," *Popular Mechanics* 52 (July 1929):8–12; and in Irving Crump and Norman Maul, *Our Airliners* (New York: Dodd, Mead, 1942). Statistics and salary figures for pilots can be found in government sources like U.S. Federal Aviation Agency, *Statistical Handbook of Aviation* (Washington, D.C.: U.S. Government Printing Office, 1960), and in contemporary texts like Charles Norcross, *Getting a Job in Aviation* (New York: McGraw-Hill, 1938).

On the evolution of cabin service, see John R. Tunis, *Million-Miler: The Story of an Air Pilot* (New York: Messner, 1942), and Charles E. Planck, *Women with Wings* (New York: Harper, 1942). Although Tunis's book discusses airline operations from the pilot's point of view, few works of this genre equal that of E. K. Gann, *Fate Is the Hunter* (New York: Simon & Schuster, 1961), which strikingly evokes the atmosphere of air transport flying during the 1930s. Bierne Lay summed up the decade's changes and the quiet professionalism of 1940s airline pilots in "The Airman," *Fortune* 23 (March 1941):122–23. Trends in marketing can be found in textbooks of the era, such as John Frederick, *Commercial Air Transportation* (Chicago: Irwin, 1942), and in Paul Peter Willis, *Your Future in the Air* (New York: Prentice-Hall, 1940), which details advertisement campaigns by American Airlines.

The Light-plane Industry

My general remarks on engines of the era come from comments in Devon Francis, *Mr. Piper and His Cubs* (Ames: Iowa State University Press, 1973) and from Peter M. Bowers, *Yesterday's Wings* (Washington, D.C.: Aircraft Owners & Pilots Association, 1974). Prices and ownership are discussed in Walter J. Boyne, "Those Anonymous Cubs," *Aviation Quarterly* 1, no. 4 (Winter 1975):252–80. Hopes for the "$700 airplane" are assessed in Tom D. Crouch, "General Aviation: The Search for a Market, 1910–1976," in *Two Hundred Years of Flight in America: A Bicentennial Survey*, ed. Eugene M. Emme, pp.108–35 (San Diego: Univelt, 1977).

The principal source for material on Piper is Francis, *Mr. Piper*. Subjectively and flamboyantly written, the book nonetheless offers many insights into the financial struggles and corporate problems of this pioneering firm in the general aviation industry. For pertinent quotations from flying reports, selected from contemporary journals of the decade, see Leighton Collins, ed., *Air Facts Reader, 1939–1941* (New York: Air Facts Press, 1974). A reliable source for Cessna is its own detailed chronology, *An Eye to the Sky* (Wichita: Cessna, 1962), replete with performance data, production figures, and snippets of the company's corporate lore. Beechcraft has produced a detailed, heavily illustrated company history in William H. Mc-Daniel, *The History of Beech: Four Decades of Aeronautical and Aerospace Achievement* (Wichita: McCormick-Armstrong, 1976).

Additional sources of information on general aviation aircraft of this period, in addition to Francis, *Mr. Piper*, and Bowers, *Yesterday's Wings*, include Paul Poberezny and S. H. Schmid, compilers, *Wings of Memory* (Hales Corners, Wis.: Experimental Aircraft Association, 1969), a series of brief descriptions of new airplanes as they were introduced in the pages of *Aero Digest* between 1937 and 1939. A valuable and concise summary is Kenneth Munson, *Private Aircraft: Business and General Purpose since 1946* (New York: Macmillan, 1967).

Other Avenues of Research

On Link flight trainers, see Lloyd L. Kelly, as told to Robert B. Parke, *The Pilot Maker* (New York: Grosset & Dunlap, 1970). The background of autogyros and helicopters can be traced in H. F. Gregory, *Anything a Horse Can Do: The Story of the Helicopter* (New York: Reynal & Hitchcock, 1944), and Kenneth Munson, *Helicopters and Other Rotocraft since 1907* (New York: Macmillan, 1969). Stratospheric research is summarized in Richard P. Hallion, *Test Pilots: The Frontiersmen of Flight* (Garden City, N.Y.: Doubleday, 1981), pp. 114–17; and appears in many previously cited studies, such as Alfred Goldberg et al., *A History of the United States Air Force, 1907–1957* (Princeton, N.J.: D. Van Nostrand, 1957).

For an outstanding survey of space explorations, see Wernher von Braun and Frederick I. Ordway III, *History of Rocketry and Space Travel* (New York: Crowell, 1975), an effectively illustrated book with a strong bibliography. On the career of Robert Goddard, see Esther C. Goddard and G. Edward Pendray, eds., *The Papers of Robert H. Goddard*, 3 vols. (New York: McGraw-Hill, 1970). See also Milton Lehman, *This High Man* (New York: Farrar, Straus, 1963). The German background is covered in detail by Frederick I. Ordway III and Mitchell Sharpe, *The Rocket Team* (New York: Crowell, 1979).

CHAPTER 4. AIR POWER AT WAR

The body of literature on World War II, already monumental, continues to grow, and the roster of books on the air war follows the trend. Specialized publications on combat theaters, particular units, or specific aircraft types—the special province of popular, or "buff," literature—frequently appear to have the redundancy of clones. At the same time, many of these narrowly focused studies provide useful details and oddments of information. More often than in other chapters, the popular titles included below are the result of personal choice from the vast field of air war publications that are available.

My interpretations and conclusions regarding the air war have strongly been influenced by three studies. Robin Higham, *Air Power: A Concise History* (New York: St. Martin's, 1972) represents the author's many years as a military and aeronautical scholar. R. J. Overy, *The Air War, 1939–1945* (New York: Stein & Day, 1981), is an absorbing analysis by a British military historian. Kent Roberts Greenfield, *American Strategy in World War II: A Reconsideration* (Baltimore: Johns Hopkins Press, 1963), draws from the author's experience as Chief Historian, Department of the Army (1946–58) and includes a stimulating essay, "Air Power and Strategy."

Military Aviation between the Wars

The NACA's research projects of the late 1930s and the war years are summarized in David A. Anderton, *Sixty Years of Aeronautical Research, 1917–1977* (Washington, D.C.: U.S. Government Printing Office, 1977). Alex Roland, of Duke University, is preparing an administrative history of the NACA, 1915–58; my comments on the NACA in the 1930s are drawn from his unpublished but revealing essay "Tunnel Vision" (1977), in the files of the NASA History Office.

Army Air Corps doctrine in the 1930s is discussed in Alfred Goldberg et al., *A History of the United States Air Force, 1907–1957* (Princeton, N.J.: D. Van Nostrand, 1957), and my quotes from FDR and the comments on overemphasis for bombers are from p. 44. Henry H. ("Hap") Arnold, *Global Mission* (New York: Harper, 1949), has the authenticity of a first-hand observer. Air Corps efforts to carry the mail in 1934 are assessed in Arthur M. Schlesinger, Jr., *The Coming of the New Deal* (Boston: Houghton Mifflin, 1958); this book is the source of the Rickenbacker quote about "legalized murder." See also Richard Patterson, "The Army Flies the U.S. Air Mail," *Aviation Quarterly* 2, no. 4 (1976):454–63. David Nevin, *Architects of Air Power* (Alexandria, Va.: Time-Life, 1981), describes the interwar years through the Spanish Civil War of 1939. Two older books that remain standard sources on naval aviation in the 1930s and on Pearl Harbor, respectively, are Archibald D. Turnbull and Clifford L. Lord, *History of United States Naval Aviation* (New Haven: Yale University Press, 1949), and Samuel Eliot Morison, *The Rising Sun in the Pacific* (Boston: Little, Brown, 1948). Turnbull and Lord, *History*, p. 317, are the source of the quote about the sergeant's guard. Details of the era's military aircraft come from Kenneth Munson, *Fighters between the Wars, 1919–39* (New York: Macmillan, 1970) and his *Bombers between the Wars, 1919–39* (New York: Macmillan, 1970).

Air War over Europe

As a general assessment of the background and conduct of World War II, see Peter Calvocoressi and Guy Wint, *Total War* (New York: Penguin, 1979). Samuel Eliot Morison, *Strategy and Compromise: A Reappraisal of the Crucial Decisions Confronting the Allies, 1940–1945* (Boston: Little, Brown, 1958) is a model of elegant brevity and clarity that sets out the rationale that influenced strategy in the principal theaters of the war.

Overy, *The Air War*, not only covers Allied, German, and Japanese theory and strategy but includes economics and innovation. Richard Humble, *The War in the Air, 1939–1945* (London: Salamander, 1975), is a concise, popularly written, and colorfully illustrated coverage of major aerial engagements. David A. Anderton, *The History of the U.S. Air Force* (New York: Crescent, 1981), a lavishly illustrated work, covers events from the Civil War through Vietnam; about one-third is devoted to World War II. Edward Jablonski, *America in the Air War* (Alexandria, Va.: Time-Life, 1982), is a dramatic narrative, with special attention given to personalities and combat stories. It is an effective work, backed by the author's many prior aviation books, including his four-volume *Airwar* (Garden City, N.Y.: Doubleday, 1971).

The official history of the U.S. Army effort is Wesley F. Craven and James L. Cate, eds., *The Army Air Forces in World War II*, 7. vols. (Chicago: University of Chicago Press, 1948–50). In its coverage of World War II, Goldberg, *History*, represents a synthesis of this work.

Arnold, *Global Mission*, includes valuable background on the problems of gathering forces and mounting combat missions in all theaters. See also Lewis H. Brereton, *The Brereton Diaries: The War in the Pacific, Middle East, and Europe* (New York: Morrow, 1946), by a top-ranking officer who eventually commanded the Ninth Air Force in Europe. For more detailed reviews of two major air raids, see James Dugan and Carroll Stewart, *Ploesti: The Great Ground-Air Battle of August 1943* (New York: Random House, 1962), and Martin Caidin, *Black Thursday* (New York: Dutton, 1960), which chronicles the Schweinfurt raid. Caidin also includes informative discussion of bomber tactics like the "combat boxes," as well as the ploys of German fighter pilots. Alan Osur, *Blacks in the Army Air Forces during World War II* (Washington: U.S. Government Printing Office, 1977), is a story of prejudice as well as valor.

A fascinating view of the war from the German perspective is Adolph Galland, *The First and the Last* (New York: Holt, 1954). Galland, chief of the German fighter squadrons, offers a number of insights into the collapse of Germany's air defenses, including the failure of the ME-262. Likewise, the overall British perspective is well served by J. E. Johnson, the Allies' top fighter ace with 38 victories and later Air Vice-Marshall, *Full Circle* (New York: Bantam, 1980).

For specifics and informed comments on aircraft, general works include John W. R. Taylor, *Combat Aircraft of the World* (New York: Putnam's, 1969), and Ray Wagner, *American Combat Planes*, 3d ed. (Garden City, N.Y.: Doubleday, 1982). Details of the principal military aircraft of 1939–45 are conveniently reviewed in three guides by the well-known British expert Bill Gunston: *Allied Fighters of World War II; German, Italian, and Japanese Fighters of World War II;* and *Bombers of World War II* (New York: Arco, 1981). Each is heavily illustrated in both color and black and white and includes three-view drawings of each plane. For a sample of popularly written literature on one of the outstanding fighters of the European Theater, see Jeffrey Ethell, *Mustang: A Documentary History* (New York: Jane's, 1981), thoroughly illustrated with a carefully written narrative.

Air War in the Pacific

The Navy's official chronicle is Samuel Eliot Morison, *History of United States Naval Operations in World War II*, 15 vols. (Boston: Little, Brown, 1948–62). A one-volume synthesis is Morison's, *The Two-Ocean War* (Boston: Little, Brown, 1963). Robert Sherrod, *History of United States Marine Corps Aviation in World War Two* (Washington, D.C.: Combat Forces, 1952), is a thorough study. An authoritative account of carrier warfare is Clarke G. Reynolds, *The Fast Carriers: The Forging of an Air Navy* (New York: McGraw-Hill, 1968). Reynolds also authored *The Carrier War* (Alexandria, Va.: Time-Life, 1982). Norman Polmar, *Aircraft Carriers: A Graphic History of Carrier Aviation and Its Influence on World Events* (Garden City, N.Y.: Doubleday, 1969), is a detailed and valuable reference. Among the many histories of major aerial engagements in the Pacific, see the dramatic story of Midway by Walter Lord, *Incredible Victory* (New York: Harper & Row, 1957).

The story of Army aviation in the Pacific is covered in the official and general histories of AAF cited above. A memoir of particular interest to Pacific operations is Curtis E. LeMay, with MacKinlay Kantor, *My Story: Mission with LeMay* (Garden City, N.Y.: Doubleday, 1965), which covers B-29 operations over the Japanese home islands. A skillful assessment of air combat from the Japanese perspective, including comments from many Japanese fighter pilots is Martin Caidin, *Zero Fighter* (New York: Ballantine, 1969). Background and data about aircraft flown in the Pacific Theater are included in the surveys by Taylor, *Combat Aircraft*, Wagner, *American Combat Planes*, and Gunston. For details of an outstanding fighter of the Pacific war, see Richard Abrams, *F4U Corsair at War* (New York: Scribner's, 1981), a popular review with many illustrations.

Components for a Juggernaut

For a convenient summary of the Battle of the Atlantic, see the synthesis by Morison, *The Two-Ocean War*. The remarkable story of U.S. aircraft production is the subject of the definitive study by Irving Brinton Holley, *Buying Aircraft: Materiel Procurement for the Army Air Forces* (Washington, D.C.: U.S. Government Printing Office, 1964). The equally remarkable accomplishments of the Air Transport Command on a global scale are celebrated in the lively account by Oliver LaFarge, *The Eagle in the Egg* (Boston: Houghton, 1949). The role of light aircraft in liaison and ancillary roles, and the impact of light aircraft in flight instruction can be followed in Devon Francis, *Mr. Piper and His Cubs* (Ames: Iowa State University Press, 1973), and Patricia Strickland, *The Putt-Putt Air Force: The Story of the Civilian Pilot Training Program and the War Training Service* (Washington, D.C.: U.S. Federal Aviation Administration, 1970). The contributions of women as flight instructors and as indefatigable ferry pilots is the subject of Sally Van Wagenen Keil, *Those Wonderful Women in Their Flying Machines* (New York: Rawson Wade, 1979). The standard reference on American science during the war, including innovations relating to aviation, is James Phinney Baxter III, *Scientists against Time* (Boston: Little, Brown, 1946). A sufficient summary of guided weapons is Roger Beaumont, "Rapiers versus Clubs: The Fitful History of Smart Bombs," *Journal of the Royal United Services Institute for Defence Studies* 126 (September 1981):45–50. Von Braun's own recollections of German rocket development and the search for Allied contacts are noted in Wernher von Braun and Frederick I. Ordway III, *History of Rocketry and Space Travel* (New York: Crowell, 1975). The American scheme to round up leading German scientists, including aeronautical and astronautical experts, is the subject of Clarence Lasby, *Operation Paperclip* (New York: Antheneum, 1971).

CHAPTER 5. AIR-AGE REALITIES

New Air-Age Concepts

During the early years of World War II, many groups had already begun to consider the global role of aviation. Representative of the trend was the Aviation Education Research Group of Teachers College, Columbia University. A number of the group's personnel prepared titles in a fascinating series of compact little books sponsored by the Institute of the Aeronautical Sciences, Air-Age Education Series, 20 vols. (New York: Macmillan, 1942–43). Topics ran the gamut from pure pedagogy, industrial arts, and biology to concise anthologies. Hall Bartlett prepared *Social Studies for the Air Age* (1942) and George T. Renner wrote *Human Geography in the Air Age* (1943). J. Parker Van Zandt, World War I flier and aviation policy consultant, wrote an informative contemporary treatment of the air age, the two-volume *America Faces the Air Age:* vol. 1, *The Geography of World Air Transport* (Washington, D.C.: Brookings Institution, 1944), and vol. 2, *Civil Aviation and Peace* (Washington, D.C.: Brookings Institution, 1944). These books as well as the Air-Age Education Series include fascinating bibliographies. A particularly interesting assessment is that of the administrator of the CAA, Theodore P. Wright, "Aviation's Place in Civilization," taken from the 33d Wright Memorial Lecture in London in 1945, and issued as a U.S. government document (copy in the NASM Archives). Although William F. Ogburn wrote *The Social Effects of Aviation* (New York: Houghton Mifflin, 1946), as an exercise in forecasting, the book contains much contemporary information of considerable interest.

Scheduled Air Transport

For a balanced and informed treatment of postwar air route politics, the origins of the ICAO, and related events, see Henry Ladd Smith, *Airways Abroad: The Story of American World Air Routes* (Madison: University of Wisconsin Press, 1950); the quote from Claire Boothe Luce is cited there, p. 129. Kenneth Munson, *Airliners since 1946* (New York: Macmillan, 1975), contains a wealth of pertinent information on selected aircraft. Technological progress and innovations are covered in Ronald Miller and David Sawers, *The Technical Development of Modern Aviation* (New York: Praeger, 1970), and Peter W. Brooks, *The Modern Airliner: Its Origins and Development* (London: Putnam, 1961). A detailed account of one of the era's standard airliners is M. J. Hardy, *The Lockheed Constellation* (New York: Arco, 1973). Changing patterns in passenger amenities, airports, and air travel generally are discussed in Kenneth Hudson, *Air Travel: A Social History* (Totowa, N.J.: Rowman & Littlefield, 1972), the source of my quote about stewards (p. 119). Martin Greif, *The Airport Book* (New York: Mayflower, 1979), is heavy on illustrations. For trends in government policy, see Stuart Rochester, *Takeoff at Mid-Century: Federal Civil Aviation Policy in the Eisenhower Years, 1953–1961* (Washington, D.C.: U.S. Government Printing Office, 1977). The evolution of presidential aircraft is described in J. F. terHorst and Ralph D. Albertazzie, *The Flying White House: The Story of Air Force One* (New York: Coward, McCann & Geohegan), 1979. Mr. terHorst has been a long-time member of the White House press corps; Albertazzie, an air force colonel, was the pilot of *Air Force One* during the Nixon years.

Military Aeronautics

For a fascinating memoir on the origins of the turbojet, see Sir Frank Whittle, *Jet: The Story of a Pioneer* (New York: Philosophical Library, 1954). Brief reminiscences by Whittle, Hans von Ohain, and others are included in the useful survey by Walter J. Boyne and Donald S. Lopez, eds., *The Jet Age: Forty Years of Jet Aviation* (Washington, D.C.: Smithsonian Institution Press, 1979). Edward W. Constant II, *The Origins of the Turbojet Revolution* (Baltimore: Johns Hopkins University Press, 1980), is a scholarly treatment with informative analysis and interpretation. See also Charles D. Bright, *The Jet Makers: The Aerospace Industry from 1945 to 1972* (Lawrence: Regents Press of Kansas, 1978).

The most thorough roster of postwar military jets, as well as piston-engine aircraft like the B-36, is Ray Wagner, *American Combat Planes*, 3d ed. (Garden City, N.Y.: Doubleday, 1982). Bill Gunston, *Bombers of the West* (New York: Scribner's, 1973), presents informed discussions of jet bombers of both the United States and Britain, noting how they evolved and became modified against the shifting background of the Cold War's economic and geopolitical influences. Richard P. Hallion, *Test Pilots: The Frontiersman of Flight* (Garden City, N.Y.: Doubleday, 1981) analyzes the new generation of high-speed aircraft, like the F-86, along with the aerodynamic challenges they presented. On the X-1 and other supersonic projects, see Richard P. Hallion, *Supersonic Flight: Breaking the Sound Barrier and Beyond* (New York: Macmillan, 1972).

Cold War and Hot War

Alfred Goldberg et al., *A History of the United States Air Force, 1907–1957* (Princeton, N.J.: D. Van Nostrand, 1957), with its strong emphasis on postwar organization and policy since 1945, is still an excellent source. Contemporary air policy ideas are expressed in Thomas K. Finletter, *Survival in the Air Age* (Washington, D.C.: President's Air Policy Commission,

1948). For a sample of contemporary opinions by defense analysts, Congress, the air force, and others, see Eugene M. Emme, ed., *The Impact of Air Power: National Security and World Politics* (New York: D. Van Nostrand, 1959). Robert F. Futrell, *Ideas, Concepts, Doctrine: A History of Basic Thinking in the United States Air Force, 1907–1964*, 2 vols. (Maxwell AFB, Ala.: Air University, 1971), is an informed assessment of military thought on the aerial weapon. Futrell's monograph, *The United States Air Force in Korea, 1950–1953* (New York: Duell, 1961), covers combat and weapons, as well as strategy and foreign relations. For photographic impressions of both air force and naval forces in combat, see David A. Anderton, *The History of the U.S. Air Force* (New York: Crescent Books, 1981), and Martin Caidin, *Golden Wings: A Pictorial History of the United States and Marine Corps in the Air* (New York: Random House/Arno, 1974).

Hurricanes, Blizzards, and Blockades

Information on hurricanes has been culled from U.S. Air Force, Office of Public Affairs, *A Chronology of American Aerospace Events* (Washington, D.C.: Office of Information Services, July 1, 1959), and from Clark Van Fleet et al., *U.S. Naval Aviation, 1910–1970* (Washington, D.C.: Naval Air Systems Command, 1970). See also Clayton Knight, *Lifeline in the Sky* (New York: Morrow, 1957), which includes information on weather reconnaissance squadrons. Information on the "Blizzard of '49" is based on the author's recollections and on scrapbooks in the Grant County (Nebraska) Historical Museum. Walter P. Davidson, *The Berlin Blockade: A Study in Cold War Politics* (Princeton: Princeton University Press, 1958), is an authoritative analysis of a critical Cold War confrontation. See also Knight, *Lifeline*. I have also drawn information from a useful collection of contemporary materials in the NASM Archives, "Miscellaneous Files: Berlin Airlift."

The General Aviation Sector

Miscellaneous statistics are summarized by Tom D. Crouch, "General Aviation: The Search for a Market, 1910–1976," in *Two Hundred Years of Flight in America: A Bicentennial Survey*, ed. Eugene Emme (San Diego: Univelt, 1977). A witty, authoritative summary of the early postwar era is Frank Kingston Smith, "The Turbulent Decade," in *Flying: Fiftieth Anniversary Issue*, which is the source of the quotes on flying in Texas and on airplanes for Everyman. The contemporary article, "New Planes for Popular Flying" appeared in *Fortune* 33 (February 1946):124–29. Developments affecting Piper, Cessna, and Beech, respectively, are covered in Devon Francis, *Mr. Piper and His Cubs* (Ames: Iowa State University Press, 1973), William H. McDaniel, *The History of Beech: Four Decades of Aeronautical and Aerospace Achievement* (Wichita: McCormick-Armstrong, 1976), and Cessna Corporation, *An Eye to the Sky* (Wichita: Cessna, 1962). Kenneth Munson, *Private Aircraft: Business and General Purpose since 1946* (New York: Macmillan, 1976), includes informative commentary along with performance and specifications. Larry A. Ball, *Those Imcomparable Bonanzas* (Wichita: McCormick-Armstrong, 1971) stresses technical data but includes many useful details. See also Ogburn, *Social Effects*, which enumerates many typical general aviation uses.

CHAPTER 6. HIGHER HORIZONS

Missiles and Rockets

Because both authors were associated with space activities of the period, the book by Wernher von Braun and Frederick I. Ordway III, *History of Rocketry and Space Travel* (New

York: Crowell, 1975), is still an authentic source. A recent, comprehensive study is David Baker, *The Rocket: The History and Development of Rocket and Missile Technology* (New York: Crown, 1978). Instructive essays on the Polaris, Thor, Atlas, Redstone, and other military missiles are included in Eugene M. Emme, ed., *The History of Rocket Technology: Essays on Research, Development, and Utility* (Detroit: Wayne State University Press, 1964). See also Ernest G. Schwiebert, ed., *A History of the U.S. Air Force Ballistic Missiles* (New York: Praeger, 1965). Indicative of the struggles for major roles in the new arena of military rocketry is Michael H. Armacost, *The Politics of Weapons Innovation: The Thor-Jupiter Controversy* (New York: Columbia University Press, 1969). Other details of the period are summarized in Roger E. Bilstein, *Stages to Saturn: A Technological History of the Apollo/Saturn Launch Vehicles* (Washington, D.C.: U.S. Government Printing Office, 1981).

Space Research

For the trials and tribulations of Vanguard and the impact of Sputnik. see Constance Green and Milton Lomask, *Vanguard: A History* (Washington, D.C.: Smithsonian Institution Press, 1971). Controversy regarding and reorganization of America's space program in the wake of Sputnik is covered in Robert L. Rosholt, *An Administrative History of NASA, 1958–1963* (Washington, D.C.: U.S. Government Printing Office, 1966). Discussion of relevant rocket probes and space science is included in Cargill Hall, *Lunar Impact: A History of Project Ranger* (Washington, D.C.: U.S. Government Printing Office, 1977). The growing number of un-manned space projects from the United States and several other countries are catalogued in Kenneth Gatland, *Missiles and Rockets* (New York: Macmillan, 1975). Projects of particular significance to NASA are gracefully summarized in Frank Anderson, Jr., *Orders of Magnitude: A History of NACA and NASA, 1915–1980* (Washington, D.C.: U.S. Government Printing Office, 1981). On the X-15 series of research aircraft, see Richard P. Hallion, *Test Pilots: The Frontiersmen of Flight* (Garden City, N.Y.: Doubleday, 1981). John M. Logsdon, *The Decision to Go to the Moon: Project Apollo and the National Interest* (Cambridge, Mass.: MIT Press, 1970), discusses political events leading to the decision for manned lunar missions, and Loyd S. Swenson, Jr., James M. Grimwood, and Charles C. Alexander, *This New Ocean: A History of Project Mercury* (Washington, D.C.: U.S. Government Printing Office, 1966) details both policy and hardware development. The quotes from the *New York Times* and from Wernher von Braun are cited in Leonard C. Bruno, *We Have a Sporting Chance: The Decision to Go to the Moon* (Washington, D.C.: Library of Congress, 1979), a catalog prepared for a Library of Congress exhibit honoring the 10th anniversary of the manned lunar landing.

Military Aviation

See the sources previously cited for the section on "Military Aeronautics" for chapter 5. Gunston, *Bombers*, discusses the B-52. The air force's high-altitude balloon flights and the sled tests are assessed in Hallion, *Test Pilots*. Many publications of the period were fascinated by the air force's shiny, scientific-technological aura and global presence. See, for example, Beverly M. Bowie, "MATS: America's Long Arm of the Air," *National Geographic* 112 (March 1959), and a special double section, "United States Air Force," *National Geographic* 128 (September 1965).

The Airlines

On the evolution of jets and turbojets, see the sources cited in the section on "Scheduled Air Transport," chapter 5. Robert J. Serling, *The Jet Age* (Alexandria, Va.: Time-Life, 1982), is a well-written and well-illustrated treatment of a historic era of transition into the jet age. Martin Caidin, *Boeing 707* (New York: Ballantine, 1959), catches the contemporary excitement and awe of the new jet-powered giants, commenting on the startling speed of jet travel, maintenance procedures, the role of stewardesses, etc. In addition to Carl Solberg, *Conquest of the Skies: A History of Commercial Aviation in America* (Boston: Little, Brown, 1979), and Davies, *Airlines of the United States,* in particular, see W. David Lewis and Wesley P. Newton, *Delta: The History of an Airline* (Athens: University of Georgia Press, 1979), an outstanding scholarly study of a major air carrier. Increasing awareness of the U.S. aircraft manufacturing sector and its growing influence, both domestically and internationally, is inherent in the series of annual reports, *Aerospace Facts and Figures,* (Washington, D.C.: Aerospace Industries Association of America, 1955–82). The same association also publishes a quarterly digest of events, *Aerospace,* with an informative narrative review and forecast at the end of each year.

New Styles in Postwar Air Travel

In addition to my own recollection of the state of air travel during this period, I have relied on the sources above, especially the books by Solberg, *Conquest,* and R. E. G. Davies, *Airlines of the United States since 1914* (New York: Rowman, 1972). Solberg, formerly a senior editor at *Time,* recounts the Marquis Childs anecdote on the effect of intercontinental flying during Dulles's travels as secretary of State. Daniel J. Boorstin, *The Americans: The Democratic Experience* (New York: Random House, 1973), includes commentary on the phenomenon of migratory teams in professional sports, as well as the effect of air transportation on business and the democratization of vacation flights. I also drew heavily on a series of articles that appeared in a special edition of the *Saturday Review* (April 16, 1977), which honored the 50th anniversary of Lindbergh's trans-Atlantic flight. Particularly useful were: Caskie Stinett, "How Mombasa Became the New Place," pp. 11–14; Wayne Parrish, "A Ticket to Anywhere," pp. 16–17 and Irving Kolodin, "Headphone Hunting at 30,000 Feet," pp. 20–22.

General Aviation

See the sources previously cited for the section on "The General Aviation Sector" for chapter 5. The significance of Piper's introduction of agricultural aircraft like the Pawnee is discussed in Devon Francis, *Mr. Piper and His Cubs* (Ames: Iowa State University Press, 1973). A useful summary of William Lear and his speedy jet is Robert J. Serling, *Little Giant: The Story of Gates Learjet* (Potomac, Md.. R. J. Serling, 1974). During 1978, when I helped to coordinate an oral history program at the National Air and Space Museum, I had the opportunity to record interviews with 15 senior executives from Beech, Cessna, and Learjet; these discussions helped clarify many trends in the industry. The tapes are filed in the NASM Archives.

CHAPTER 7. FROM THE EARTH TO THE MOON

Military Aviation

For the background of MAC, see the sources cited in chapter 5. Pertinent background and statistics for principal American transports, as well as those of several foreign nations, can be found in Kenneth Munson, *Bombers in Service: Patrol and Transport Aircraft since 1960* (New York: Macmillan, 1975).

On Vietnam, Robin Higham's summary in *Air Power: A Concise History* (New York: St. Martin's, 1972), represents a careful and thoughtful analysis. David A. Anderton's chapters on

Vietnam in *The History of the U.S. Air Force* (New York: Crescent, 1981), though weighted toward operational analysis, also record the dilemma of escalating costs. Carl Berger, ed., *The United States Air Force in Southeast Asia, 1961–1973* (Washington, D.C.: U.S. Government Printing Office, 1977) stresses combat narratives and features numerous illustrations. A semiofficial survey, based on U.S. Air Force reports, was released as *Air War: Vietnam*, then reprinted and commercially distributed (New York: Arno, 1978), covers both air force and naval operations; despite its bureaucratic, chauvinistic tone it provides an interesting background. The Office of Air Force History is currently undertaking a multivolume series of scholarly quality. See, for example, Robert F. Futrell and Martin Blumenson, *The United States Air Force in Southeast Asia: The Advisory Years to 1965* (Washington, D.C.: U.S. Government Printing Office, 1981).

For a sample of profusely illustrated and highly detailed narratives representing the genre favored by air power enthusiasts, see J. C. Scutts, *F-105 Thunderchief* (New York: Scribner's, 1981). A similar study by Jeffrey Ethell and Joe Christy, *B-52 Strafortress* (New York: Scribner's, 1981), is more judicious and analytical. Both place this well-known plane of the Vietnam war in historical perspective and trace its prewar and postwar deployment. Bernard C. Nalty, George M. Watson, and Jacob Neufeld, *The Air War over Vietnam: Aircraft of the Southeast Asia Conflict* (New York: Arco, 1981) is a useful handbook that encompasses vintage retreads like Douglas A-26 Invaders, DC-3 transports converted to gunships, helicopters, North Vietnamese fighters, and others. The book depicts the manner in which the skies over Vietnam became a deadly proving area for modified and updated weaponry.

Civil Aviation

For cumulative financial data and production statistics, see the elaborate series of tables in Aerospace Industries Association of America, *Aerospace Facts and Figures, 1977–78* (Washington, D.C.: AIAA, 1978). Remarks on the general aviation sector, including the range of price tags, have been synthesized from relevant editions of *Flying Annual and Pilots' Buying Guide*, a yearly newsstand compendium issued by Ziff-Davis, publishers of the popular magazine *Flying*.

The striking technological improvements in jet airliners are fully assessed in Ronald Miller and David Sawers, *The Technical Development of Modern Aviation* (New York: Praeger, 1970), and other sources cited in chapter 6. Appearance of the first jumbo jet is thoroughly covered by Douglas J. Ingells, *747: Story of the Boeing Super Jet* (Fallbrook, Calif.: Aero, 1970). In a similar vein, see the author's *L-1011 TriStar and the Lockheed Story* (Fallbrook, Calif.: Aero, 1973). Both of Ingells's books cover the origins and prior aircraft programs of the companies. The SST competition is summarized in Robert J. Serling, *The Jet Age* (Alexandria, Va.: Time-Life, 1982), although most of my remarks are based on editorial comment and news stories from my own files of clippings of the period, and on Mary E. Ames, *Outcome Uncertain: Science and the Political Process* (Washington, D.C.: Communications, 1978), which includes an impressive chapter, "The Case of the U.S. SST: Disenchantment with Technology" (pp. 49–82). For a useful reference on contemporary aircraft, as well as early postwar types still in service somewhere in the world, see David Mondey, comp., *Encyclopedia of the World's Commercial and Private Aircraft* (New York: Crescent, 1981).

Space Exploration: Apollo-Saturn

In addition to books cited in chapter 6, see especially Courtney Brooks, James M. Grimwood, and Loyd S. Swenson, Jr., *Chariots for Apollo: A History of Manned Lunar Spacecraft*

(Washington, D.C.: U.S. Government Printing Office, 1979); Roger E. Bilstein, *Stages to Saturn: A Technological History of the Apollo/Saturn Launch Vehicles* (Washington, D.C.: U.S. Government Printing Office, 1981); and Edward C. Ezell and Linda N. Ezell, *The Partnership: A History of the Apollo-Soyuz Test Project* (Washington, D.C.: U.S. Government Printing Office, 1978). The detailed, official NASA history of Skylab is in progress; in the meantime, consult Frank W. Anderson, Jr., *Orders of Magnitude: A History of NACA and NASA, 1915–1980* (Washington, D.C.: U.S. Government Printing Office, 1981).

The review of the *Saturn V* is essentially that which appeared as my essay, "The Saturn Launch Vehicle Family," in *Apollo: Ten Years since Tranquillity Base*, eds. Richard P. Hallion and Tom D. Crouch (Washington, D.C.: Smithsonian Institution Press, 1979). This book contains other useful essays that place the technological aspects of the Apollo program in historical perspective. Edgar M. Cortright, ed., *Apollo Expeditions to the Moon* (Washington, D.C.: U.S. Government Printing Office, 1975) is a particularly interesting volume, since the various sections on astronaut preparations, launch facilities, management, and other topics were contributed by contemporary executives and participants.

Giant Leap: Manned Missions

Norman Mailer, *Of a Fire on the Moon* (Boston: Little, Brown, 1969) is an intensely personal but compelling view of the Apollo program and the *Apollo 11* mission. Richard S. Lewis, *The Voyages of Apollo: The Exploration of the Moon* (New York: Quadrangle, 1974) represents a knowledgeable assessment by a highly skilled scientific journalist. Tom Wolfe, *The Right Stuff* (New York: Farrar, Straus & Giroux, 1979), is full of dazzling prose and vignettes that focus on the lives of the test pilots who became astronauts. It should be balanced with Michael Collins, *Carrying the Fire: An Astronaut's Journeys* (New York: Farrar, Straus & Giroux, 1974), a memoir that is outstanding for its lucidity, candor, and modesty.

CHAPTER 8. AEROSPACE PERSPECTIVES

Civil Aviation

In honor of the 75th anniversary of powered flight, the Air Transport Association issued an informative brochure, written by Robert J. Serling, *Wrights to Wide-Bodies: The First Seventy-five Years* (1978), full of useful comparisons and data. The ATA also publishes an annual statistical summary, *Air Transport,* which tabulates a considerable variety of data on finances, fleet sizes, and highlights of the past decade. The Aerospace Industries Association of America (AIAA) publishes its own statistical summaries, cited in chapter 7, in addition to annual highlights in its *Aerospace Review and Forecast, 1981–82*, which was a source for several assessments in this chapter.

Aerospace Industry Trends

Many of my comments on aerospace trends are drawn from industry publications and trade magazines used to organize an annual essay, "Aerospace" which I prepared for *Collier's Year Book* during the years 1978–83 (New York: Macmillan, 1978–84). Journals such as *Aviation Week and Space Technology, Interavia,* and *Air International,* among others, have been especially useful. David Mondey, comp., *Encyclopedia of the World's Commercial and Private Aircraft* (New York: Crescent, 1981), is full of details and informative commentary.

All of the sources cited above include relevant information about general aviation. The General Aviation Manufacturers Association issues a well-illustrated and instructive brochure, *The General Aviation Story* (Washington, D.C.: GAMA, 1980), an updated and revised version of prior titles. A particularly valuable report is Frank Kingston Smith, "An Appreciation of the Social, Economic, and Political Issues of General Aviation" (Washington, D.C., 1977), available from the Federal Aviation Agency, Office of General Aviation. For examples of the business community's growing interest in the potential of executive aircraft, see "Corporate Flying: Changing the Way Companies Do Business," *Business Week,* February 6, 1978, pp. 62–67, and Thomas J. Murray, "Corporate Flying Boom," *Dun's Review,* February 1981. My comments on aerial ambulance service and police operations in Houston are based on personal interviews and locally supplied information.

Aerospace Weapons

The annual industry reviews and periodicals used to prepare my essays in *Collier's Year Book,* cited above, also furnished the basis for this section. Many of the books cited in chapter 6 are also relevant. For reliable and convenient surveys of international military aircraft, see Mark Hewish et al., *Air Forces of the World* (New York: Simon & Schuster, 1979), and also William Green and Gordon Swanborough, *Observer's Directory of Military Aircraft* (New York: Arco, 1982). I have also relied on a yearly compendium published by *Air Force Magazine,* the Annual Air Force Almanac issue (1975–83).

Folklore, Fantasy, and Artifacts

This sampling of examples from literature, film, and other media is obviously a personal one. Many of my ideas have been shaped by earlier research done for my book *Flight Patterns,* and by exchanges of aeronautical trivia with colleagues at the National Air and Space Museum. In terms of scholarly assessment of aerospace themes, little seems to have been done in this field, although Dominick A. Pisano of the National Air and Space Museum is at work on a comprehensive manuscript that promises to be a valuable study.

"Darius Green and His Flying Machine" has appeared in many anthologies, such as Hazel Felaman, *Poems that Live Forever* (New York: Macmillan, 1965). Sinclair Lewis's career is thoroughly analyzed in Mark Schorer, *Sinclair Lewis: An American Life* (New York: McGraw-Hill, 1961). The Book-of-the-Month-Club titles on various aspects of aviation and space have been drawn from an anniversary brochure issued by the BOMC in 1976, in which all its selections to date are listed. Among anthologies of poetry and literature having flight as the central theme, see Selden Rodman, ed., *The Poetry of Flight: An Anthology* (New York: Duell, Sloan & Pearce, 1941); Joseph B. Roberts and Paul Briand, ed., *The Sound of Wings: Readings for the Air Age* (New York: Holt, 1957).

On music, see John Vinton, ed., *Dictionary of Contemporary Music* (New York: Dutton, 1974), although most of my comments are based on a review of materials in the Bella C. Landauer Collection of Aeronautical Music, a remarkable collection housed in the National Air and Space Museum. The collection includes scores from Antheil and Blitzstein, as well as programs and news clippings. The material on comic strips is largely based on a lifetime's avid reading. There is a helpful sampling in an article by James Silke, "Aviation Comic Strips," *Air Progress/Aviation Review,* August 1980, pp. 18–23, 68. The National Air and Space Museum also has some samples of cartoon strips in its archival files, where I found a particularly informative article by Louis Kraar, "Funny-Paper Lobbyists Help Air Force, Navy Seek Funds, Recruits," which appeared, of all places, in the *Wall Street Journal,* August 4, 1960.

Many of my statements about motion pictures and television, like those about comics, are based on personal observation. The National Air and Space Museum possesses an impressive collection of feature films, and has shown many older ones as part of museum programs and exhibits. Among many useful references, I most frequently turned to Leslie Halliwell, *The Filmgoer's Companion* (New York: Hill & Wang, 1977); also, K. W. Munden and Richard P. Krafsur, eds., *The American Film Institute Catalog of Motion Pictures: Feature Films, 1911–1970,* 6 vols. (New York: Bowker, 1970–76). Additional insights into this genre of entertainment came from Judith Crist, ''The Chairborne Aviator at the Flicks,'' *Saturday Review,* April 16, 1977, pp. 24–29, and from George C. Larson, ''Breaking the Movie Barrier,'' *Flying,* October 1978, pp. 62–66. A handy guide to television programs is Les Brown, ed., *Les Brown's Encyclopedia of Television* (New York: Zoetrope, 1982).

Arshile Gorky's aviation murals appear to have been a significant element in his career; the story of the murals' accidental disclosure in 1972, leading to their removal and restoration in 1976, is described in Ruth Bowman, comp., *Murals without Walls: Arshile Gorky's Aviation Murals Rediscovered* (Newark, N.J.: The Newark Museum, 1978). The airplane as an aesthetic and cultural force in American life, including industrial design, has possibly been more influential than some writers have thought. As noted earlier, Dominick A. Pisano is preparing a new study that will assess aviation and industrial design, as well as other aspects. I first came across the ''His and Hers'' airplanes in the book by Stanley Marcus, *His and Hers: The Fantasy World of the Nieman-Marcus Catalogue* (New York: Viking, 1982). John Yardley's remark about gremlins is quoted in Richard P. Hallion, *Test Pilots: The Frontiersmen of Flight* (New York: Doubleday, 1981). My interest in gremlins dates to childhood reading during World War II. Through the fantasy world of my own children, Paula and Alex, I rediscovered gremlins in one of Roald Dahl's books, *The Wonderful Story of Henry Sugar and Six More* (New York: Knopf, 1977), in which Dahl explains his start as a writer and the origins of these irascible little creatures.

Astronautics

Again, the AIAA illustrated annual summary of highlights, *Aerospace Review and Forecast,* is useful, as is NASA's own yearly review, *Spinoff 1982* (see also previous years). Frank W. Anderson, Jr., *Orders of Magnitude: A History of NACA and NASA, 1915–1980* (Washington, D.C.: U.S. Government Printing Office, 1981), includes a succinct review of manned and unmanned programs of the seventies. See also Howard Allaway, *The Space Shuttle at Work* (Washington, D.C.: U.S. Government Printing Office, 1979), which includes comments on the application of earth resource satellites. Recent NASA press releases and news articles have given thorough coverage to the various shuttle flights. The missions of the Pioneer, Mariner, Viking, and Voyager spacecraft, along with other satellites, are summarized in a wide variety of NASA publications. See, for example, *Voyager Encounters Jupiter* (Pasadena, Calif.: NASA/ Jet Propulsion Laboratory, 1979).

CHAPTER 9. TURMOIL AND TRANSITION

Global Reach—Global Power

For a readable summary of American as well as Soviet and European atomic weaponry of the 1980s, edited by an expert from the authoritative Jane's organization in England, see Christopher Campbell, *Nuclear Weapons Fact Book* (Novata, Calif.: Presidio Press, 1984). A recent study of the Strategic Defense Initiative is Donald R. Baucom, *The Origins of SDI, 1944–1983* (Lawrence: University Press of Kansas, 1993), which covers the postwar history of ballistic missile defense and related security issues.

An informed review of contemporary air power from a global perspective is Michael J. Gething, *Air Power 2000* (London: Arms and Armour Press, 1992). Michael Brown, *Flying Blind: The Politics of the U.S. Strategic Bomber Program* (Ithaca: Cornell University Press, 1992), provides an instructive review of strategic and tactical assumptions that guided bomber design and contracts since World War II, including recent B-1 and B-2 programs.

For an authoritative analysis of Desert Storm, see Richard P. Hallion, *Storm over Iraq: Air Power and the Gulf War* (Washington, D.C.: Smithsonian Institution Press, 1992). This superb study not only covers the war but also discusses the evolution of Air Force doctrine during the 1970s and 1980s. A fascinating photographic survey of the Gulf War has been assembled by the Editors, Time-Life Books, *Air Strike* (Alexandria, Va.: Time-Life Books, 1991). The F-117 stealth fighter has sparked a variety of popular studies; for a summary that includes combat operations, see Steve Pace, *The F-117A Stealth Fighter* (Blue Ridge Summit, Pa.: TAB Books, 1992).

My quotation from Secretary of the Air Force Donald Rice comes from Department of Defense, *Report of the Secretary of Defense to the President and the Congress: January 1993* (Washington, D.C.: Government Printing Office, 1993), p. 136; a document that also provides insights into current thinking about Air Force and armed forces roles in the future. For a more specific analysis of Air Force doctrine and roles likely to evolve in the future, see Richard H. Shultz, Jr., and Robert L. Pfaltzgraff, Jr., eds., *The Future of Air Power in the Aftermath of the Gulf War* (Maxwell Air Force Base, Ala.: Air University Press, 1992).

The Adversities of Air Travel

Reacting to the rash of terrorism in the air, the aviation press and other periodicals produced a wide variety of articles and commentary. One example was William Triplett, "An Industry Held Hostage," *Air and Space* 7 (February/March 1993): 26–38, surveying selected events and summarizing contemporary procedures to thwart terrorist actions. On deregulation, scholarly studies include Anthony E. Brown, *The Politics of Airline Deregulation* (Knoxville: University of Tennessee Press, 1987) and informative journalistic articles such as Christopher Power and others, "The Frenzied Skies," *Business Week* (December 19, 1988), pp. 70–80, featured as the magazine's cover story.

The Airline Upheaval

For a fascinating and informative review of airlines in the United States, including profiles of major leaders, see William M. Leary, ed., *Encyclopedia of American Business History and Biography: The Airline Industry* (New York: Facts on File, 1992), a reference that touches on many contemporaneous trends, including commuter operations as well as the major carriers. John Newhouse, *The Sporty Game* (New York: Knopf, 1983) primarily analyzes the airframe manufacturers while offering incisive commentary on the airlines as well. Useful studies of major airlines that encompass recent years include Robert Serling, *Eagle: The History of American Airlines* (New York: St. Martin's Press, 1985) and R.E.G. Davies, *Delta: An Airline and Its Aircraft* (Miami: Paladwr Press, 1990).

On the career of the controversial Frank Lorenzo, airline expert R.E.G. Davies, *Rebels and Reformers of the Airways* (Washingon, D.C.: Smithsonian Institution Press, 1987), includes a favorable argument on Lorenzo's efforts to bring fat wage agreements and other costs into line with operating realities. Aaron Bernstein, *Grounded: Frank Lorenzo and the Destruction of Eastern Air Lines* (New York: Simon & Schuster, 1990), takes an opposite view. Edmond Preston, "Frank Lorenzo," in Leary, ed., *The Airline Industry* represents a cautionary synthesis.

The issue of airlines and global mergers received considerable attention during the negotiations between British Airways and USAir during 1992–93. A knowledgeable business journalist, John Newhouse, published "The Battle of the Bailout," in the *New Yorker* (January 18, 1993), pp. 42–51. Also see Richard Weintraub, "Rebuilding USAir: Investment by British Airways Gives Carrier a Big Lift," in the *Washington Post* (March 22, 1993), pp. 19–21, which provides the source of the "food chain" analogy.

Readjustments in the Aerospace Industry

For an indispensable background about the matrix of technology, economics, politics, marketing, and sheer gambling instincts that characterize the modern airline manufacturing industry in America and Europe, read John Newhouse, *The Sporty Game* (New York: Knopf, 1983). Its title is richly descriptive. A useful analysis from the American perspective is Virginia C. Lopez and Loren Yager, *The U.S. Aerospace Industry and the Trend toward Internationalization* (Washington, D.C.: Aerospace Industries Association of America, 1988), with commentary on military developments as well as civil trends. Aspects of location, economic impact, and political factors of many aerospace activities are discussed in Ann Markusen et al., *The Rise of the Gunbelt: The Military Remapping of Industrial America* (New York: Oxford University Press, 1991), a study with a rich bibliography. Problems of adjustment to the end of the cold war and slashed budgets in the wake of the Gulf War are discussed in "America's Arsenal: Defense Contractors and Their High-Tech Weapons Face the Toughest Test Yet," a special report appearing in *Fortune* 123 (February 25, 1991): 34–64. Aerospace contractors were the subject of several articles in this particular issue.

Space: A Star-Crossed Saga

The Pulitzer–prize winning study by Walter McDougal, *The Heavens and the Earth: A Political History of the Space Age* (New York: Basic Books, 1985), provides a stimulating perspective for any study of American-Soviet space efforts and the ethos of NASA. Various NASA programs are discussed in special press releases from Washington headquarters; an annual summary of principal launches and initiatives appears each December in *NASA News*. Through the Government Printing Office, NASA also publishes a slick, magazine-style annual report, *Spinoff*, with color photos and informative descriptions of programs under way or planned.

The destruction of the *Challenger* space shuttle is officially assessed in "Report of the President's Commission on the Space Shuttle *Challenger*" (Washington, D.C.: Government Printing Office, 1986). The agency became the subject of numerous critical periodical articles and books, such as Joseph J. Trento, *Prescription for Disaster: From the Glory of Apollo to the Betrayal of the Shuttle* (New York: Crown Publishers, 1987). The prolonged evolution of the space station program is treated by Howard E. McCurdy, *The Space Station Decision: Incremental Politics and Technological Choice* (Baltimore: Johns Hopkins University Press, 1990). The same author also provides a historical analysis of NASA's bureaucratic imbroglios, including the Hubble telescope, in *Inside NASA: High Technology and Organizational Change in the American Space Program* (Baltimore: Johns Hopkins University Press, 1993). Regarding Hubble, the *Newsweek* quote appears in "Heaven Can Wait" (July 9, 1990), p. 49; the astrophysicist is quoted in Dick Thompson, "Big Gamble in Space," *Time* (March 22, 1993), p. 63.

Notes for the Third Edition

For coverage of recent events described in the third edition, I have relied on the authoritative reporting in periodicals such as *Aviation Week and Space Technology* along with the professional journal of the American Institute of Aeronautics and Astronautics, *Aerospace America,* and its British counterpart, *Aerospace International,* issued by the Royal Aeronautical Society, as well as other contemporary periodicals. I also relied on NASA's annual summary, *Spinoff,* as well as *Aviation Week and Space Technology*'s invaluable annual compendium, the *Aerospace Source Book 2000* (January 17, 2000), as well as prior numbers of this publication, released as a special edition each January. The Air Transport Association hosts a highly informative e-mail site, *ATA SmartBrief,* available at smartbrief@smartbrief.rsvp0.net, which compiles noteworthy news stories with links to American and foreign sources. See also William E. Burrows, *This New Ocean: The Story of the First Space Age* (New York: Random House, 1998), an engrossing account by a skilled journalist, which includes Space Shuttle operations, planetary probes, and the International Space Station. Roger Launuis, ed., *Innovation and the Development of Flight* (College Station: Texas A&M University Press, 1999), embodies a series of outstanding essays on historical developments that are relevant to this new edition. Additional sources for two new subsections in Chapter 9 include the following:

Rethinking Airpower. For a thoughtful analysis of recent programs for combat planes and issues of international trade, see James W. Canan, "Fighters Vie for Future Markets," *Aerospace America* (January 1998), 16–33. For an interpretive summary of recent trends in both the aviation and the space industry, including mergers, see Roger E. Bilstein, *The American Aerospace Industry: From Workshop to Global Enterprise* (New York: Twayne Publishers / Simon & Schuster Macmillan, 1996). See also Donald Pattillo, *Pushing the Envelope: The American Aircraft Industry* (Ann Arbor: University of Michigan Press, 1998), which describes airframe manufacturers exclusive of the light plane sector and the space industry. Benjamin S. Lambeth, *The Transformation of American Air Power* (Ithaca: Cornell University Press, 2000), surveys the contemporary scene from the viewpoint of an informed flier and knowledgeable military analyst.

General Aviation: Dark Clouds and Silver Linings. Roger Bilstein, *The American Aerospace Industry,* cited above, includes descriptive commentary on trends in this colorful sector of aviation. Donald Pattillo, *A History in the Making: Eighty Turbulent Years in the American General Aviation Industry* (New York: McGraw Hill, 1998), features details of numerous manufacturers. In their engaging book, Frank Joseph Rowe and Craig Miner, *Borne on the Wind: A Century of Aviation in Kansas* (Wichita: Wichita Eagle & Beacon Publishing Co., 1994), discuss general aviation at length, since Kansas is home to major suppliers like Beechcraft, Cessna, and others. My quotations on aerial fire fighting come from Edwin Kiester Jr., "Battling the Orange Monster," *Smithsonian Magazine* (July 2000), 33–42.

Index

Italic page numbers indicate illustrations.

ABMA (Army Ballistic Missile Agency), 208–11, 269–70
Accidents, 22, 24, 25–26, 89, 98, 232
Adams, Henry Brooks, 13
Advertising: airline, 104, *105, 177;* general aviation as subject of, 317, *318*
Aerial Experiment Association, 16, 28
Aero Commander, 200–201, *201*
Aeromarine West Indies Airways, 55–56
Aeronautical Chamber of Commerce, 74, 99
Aeronautical engineering, 37–38, 70–74, 86–96. *See also* Manufacturing
Aeronautical Society of Great Britain, 5, 8
Aerospace industry, 350–58, 362–69; origin of term, 209; revenues, 256. *See also* Manufacturing
Aerospace medicine, 226–27. *See also* Aviation medicine
Agricultural aviation, 65–66, 201–2, 240–41, 298, *299*
Air cargo, 26, 58–59, 260–61, *261,* 263
Air Commerce Act of 1926, 51, 75, 97
Air Corps Act of 1926, 43
Airbus Industrie, 351, 352–55
Aircraft carrier, 47–48, 129–33, 145–58, 159, 220–21, *306*
Aircraft Owners and Pilots Association, 245
Air Line Pilots Association, 53, 99
Airlines: before WWII, 26–28, 51–59, 85–108; after WWII, 169–78, 227–39, 286–92; alliance system, 349–50; first jet transports, 227–31; trends in postwar service, 232–39, 345–50, 351, 356; and social changes, 234–39; deregulation of, 291–92, 343, 346. *See also* Mergers
Airmail, 16, 26, 50–53, *51, 54;* carried by Air Corps, 127–28
Airmail Act of 1925, 51
Airports, 50, 61, 74–75, 100–101, 104, 233–34, 237–38, *293,* 343–45, 347
Air races, 84
Air traffic control, 98, 100–101, 231–32, 286–88
Air Transport Association, 99
ALCM (air-launched cruise missile), 300–301. *See also* Cruise missiles
Aldrin, Edwin, 278, 279
American Airlines, 51, 103, 104; advertising campaign, *105. See also* Airlines
American Eagle (biplane), *62,* 63
Anti-drug campaign, 341
Antisubmarine patrol, WWII, 159
Apollo missions, 275–79; *Apollo 11,* 276–79; Apollo-Saturn V space vehicle, *277; Apollo 15, 280. See also* NASA, Apollo-Saturn program
Area rule fuselage, 219, *223*
Armstrong, Neil, 278, *279*
Army Reorganization Act of 1920, 42
Arnold, Henry H. (Hap), 17, 24, 45, 129, 157, 182, 184
Art, 316
Astronauts. *See* Apollo missions; NASA, first manned missions; Skylab; Space Shuttle

Atomic bomb, 158
Attitudes: toward aviation, 13–15, 17, 18–19, 22–26, 39, 76–77, 107–8, 167–69, 185–86, 189, 307–18, 327–29, 341; toward space exploration, 218, 363–65, 369
Autogyro, 117
Aviation medicine, 73, 106, 226–27
AWACS (airborne warning and control system), 300

Bacon, Roger, 4
Balkan conflict, 339–40
Balloons, 4–8, 327; in the Civil War, 6, 7; in WWI, 34; stratospheric flights of, 117–18, 226. *See also* Goodyear airship
Barnstormers, 60–62
Beachey, Lincoln, 7, 20–21
Beech, Walter, 111–12, 115
Beech Aircraft Corp., 112–16, 197–200, 241, 242, 293–94, 336, 357
Beechcraft (airplanes): Bonanza, 198–200, *199, 318;* King Air, 242, *243,* 260; Model 17, 112–13; Model 18, 115–16, 241, 257, *318. See also* General aviation
Bell, Alexander Graham, 13, 16, 29, 31
Bell (aircraft): Model 47, *190;* UH-1 Huey, 251, *252;* X-1, 182, 184, 185; X-1A, *185;* X-5, *185;* 206 Longranger, *298*
Berle, Adolph A., 169–70
Berlin airlift, 192–95
Bierce, Ambrose, 14
Black Americans: as aviators, 138, 368; as astronauts, 322
Boeing (aircraft): B-17, *141;* B-29, 156–58, *157;* B-47, 181–82; B-52, 223–26, *224,* 253–54, 299–300; P-26, *126;* Clipper, 93–94, *94, 95,* 173; F2B-1, *49;* Monomail, 87–88; Stratocruiser, 171–72; Model 40, 52, 56; Model 247, 51, 85–89, *102,* 104; Model 707, 228–30, *229, 235,* 261, *263;* Model 727, 230, *261,* 263; Model 737, 257, 350–52; Model 747, 262, *263,* 351, *355;* Models 757, 767, and 777, 289–90, 352, 366. *See also* Aerospace industry; Military aviation, in Korean War, in Vietnam, WWII
Boeing Company, 89, 228–29, 261–63, 327, 336, 351–55, 358, 368; McDonnell Douglas merger, 353
Bombing, 32, 34; trials of 1921, 43, *44, 48;* dive-bombing, 45–46; strategic theory of, 126–28; in WWII, 158–64; in Korean War, 186–89; in Vietnam, 249–75; in Persian Gulf, 329, 331–35
Books: about aviation, 19, 76, 307–9, 330; commenting on aviation, 316–17
Braniff, 51, 169, 266, 292. *See also* Airlines
Buck Rogers (comic strip), 118, 310

Bureau of Air Commerce, 97–98, 106
Bush, George, 328
Business flying. *See* General aviation
Byrd, Richard E., 80

Cabin attendants, 234, *258. See also* Stewardesses; Stewards
Cayley, George, 8
Cessna (aircraft), AGtruck, *299;* C-34 Airmaster, 112, 113; model 185, *240. See also* General aviation
Cessna Aircraft Company, 111–12, 116, 197, 239–40, 293, 294, 357
Chanute, Octave, 9, 14
Christofferson, Silas, 26–27
Civil Aeronautics Act of 1938, 98
Civil Aeronautics Administration, 99
Civil Aeronautics Board, 99, 291, 292
Cochran, Jacqueline, 84, 163
Collins, Michael, 278, *279*
Commuter airlines. *See* Airlines, trends
Comic strips, 118, 310–11
Consolidated (aircraft): B-24, *139;* B-36, 180–81, *182,* 184–85; PBY, 69, 129, *130,* 159
Continental Airlines, 345–46
Convair (aircraft): B-58, 222–23; F-106, 222, *223;* XF-92A, *185;* 240/340/440 series, 171, 175, 257
Coral Sea, battle of, 148
Corporate flying. *See* General aviation
Cowling, 70, 86–87, 89, 90
Cruise missiles, 329, 332, *333*
Curtiss, Glenn, 15–18, 25, 28
Curtiss (aircraft): JN-4, 36–37, *38,* 50, *51,* 60; P-6, 45, *46;* SB2C, *152;* model 75 flying boat, 55, *56*

Daniel Guggenheim Fund for the Promotion of Aeronautics, 73–74, 76, 81. *See also* GALCIT
de Havilland (aircraft): Comet, 228; DH-4, 37, *44,* 45–46, 50, 52, 59, 65; DHC-6 Twin Otter, 257–58, *259*
Delta Air Lines, 348–50. *See also* Airlines
de Rozier, Pilatre, 4–5
Desert Shield/Desert Storm, 329, 331–35
DEW (Distant Early Warning) line, 218–19
Dirigibles, 5–8; in WWI, 34; U.S. Navy, 45, 46–47, 93, 129
Doolittle, James H. ("Jimmy"), 73, 84, *135,* 136, 146–47
Douglas (aircraft): C-47, *155,* 252; C-54, *194;* D-558-I, *185;* D-558-II, *185;* DC-3, 51, 85–86, 89–92, *91, 103,* 104, 107, 257; DC-4, 94–96; DC-6, 171, 174, *175;* DC-8, 229, 261–62; DC-9, 257, 263–64; DC-10, 264, *266, 289;* SBD Dauntless, *132;* X-3, *185. See*

also Military aviation, WWII; McDonnell Douglas

Douglas, Donald, 38, 49, 59, 89

Eaker, Ira C., 134, 136, 139
Earhart, Amelia, 76, 84, *85*
Eastern Airlines. *See* Airlines; Lorenzo, Frank
Eilmer of Malmesbury, 4
Electronics: and WWII aircraft, 164; civil aerospace, 178, 200–201, 231–32, 286–88, 342; military aerospace, 194, 218–19, 222, 223–25, 299–301, 329, 334, 337, 338; and aircraft in Vietnam, 251, 255–66. *See also* Satellites
Ely, Eugene, 32
Emergency services, 61, 191–92, 296–98, *298*
Engines, piston, 5, 10, 12, 37, 47, 52, 71–72, 84–85, 88–89, 108–9. *See also* Jet propulsion
European Aeronautic Defense and Space Company, 354
European influence: on early aeronautics, 4–6, 8–10; in WWI, 32, 33–34, 46, 47; in 1920s, 55, 62, 68, 70–73, 77–78, 80–81; in 1930s, 86, 88, 93, 117; and rocketry, 118–23, 164–65; in WWII, 125–26, 128–29, 134, 136, 145, 159, 164–65; in postwar era, 170–71, 182, 220, 230, 301, 327, 336–40, 353–56, 365, 367–78; on development of jets, 179, 227–28, 264, 288–90, 291, 303–5. *See also* Gremlins; Spacelab; SST; V-2; *VfR;* von Braun, Wernher
Explorer I, 211, *212*

Fairchild (aircraft), 63; FC-2, *67;* Fairchild-Saab 340, 295–96
Fairchild Aerial Surveys, 68
Fighter-bombers. *See* Military aviation, WWII
Finletter report, 185–86
Fixed-base operators, 61, 116
Flying Tigers, 153–54˙
Food service, 93, *95,* 96, *102,* 177–78, 262
Ford trimotor, 57, *58,* 63, *64,* 71
Forest-fire patrol, 61, 66
Foulois, Benjamin D., 33, 127
Fuels, 72, 85, 175

GALCIT (Guggenheim Aeronautical Laboratory of the California Institute of Technology), 87, 90; and rocketry, 122
Gann, E. K., 91, 101, 107
General aviation: in 1920s, 59–69; in 1930s, 108–16; in WWII, 161; after WWII, 191–92, 195–203, 239–44, 257, 292–98, 356–60; limited liability, 357; and use of jets, 242–45, 259–60, 294–95, 357–59
General Aviation Manufacturers Association, 245
General Dynamics (aircraft): F-16, 303–4, *304,*
353; F-111, 254–55. *See also* Aerospace industry; Consolidated; Convair; Lockheed
Glenn, John H., 218, 219, 365
Gliders, 8–10
Goddard, Robert, 119–20, *121,* 121–22
Goodyear airship (blimp), *202*
Gremlins, 317–18
Grumman (aerospace vehicle manufacturer), 353. *See also* NASA; Northrop Grumman
Grumman (aircraft): A-6, 334; F6F Hellcat, *147, 152;* F9F Panther, 180, *188;* F-14 Tomcat, 304–5; Gulfstream I, 241–42, 260, *297;* TBF Avenger, *152. See also* Aerospace industry; Military aviation: in Korean War, WWII
Gunships: in Vietnam, 251–52, 329; in Persian Gulf, 332
Guppy (aircraft), 273–74; Super Guppy, *274*

Helicopters, 117–18, 190, 192, 202, 260, 296–98, 305, 307, 329; in WWII, 117; in Korea, 189; in Vietnam, 249, 251–52
Hostesses, airline. *See* Stewardesses
Hoxsey, Arch, 17, 18, *19,* 21
Huff-Daland Company, 66
Hughes AH-64, 305, 307
Hunsaker, Jerome C., 31, 37–38
Hurricane-hunter squadrons, 191

IATA (International Air Transport Association), 169–71, 178
ICAO (International Civil Aviation Organization), 169–71, 178
ICBM (intercontinental ballistic missile), 206, 209, *214,* 300, 328, 340. *See also* Missiles
Instrument flying, 21, 73–74
International relations, 76–80, 167–70, 234–39, 327, 340. *See also* Attitudes

Jet lag (circadian rhythm), 237
Jet propulsion, 178–80, 227–39. *See also* Messerschmitt Me-262; Military aviation, after WWII; Turboprop engines
Jet set, 238
Joint Strike Fighter (JSF), 338–39
Jumbo jets. *See* Wide-body transports

Kamikaze Corps, 152–53
Kellogg-Briand Pact of 1928, 77
Kelly Bill, 53, 56, 74. *See also* Airmail Act of 1925
Kitty Hawk, North Carolina, 3, 10
Knabenshue, Roy, 7, 17, 26
Knight, Jack, 50
Korean War. *See* Military aviation

Lafayette Flying Corps, 33
Langley, Samuel Pierpont, 9–10, *11,* 13–14, 31

Law, Ruth, 22
Lear, William (Bill), 243–44; and Learjet, *244, 357*
LeMay, Curtis E., 157, 185
Lend-Lease Program, 133–34
Lexington, 47, *49*
Lilienthal, Otto, 8, *9*
Lindbergh, 49, 53, 61, 69, 77, *78;* and Robert Goddard, 120
Lockheed (aircraft): Constellation, 94, 171–72, *172, 173,* 174; Electra, 92, 257; F-117, 331, 334–35, 339; L-1011 TriStar, 264, *265,* 289; P-38 Lightning, *143;* P-80 Shooting Star, 179, *187;* SR-71 Blackbird, 301, *302;* U-2, 225, 301; Vega, 86, 87. *See also* Aerospace industry; Balkan conflict; Military aviation: in Korea, in WWII
Lockheed Martin (company): mergers, 327, 353; F-16, 339; F-22, 337–38. *See also* Joint Strike Fighter
Lorenzo, Frank, 345–46
Lowe, Thaddeus S. C., 6
Lunar landing, early plans, 218. *See also* Apollo missions; NASA
Luscombe (aircraft), 113

McDonnell Douglas (aircraft): AV-8B Harrier, 305; F-4 Phantom, 219–20, *221,* 222; F-15 Eagle, 303, 332; F-18, 305. *See also* Balkan conflict; Douglas (aircraft); Military aviation, in Vietnam
McDonnell Douglas Corporation, merger of, 264n, 328, 336–37, 351–53. *See also* Aerospace industry; NASA
Machine guns, WWI, 31–32
Manly, Charles, 10, *11*
Manufacturing, 29, 30–31, 34–35, 74–76, 327, 336; in WWII, 159–60; and risk-sharing, 264, 289–90. *See also* Aerospace industry; Mergers
Maps, 79–80
Marine Corps. *See* Bombing, dive-bombing; Military aviation
Marketing, 103–4, 291–92
Martin, Glenn, 37–38
Martin (aircraft), M-130 China Clipper, 92–93; MB-2, 43, *48*
Martin Marietta Corporation, 350
MASH (Mobile Army Surgical Hospital), in Korea, 189
Massachusetts Institute of Technology, 12, 31, 37
MATS (Military Air Transport Service), 248–49
Mergers: of manufacturers, 290–92, 294, 327, 351–55; of airlines, 291–92, 345–50
Messerschmitt Me-262, 145
Mexican Border Patrol, 42
Midway, battle of, 149–50
Military aviation: in Civil War, 6–7; in WWI,

31–39; from 1920 to 1941, 41–48, 125–33, 158–64; after WWII, 178–90, 218–27, 299–307, 329; in Korean War, 186–89; in Vietnam, 249–75; post-Vietnam, 327–31, 335–41; in Persian Gulf, 329, 331–35
—WWII, 158–64; in Europe, 133–45; in Pacific, 132–33, 145–58
Missiles, military, 206, 209, 299–301, 327–29. *See also* ALCM; ICBM; Rocketry; V-2
Mitchell, William (Billy), 34, 41, 42, *44*
Model airplanes, 19
Moffett, William A., 46–47
Montgolfier brothers, 4, 8
Mooney (aircraft), 241, 295
Motion pictures, 20, 311–16; as in-flight entertainment, 234; *Wings, 313; Star Wars, 315*
Music, 19, 76, 309–10, 314; as in-flight entertainment, 234

NACA (National Advisory Committee for Aeronautics), 31, 47, 69–71, 86–87, 89, 182–84, 219. *See also* NASA
NASA: origins of, 211, 267; and aeronautics, 211–18, 366–67. *See also* Apollo missions; NACA; North American (aircraft), X-15
—and astronautics, 211–18, 360–67; first manned missions, 215, 217–18, 267; Apollo-Saturn program, 267–78; Skylab, 278–82, 322–23; Soviet joint mission, 282–83; Space Shuttle, 318–22, *320,* 360–63; satellites, 322–24; planetary probes, 324–25, 364–65; Hubble Telescope, 362–64; space station, 362–63, *363,* 365–66, 368
National Business Aircraft Association, 245
National Security Act of 1947, 184, 205
Navaids. *See* Navigational aids
Navigational aids: beacons, 50–51; early instrumentation, 73, 97–98. *See also* Air traffic control; Electronics; VOR
Newcomb, Simon, 14
Nichols, Ruth, *87*
Ninety-nines, *85*
NORAD (North American Air Defense Command), 219
North American (aerospace vehicle manufacturer), rocket hardware. *See* NASA
North American (aircraft): B-25 Mitchell, 146; F-86 Sabre, 179, *181;* F-100 Super Sabre, 221–22; P-51 Mustang, 129, 142–43, 144; Sabreliner, 242; X-15, 84, 214–17, *217. See also* Military aviation: in Korea, in Vietnam, WWII
Northrop (aircraft): Alpha, 87–88; T-38 Talon, *323;* X-4, *185*
Northrop Grumman, 337; merger, 353; B-2, 336–37, 339–40
Northwest Airlines, 15, 99, 169

Oberth, Hermann, 120–21, *208*
Oldfield, Barney, 20
Operation Paperclip, 164

Painting, 76, 316
Pan American World Airways, 78–79, 91–92, 169–72, 234, 342–43, 345, 347–48; advertisement, *177. See also* Airlines
Paratroops, 137, 189
Pearl Harbor (Hawaii), 133, 145–46
Pershing, John, 33
Photography, 34, 66–69. *See also* Satellites
Piper, William T., 110
Piper (aircraft): Archer II, *295;* J-3 Cub, 109, 110–11; Pawnee, 240–41. *See also* General aviation
Piper Aircraft Corporation, 110, 196, 240–41, 292–93, 294, 295, 357
Planetary space probes, 324–25. *See also* NASA
Ploesti, 136, 138–40
Popular culture. *See* Attitudes
Post, Wiley, 84
Presidential flying, 96–97, 172–74, 201, 229–30; *Air Force One,* 230
Pressurization, 94, 175; in general aviation, 242, 260, 294, 295
Propellers, 12, 72, 88

Quimby, Harriet, 22

Radar, 164, 286–87. *See also* DEW line; Electronics; NORAD
Radio range beacon. *See* Navigational aids
Radio shows, 311; as airline entertainment, 234
RAF. *See* Royal Air Force
Reagan, Ronald, 327, 330, 361
Regional airlines. *See* Airlines, trends
Republic (aircraft): F-84 Thunderjet, 179; F-105 Thunderchief, 222, 250–53, *250;* P-47 Thunderbolt, *137. See also* Military aviation: in Korea, in Vietnam, WWII
Rickenbacker, Eddie, 37
Ride, Sally, 322, *323*
Rocketdyne (aerospace vehicle manufacturer), rocket hardware. *See* NASA
Rocketry, 118–22, 164–65; postwar rocketry, 205–18. *See also* NASA
Rockwell (aerospace vehicle manufacturer), rocket hardware. *See* NASA
Rockwell B-1, 299, 301
Rodgers, Calbraith P., 25
Roosevelt, Franklin Delano, 96–97, 172–74; decision to cancel air-mail contracts, 127–28
Roosevelt, Theodore, 9, 18, 19
Root, Amos I., 15
"Rosie the Riveter," *160*

Royal Aeronautical Society. *See* Aeronautical Society of Great Britain
Royal Air Force, 128, 133–34, 136, 139–40, 142, 145, 159, 163, 179, 194, 220, 337, 338

SAC (Strategic Air Command), 184–86, 247–48, 340
St. Petersburg-Tampa Airboat Line, 27–28
SAM (surface-to-air missiles), 251, 256, 300
Santos-Dumont, Alberto, 6
Saratoga, 47–48, 49
Satellites, 211–15, 301, 322–25, 329, 334; commercial satellites, 367–68. *See also* NASA; Rocketry
Saturn launch vehicles, 267–75; Saturn I, *216, 270;* Apollo-Saturn V, *277. See also* NASA
Sikorsky, Igor, 117–18, *118*
Sikorsky (aircraft): helicopters, 117, *118;* S-42, 92–93
Silverstein Committee, 271
Skylab, 278–82, *281,* 322. *See also* NASA
Smithsonian Institution, 9, 10, 13, 31
Société d'Aviation, 5, 8
Southwest Airlines, 350
Spaatz, Carl A., 42, 134, *135,* 136
Spacelab, 219, 360. *See also* NASA
Space race, 215, 217–18
Space research, early postwar, 210–18, 360–66. *See also* NASA; Rocketry; Satellites
Space Shuttle, 318–22, *320, 327,* 360–63. *See also* NASA
Space station, 362–63, *363,* 365–66, 368
Spartan Executive, 113–14, *114*
SST (supersonic transport), 264–66
START (Strategic Arms Reduction Talks treaty), 328–29
Stealth aircraft, 331, 332, 339
Stewardesses, 101–3, *236, 258. See also* Cabin attendants
Stewards, *95,* 101, 176. *See also* Cabin attendants
Stinson, Katherine, 22
Stinson (aircraft), 63
Strategic Defense Initiative, 327–28
Stressed-skin construction, 86–87
Stuhlinger, Ernst, *208*
Sullivan, Mark, 13
Supersonic flight, 182–84
Symbolism. *See* Attitudes

Takeoff and landing, from ship, 32
Tarbell, Ida, 22–23
Television, 189; "Sky King" (program), *311,* 314–15
Terrorism, 330, 341–43
Toftoy, H. N., *208*
Training, 21, 161–62

Trimotors, 57, 77, 88, 95, 101. *See also* Ford trimotor
Trippe, Juan, 78–79. *See also* Pan American World Airways
Tsiolkovsky, Konstantin, 119
Turboprop engines, 227–28, 242, 248, 257, 259–60, 295–96
TWA (Trans World Airlines), 51, 169, 171–72, 234, 341–42, 344, 348. *See also* Airlines

United Airlines, 51, 101, 346–48, 350. *See also* Airlines
Unmanned Aerial Vehicle (UAV), 333–36
USAF (United States Air Force), created, 184

V-2 (rocket), 164–65, 206, *207*. *See also* VfR; von Braun, Wernher
Variable-sweep wing, 185, 254, *255*
Vera Cruz, Mexico, 32–33
VfR (Verein für Raumschiffart), 121–23
Vietnam. *See* Military aviation
Villa, Pancho, 33
Vinci, Leonardo da, 4
von Braun, Wernher, 121–23, 164–65, 205–8, *208*, 211, *216*, 269, 271, 272, 274
von Karman, Theodore, 73
von Ohain, Hans, 179
VOR (very high frequency omnidirectional radio range), 178, 200. *See also* Electronics

Vought (aircraft): A-7 Corsair, 254; F4U Corsair, *149*; F-8 Crusader, 227, 254. *See also* Military aviation: in Vietnam, WWII
Voyager (aircraft), 356

Waco biplane, *60, 63*
Walcott, Charles D., 31
Warner, E. P., 70, 170
WASPs (Women Airforce Service Pilots), 163
Whittle, Frank, 179
Wide-body transports, introduction of, 262, 264
Wilson, Woodrow, 31, 32–33, 50
Wind tunnels, 70–71, 86
Women, 22, 76, 84–85, 368–69; in wartime production, 159–60, *160;* as instructors, 163; as astronauts, 322–23, *323*. *See also* Stewardesses
World War II, bombers and fighters in. *See* Military aviation, WWII
Wright (aircraft): first flight, 12, *13;* Military Flyer, *16*
Wright brothers, 3, 4, 10, *13*, 14–15, 17, 29, *30*, 38–39; first flight, 12
Wright Company, 29, 30
Wright Exhibition Company, 17, 24

Yeager, Charles E., 182

Zahm, Albert F., 31

About the Author

Dr. Roger Bilstein, Emeritus Professor of History at the University of Houston-Clear Lake, has written dozens of articles and eight books on the aviation business, the aerospace industry, and space exploration. His honors include appointment to the Charles Lindbergh Chair of Aerospace History at the Smithsonian's National Air and Space Museum and as visiting professor of history at the U.S. Air Force Air War College. Dr. Bilstein has also served as a consultant and guest curator for several U.S. museums. Since his retirement from the classroom, he has continued activities as consultant and author at his home in the Texas Hill Country near Austin, Texas.